T0325234

AI-Aided IoT Technologies and Applications for Smart Business and Production

This book covers the need for Internet of Things (IoT) technologies and artificial intelligence (AI)–aided IoT solutions for business and production. It shows how IoT-based technology uses algorithms and AI models to bring out the desired results. *AI-Aided IoT Technologies and Applications for Smart Business and Production* shows how a variety of IoT technologies can be used toward integrating data fabric solutions and how intelligent applications can be used to greater effect in business and production operations. The book also covers the integration of IoT data-driven financial technology (fintech) applications to fulfill the goals of trusted AI-aided IoT solutions. Next, the authors show how IoT-based technology uses algorithms and AI models to bring out the desired results across various industries including smart cities, buildings, hospitals, hotels, homes, factories, agriculture, transportation, and more. The last part focuses on AI-aided IoT techniques, data analytics, and visualization tools.

This book targets a mixed audience of specialists, analysts, engineers, scholars, researchers, academics, and professionals. It will be useful to engineering officers, IoT and AI engineers, engineering and industrial management students, and research scholars looking for new ideas, methodologies, technologies, models, frameworks, theories, and practices to resolve the challenging issues associated with leveraging IoT technologies, data-driven analytics, AI-aided models, IoT cybersecurity, 5G, sensors, and augmented and virtual reality techniques for developing smart systems in the era of Industrial Revolution 4.0.

AI-Aided IoT Technologies and Applications for Smart Business and Production

Edited by
Alex Khang, Anuradha Misra,
Shashi Kant Gupta, and Vrushank Shah

CRC Press
Taylor & Francis Group
Boca Raton London New York

CRC Press is an imprint of the
Taylor & Francis Group, an **informa** business

First edition published 2024
by CRC Press
6000 Broken Sound Parkway NW, Suite 300, Boca Raton, FL 33487–2742

and by CRC Press
4 Park Square, Milton Park, Abingdon, Oxon, OX14 4RN

CRC Press is an imprint of Taylor & Francis Group, LLC

Library of Congress Cataloging-in-Publication Data
Names: Khang, Alex, editor.
Title: AI-aided IoT technologies and applications for smart business and production /
    edited by Alex Khang, Anuradha Misra, Shashi K. Gupta and Vrushank Shah.
Description: Boca Raton : CRC Press, 2024. | Includes bibliographical references and
    index.
Identifiers: LCCN 2023032031 (print) | LCCN 2023032032 (ebook) |
    ISBN 9781032490076 (hardback) | ISBN 9781032491202 (paperback) |
    ISBN 9781003392224 (ebook)
Subjects: LCSH: Internet of things—Industrial applications. | Artificial intelligence—
    Industrial applications. | Production management. | Technological innovations—
    Management.
Classification: LCC TK5105.8857 .A399 2024 (print) | LCC TK5105.8857 (ebook) |
    DDC 658.500285/63—dc23/eng/20230812
LC record available at https://lccn.loc.gov/2023032031
LC ebook record available at https://lccn.loc.gov/2023032032

ISBN: 978-1-032-49007-6 (hbk)
ISBN: 978-1-032-49120-2 (pbk)
ISBN: 978-1-003-39222-4 (ebk)

DOI: 10.1201/9781003392224

Contents

Chapter 3   Internet of Things–Generated Data in Business and Production 28

*Alex Khang, Vugar Abdullayev, Eugenia Litvinova,
Svetlana Chumachenko, Triwiyanto, and Vusala Abuzarova*

Chapter 4   Data-Oriented Internet of Things in Intelligent Analytics
and Visualization 46

Jainam Shah, Aakansha Saxena, and Akash Gupta

Chapter 7   Ubiquitous Sensor-Based Internet of Things Platform for
Smart Farming 98

Neha Jain, Yogesh Awasthi, and Jain R. K.

Chapter 13 Smart Notice Board Using the Internet of Things–Based
NODEMCU ESP8266... 211

*Suman Turpati, Richi Sumith Raj P., Mohammed Taj S.,
Naveen Kumar S., and Ranga Reddy S. V.*

Chapter 14 Automotive Internet of Things: Accelerating the Automobile
Industry's Long-Term Sustainability in a Smart City
Development Strategy 225

Vrushank Shah, Suketu Jani, and Alex Khang

Preface

In recent years, with an ever-changing technology edge and an unpredictable world, high-tech companies are keen to leverage multifaceted Internet of Things (IoT)–based solutions and artificial intelligence (AI)–integrated techniques to deploy and deliver next-generation intelligent business and information technology (IT) applications to their clients.

In the Industrial Revolution 4.0, most of the machines, instruments, appliances, equipment, devices, intensive gadgets, and diverse resources are applied across a variety of smart business and production environments, and they have been meticulously empowered with IoT competencies.

The power and consistency of IoT competencies are being seen as a critical challenge in business and production. Specialists and engineers are insisting on the unambiguous interpretations and explanations of AI-aided IoT systems' decisions.

All kinds of connected IoT systems are collectively and/or individually enabled to be intelligent in their operations, offerings, and outputs. Precisely speaking, AI-aided IoT systems are being touted as the next-generation technology to visualize and realize a bevy of intelligent systems, networks, wireless, 5G, 6G, sensors, smart IoT devices, AI frameworks, and environments.

This book targets a mixed audience of specialists, analysts, engineers, scholars, researchers, academics, professionals, and students from different communities to share and contribute new ideas, methodologies, technologies, approaches, models, frameworks, theories, and practices to resolve the challenging issues associated with the leveraging of combating the six fields of **IoT Technologies**; **Data Science**; **Wireless, 5G/6G, and Sensors**; **AI-Aided IoT Systems**; **Cybersecurity**; and **Smart Vehicle Parking Systems** for developing smart systems in the era of Industrial Revolution 4.0.

Happy reading!

**Editors: Alex Khang, Anuradha Misra,
Shashi K. Gupta, Vrushank Shah**

Editors' Biographies

Dr. Alex Khang is a professor in Information Technology; he has over 28 years of teaching and researching information technology at universities of science and technology in Vietnam and United States. He is an AI and data scientist in the Department of AI and Data Science at the Global Research Institute of Technology and Engineering, North Carolina, United States. He is a specialist of data engineering and artificial intelligence in multinational information technology corporations and is in the contribution stage of knowledge and experience into the scope of tech talk. He is a consultant, a full-time and part-time lecturer, and an evaluator on the sides of contents and engineering for the research, an editor, Master/PhD thesis advisor and examiner for local and international institutes and schools of information technology.

Dr. Anuradha Misra is an assistant professor in the Department of Computer Science & Engineering, Amity School of Engineering and Technology, Amity University Uttar Pradesh, Lucknow Campus, India. Her main area of interest includes database security, software engineering, deep learning, data science, and machine learning.

Dr. Shashi Kant Gupta, PhD, is a researcher at the Integral University, Lucknow, Uttar Pradesh, India. He has completed his PhD in CSE from Integral University, Lucknow, Uttar Pradesh (UP), India, and worked as an assistant professor in the Department of Computer Science and Engineering, PSIT, Kanpur, UP, India. He is currently the founder and CEO of CREP, Lucknow, UP, India. He is a member of Spectrum IEEE and *Potentials Magazine IEEE* since 2019 and many more international organizations for research activities.

Dr. Vrushank Shah is the head of Electronics and Communication, Indus Institute of Technology and Engineering, Indus University, Rancharda, Via Shilaj Ahmedabad, Gujarat, India. He has more than 13 years of continuous teaching and research experience in electronics engineering, communication engineering, cybersecurity, data mining, machine learning, and the Internet of Things.

Acknowledgments

This book is based on the design and implementation of the Internet of Things (IoT), the Industrial Internet of Things, artificial intelligence (AI), machine learning, data science, big data solutions, cloud platforms, cybersecurity technology, and AI-aided IoT applications in the smart business and production.

Preparing and designing a book outline to introduce to readers across the globe is the passion and noble goal of the editorial team. To be able to turn ideas into a reality and the success of this book, the biggest reward belongs to the efforts, experiences, enthusiasm, and trust of the contributors.

To all the reviewers with whom we have had the opportunity to collaborate and monitor their hard work remotely, we acknowledge their tremendous support and valuable comments not only for the book but also for future book projects.

We also express our deep gratitude for all the pieces of advice, support, motivation, sharing, collaboration, and inspiration we received from our faculty, contributors, educators, professors, scientists, scholars, engineers, and academic colleagues.

At last but not least, we are really grateful to our publisher CRC Press (Taylor & Francis Group) for the wonderful support in making sure the timely processing of the manuscript and bringing out this book to the readers as soon as possible.

Thank you, everyone.

**Editorial team: Alex Khang, Anuradha Misra,
Shashi Kant Gupta, Vrushank Shah**

Contributors

Vugar Abdullayev
Azerbaijan State Oil and Industry
    University
Baku, Azerbaijan

Abuzarova Vusala Alyar
Azerbaijan State Oil and Industry
    University
Baku, Azerbaijan

Radha Agrawal
Amity School of Engineering and
    Technology, Amity University Uttar
    Pradesh, Lucknow Campus
Lucknow, India

Yogesh Awasthi
Department of Computer Science &
    Engineering, Shobhit University
Meerut, Uttar Pradesh, India

Ravi Chandra B.
G. Pullaiah College of Engineering and
    Technology
Kurnool, India

Mohammad Adam Baba
Department of Computer Science,
    College of Science and Arts,
    University of Bisha
Bisha, Asir Governorate, Saudi Arabia

Soumya Chandra
Amity School of Engineering and
    Technology, Amity University Uttar
    Pradesh, Lucknow Campus, India

Svetlana Chumachenko
Design Automation Department,
    Kharkiv National University of
    Radio Electronics
Kharkiv, Kharkivs'ka oblast, Ukraine

Namrata Dhanda
Department of Computer Science and
    Engineering, Amity University Uttar
    Pradesh Lucknow Campus, India

Sharmistha Dey
UIC, Chandigarh University
Punjab, India

Aryan Kumar Dubey
Kamla Nehru Institute of Technology
Sultanpur, Uttar Pradesh, India

Pankaj Kumar Dubey
Kamla Nehru Institute of Technology
Sultanpur, Uttar Pradesh, India

Radhamani E.
Sri S. Ramasamy Naidu Memorial
    College
Sattur, Viruthunagar, Tamil Nadu, India

Anuj Kumar Goel
Department of Electronics and
    Communication Engineering,
    Chandigarh University
Punjab, India

Shashi Kant Gupta
CREP
Lucknow, Uttar Pradesh, India

Olena Hrybiuk
International Science and Technology
    University, National Academy of
    Sciences
Lane Magnitogorsk 3—Kiev, Ukraine

Jyoti Jain
Sagar Institute of Research & Technology |
    Best Engineering College
Bhopal, Madhya Pradesh, India

Neha Jain
Department of Electronics &
  Communication Engineering,
  Shobhit University
Meerut, Uttar Pradesh, India

Suketu Jani
Automobile Engineering, Indus
  Institute of Technology &
  Engineering, Indus University
Rancharda, Gujarat, India

Nishant Jaiswal
Amity School of Engineering and
  Technology, Amity University
  Uttar Pradesh, Lucknow Campus
Lucknow, India

Jain R. K.
Department of Electronics &
  Communication Engineering,
  Shobhit University
Meerut, Uttar Pradesh, India

Sandip Kanase
Bharati Vidyapeeth College of
  Engineering
Navi Mumbai, India

Patan Sohail Khan
G. Pullaiah College of Engineering
  and Technology
Kurnool, India

Alex Khang
Universities of Science and
  Technology
Vietnam and United States;
Raleigh, North Carolina, United States

Anil Kumar
Amity School of Engineering and
  Technology, Amity University
  Uttar Pradesh, Lucknow Campus,
  India

Varun Kumar
Kamla Nehru Institute of
  Technology
Sultanpur, Uttar Pradesh, India

Eugenia Litvinova
Design Automation Department,
  Kharkiv National University of
  Radio Electronics
Nauky Ave, 14, Kharkiv, Kharkivs'ka
  oblast, Ukraine

Fatima M.
Sagar Institute of Research &
  Technology | Best Engineering
  College
Bhopal, Madhya Pradesh, India

Lokpriya Gaikwad M.
SIES Graduate School of Technology
Navi Mumbai, India

Karthikeyan M. P.
School of Computer Science &
  Information Technology Jain
  (Deemed-to-be University)
Bengaluru, Karnataka, India.

Anuradha Misra
Department of Computer Science &
  Engineering, Amity School of
  Engineering and Technology,
  Amity University
  Uttar Pradesh,
  Lucknow Campus, India

Praveen Kumar Misra
Dr. Shakuntala Misra National
  Rehabilitation University
Lucknow, India

Mauparna Nandan
Computer Applications,
  Techno Main Saltlake
Kolkata, India

Richi Sumith Raj P.
Department of Electronics and
  Communication Engineering,
G. Pullaiah College of Engineering
  and Technology
Kurnool, India

B. C. M. Patnaik
KIIT School of Management,
  KIIT University
Chandaka Industrial Estate,
  Bhubaneswar, Odisha, India

Mohammed Qayyum
Department of Computer Engineering,
  College of Computer Science,
  King Khalid University
Guraiger, Abha, Saudi Arabia

Jain R. K.
Department of Electronics &
  Communication Engineering,
  Shobhit University
Meerut, Uttar Pradesh, India

Ananthi S.
Department of Computer Science,
  PPG College of Arts and Science
Coimbatore, Tamil Nadu, India

Mohammed Taj S.
Department of Electronics and
  Communication Engineering,
G. Pullaiah College of Engineering
  and Technology
Kurnool, India

Naveen Kumar S.
Department of Electronics and
  Communication Engineering,
G. Pullaiah College of Engineering
  and Technology
Kurnool, India

Nidhya M. S.
School of Computer Science &
  Information Technology, Jain
  (Deemed-to-be University),
  Bengaluru
Karnataka, India

Thirupathaiah S.
G. Pullaiah College of Engineering
  and Technology
Kurnool, India

Ipseeta Satpathy
KIIT School of Management,
  KIIT University
Chandaka Industrial Estate,
  Bhubaneswar, Odisha, India

Vrushank Shah
Electronics & Communication
  Engineering, Indus Institute of
  Technology & Engineering, Indus
  University
Rancharda, Ahmedabad, Gujarat, India

Syed Shashavali
G. Pullaiah College of Engineering and
  Technology
Kurnool, India

Arvind Kumar Shukla
Department at Department of Computer
  Applications, School of Computer
  Science & Applications,
  IFTM University
Moradabad, Uttar Pradesh, India

Bindeshwar Singh
Kamla Nehru Institute of Technology
Sultanpur, Uttar Pradesh, India

Durga Prasad Singh Samanta
KIIT School of Management, Kalinga
  Institute of Industrial Technology
Bhubaneswar, Odisha, India

Thota Teja
G. Pullaiah College of Engineering and
  Technology
Kurnool, India

Triwiyanto
Department of Medical Electronics
  Technology, Poltekkes Kemenkes
  Surabaya, Jl. Pucang Jajar Tengah,
  Kertajaya, Kec. Gubeng, Surabaya,
  Jawa Timur, Indonesia

Suman Turpati
Department of Electronics and
  Communication Engineering,
  G. Pullaiah College of Engineering
  and Technology
Kurnool, India

Ranga Reddy S. V.
Department of Electronics and
  Communication Engineering,
  G. Pullaiah College of Engineering
  and Technology
Kurnool, India

Rajat Verma
Department of Computer Science
  and Engineering, Pranveer Singh
  Institute of Technology
Kanpur, Uttar Pradesh, India

Sanskar Verma
Department of Computer Science
  and Engineering, Pranveer Singh
  Institute of Technology
Kanpur, Uttar Pradesh, India

Vrinda Vishnoi
Department of Computer Science
  and Engineering, Pranveer Singh
  Institute of Technology
Kanpur, Uttar Pradesh, India

Sunil Kumar Vohra
Amity Institute of Travel and Tourism,
  Amity University
Noida, Uttar Pradesh, India

1 The Role of Internet of Things Technologies in Business and Production

Nishant Jaiswal, Anuradha Misra, Alex Khang, and Praveen Kumar Misra

1.1 INTRODUCTION

The first Internet of Things (IoT) gadget was a vending machine with remote temperature monitoring at Carnegie Mellon University in the US that kept Coke cans cold in the summertime 30 years ago (Somayya Madakam, 2015).

The IoT is widespread nowadays. The IoT is a technology that is revolutionizing business processes and generating new revenue streams, from connected cars, fridges, and virtual assistants to applications in healthcare, logistics, and retail that kept supply chains and hospitals operating during the COVID-19 pandemic (Khang et al., 2023a).

However, considering the benefits organizations may experience, adoption rates are now lower than anticipated. The purpose of this chapter is to highlight the use of the IoT in business and production and talk about potential fixes for the problems that are now preventing wider adoption.

1.1.1 WHAT IS THE IoT?

IoT refers to the interconnected network of physical objects (devices, vehicles, equipment, buildings, etc.) that are embedded with sensors, software, and network connectivity, allowing them to collect and exchange data (Jayashree et al., 2024).

The IoT allows these objects to communicate with each other and with external systems over the internet, enabling them to send and receive data and be remotely monitored and controlled.

Examples of IoT devices include smart thermostats, connected appliances, wearable fitness trackers, and industrial equipment with sensors and connectivity. These devices can collect and transmit data on a wide range of variables, such as temperature, humidity, location, and performance. The data collected by IoT devices can be used for a variety of purposes, including improving efficiency, optimizing operations, and creating new products and services (Sakthivel et al., 2022).

DOI: 10.1201/9781003392224-1

1.1.2 What Can the IoT Do in Business and Production?

In business and production, the IoT can be used to improve efficiency, reduce costs, and enhance the quality and safety of products and services (Ammar Rayes, 2022) as Figure 1.1.

Some specific ways in which the IoT can be used in business and production include the following:

1. **Monitoring and maintenance:** Sensors can be used to monitor the performance of equipment and machinery in real time, allowing businesses to identify problems before they occur and prevent downtime.
2. **Optimizing operations:** Data collected by IoT devices can be used to optimize production processes by identifying bottlenecks and inefficiencies.
3. **Improving product quality:** The IoT can be used to monitor and control the quality of products during the manufacturing process, ensuring that they meet standards and specifications.
4. **Enhancing customer service:** IoT-enabled products and services can provide customers with real-time information and support, improving their overall experience.
5. **Creating new products and services:** The data collected by IoT devices can be used to create new products and services, such as connected appliances that can be remotely monitored and controlled through a smartphone app.

1.1.3 Why Is the IoT Important?

The IoT is important because it has the potential to revolutionize the way we live and work by enabling the collection and exchange of data on a vast scale. (Sundmaeker, 2010) Some specific reasons why the IoT is important include the following:

1. **Improved efficiency:** By collecting and analyzing data from a wide range of devices and systems, the IoT can help businesses and organizations identify inefficiencies and optimize their operations.
2. **Cost savings:** The IoT can help businesses reduce costs by automating tasks, lowering energy consumption, and minimizing the need for manual labor.
3. **Enhanced safety:** The IoT can be used to monitor and control safety-critical systems, such as those found in transportation, healthcare, and manufacturing, to reduce the risk of accidents and errors.
4. **Improved decision-making:** The vast amounts of data generated by the IoT can help businesses and organizations make more informed decisions by providing real-time insights and enabling data-driven decision-making.
5. **New business opportunities:** The IoT can create new opportunities for businesses to innovate and differentiate themselves by offering connected products and services that provide customers with additional value.

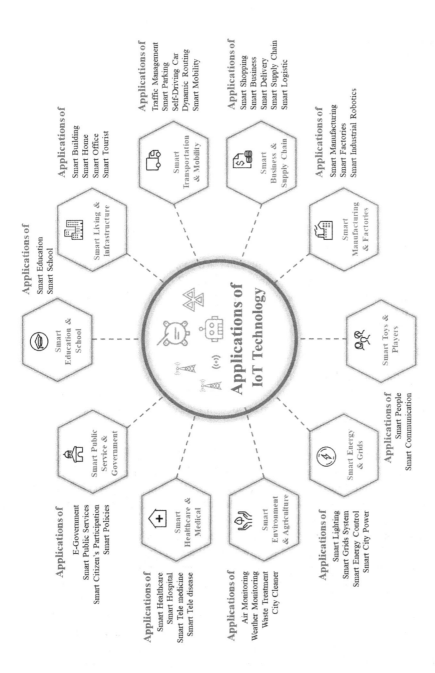

FIGURE 1.1 Applications of IoT technology in business and production.

Source: Khang (2021).

1.2   ROLE OF THE IOT IN BUSINESS AND PRODUCTION

The IoT differs from other technological improvements made over the past 20 years in that it focuses on already-existing nontechnological devices. The IoT may help organizations develop in a variety of areas, where earlier technological advancements like the release of smartphones or 4G presented prospects for rapid expansion (Khan, 2022).

We now concentrate on concrete applications used by actual firms in five sectors that are viewed as having the greatest potential for change in order to demonstrate this potential across industries:

1. Energy
2. Healthcare
3. Logistics
4. Transport
5. Retail

Across sectors, the IoT will particularly affect industries that produce, transport, or sell tangible goods, such as the pharmaceutical and electric car industries. Energy, healthcare, logistics, retail, and transportation are five of these industries that are highlighted in this portion of the research as Figure 1.2.

We connect each industry's potential transformation points with key IoT capabilities in the following subsections.

1.2.1   ENERGY

The IoT has the potential to revolutionize the energy industry by increasing the sustainability and efficiency of energy generation, delivery, and consumption (Hossein Motlagh, 2020). IoT technology may be applied in the energy industry in a variety of ways, including the following:

1. **Smart grids:** The IoT can be used to create smart grids, which are electrical grids that can monitor and control the flow of electricity in real time. Smart grids can help optimize the use of energy by balancing supply and demand and enabling the integration of renewable energy sources such as solar and wind power.
2. **Energy management:** The IoT can be used to monitor and control energy consumption in buildings and industrial facilities, enabling organizations to reduce their energy usage and costs. For example, sensors can be used to monitor the use of lighting and heating and adjust these systems to optimize energy efficiency.
3. **Renewable energy:** The IoT can be used to improve the efficiency and reliability of renewable energy sources, such as by monitoring and controlling the performance of solar panels and wind turbines.
4. **Electric vehicles:** The IoT can be used to optimize the charging and use of electric vehicles, such as by enabling the integration of electric vehicles into the grid and by optimizing charging schedules to reduce the demand on the grid during peak periods.

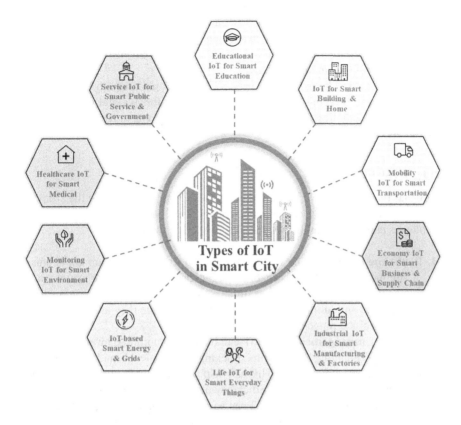

FIGURE 1.2   Types of Internet of Things technology in business and production.

Source: Khang (2021).

1.2.2 HEALTHCARE

The IoT has the potential to transform the healthcare sector by improving the quality, efficiency, and accessibility of healthcare services. Some specific ways in which IoT technologies can be used in the healthcare sector include the following:

1. **Remote patient monitoring:** Wearable devices and sensors can be used to monitor patient health remotely, allowing healthcare providers to track vital signs and detect potential problems early on. This can help improve patient outcomes and reduce the need for hospital visits (Shah et al., 2023).
2. **Electronic health records:** The IoT can be used to create electronic health records that can be accessed by healthcare providers in real time, enabling them to make more informed decisions and provide more coordinated care (Shah and Khang, 2023).
3. **Clinical decision support:** The IoT can be used to provide clinical decision support to healthcare providers, such as by integrating data from a wide range of sources and providing real-time alerts and recommendations.

4. **Medication management:** The IoT can be used to improve the accuracy and efficiency of medication management, such as by using smart pill bottles that can track medication adherence and alert patients and healthcare providers if there are any problems.
5. **Public health:** The IoT can be used to improve public health by enabling the real-time tracking and analysis of disease outbreaks and other health threats and enabling the rapid deployment of resources to respond to these threats (Khang et al., 2024).

1.2.3 LOGISTICS

The IoT has the potential to transform the logistics sector by improving the efficiency and effectiveness of supply chain management. Some specific ways in which IoT technologies can be used in the logistics sector include the following:

1. **Real-time tracking:** The IoT can be used to track the location and status of shipments in real time, enabling organizations to optimize routes and reduce the risk of delays. For example, sensors can be used to monitor the temperature and humidity of shipments to ensure that they are maintained within specified ranges.
2. **Inventory management:** The IoT can be used to optimize inventory management by enabling the real-time tracking of stock levels and the automation of restocking processes. This can help reduce the risk of stock-outs and overstocking and improve the efficiency of operations (Tailor et al., 2022).
3. **Predictive maintenance:** The IoT can be used to predict when equipment and vehicles are likely to require maintenance, enabling organizations to schedule maintenance in advance and reduce the risk of unexpected downtime.
4. **Safety and security:** The IoT can be used to enhance the safety and security of logistics operations, such as using sensors to monitor the performance of vehicles and alert drivers to potential problems.
5. **Customer service:** The IoT can be used to improve customer service by providing real-time tracking and updates to customers on the status of their shipments.

1.2.4 TRANSPORT

The IoT has the potential to transform the transportation sector by improving the efficiency, safety, and sustainability of transportation systems. Some specific ways in which IoT technologies can be used in the transportation sector include the following:

1. **Connected vehicles:** The IoT can be used to create connected vehicles that can communicate with each other and with infrastructure, such as traffic lights and road signs. This can enable the optimization of routes and the reduction of traffic congestion, as well as improve safety by enabling the detection of potential collisions and other hazards.

2. **Public transportation:** The IoT can be used to improve the efficiency and reliability of public transportation, such as using sensors to monitor the performance of buses and trains and optimize routes and schedules. The IoT can also be used to enhance the passenger experience, such as by providing real-time information and enabling the use of mobile tickets.

3. **Fleet management:** The IoT can be used to optimize the management of fleets of vehicles, such as by enabling the tracking and monitoring of vehicles in real time and automating the scheduling of maintenance and repairs.

4. **Logistics:** The IoT can be used to optimize the movement of goods, such as by enabling the tracking and monitoring of shipments in real time and automating the planning of routes and schedules (Hahanov et al., 2022).

5. **Shared mobility:** The IoT can be used to enable the creation of shared mobility services, such as car-sharing and bike-sharing, by enabling the tracking and monitoring of vehicles and optimizing their use.

1.2.5 RETAIL

The IoT has the potential to transform the retail sector by improving the efficiency and effectiveness of operations, enhancing the customer experience, and creating new business opportunities. Some specific ways in which IoT technologies can be used in the retail sector include the following:

1. **Inventory management:** The IoT can be used to optimize inventory management by tracking the movement of goods through the supply chain and enabling real-time monitoring of stock levels. This can help retailers reduce waste, reduce the risk of running out of stock, and improve the efficiency of their operations.

2. **Customer experience:** The IoT can be used to enhance the customer experience by enabling retailers to gather data on customer behavior and preferences and providing personalized recommendations and offers. For example, retailers can use IoT-enabled devices such as smart mirrors in fitting rooms to provide customers with personalized recommendations based on their past purchases and browsing history.

3. **Supply chain optimization:** The IoT can be used to optimize the efficiency of the retail supply chain by enabling real-time monitoring and control of the movement of goods through the supply chain. This can help retailers reduce costs and improve the speed and accuracy of their operations.

4. **Personalized marketing:** The IoT can be used to enable personalized marketing by gathering data on customer behavior and preferences and using these data to target marketing campaigns more effectively.

5. **Fraud detection:** The IoT can be used to detect and prevent fraud in the retail sector by enabling the real-time monitoring of transactions and the detection of unusual activity.

1.3 SMART FACTORIES USING IOT TECHNOLOGY

The IoT is a technology used in industry, and it is used to build smart factories that improve both customer and manufacturer satisfaction. This technology advances the industrial sector thanks to the revolution of the internet era (Zhifeng, 2022). It oversees all production-related divisions, including transportation, inventories, manufacturing, inspection, and quality control, among others.

With sensors, interfaces, actuators, sensors, and other specialized modules, the IoT is used to build smart factories. These devices are set up properly to offer the right data for manufacturing.

The effective operation of these gadgets aids in meeting the primary requirements of industry. For an IoT device to be installed in a production workhouse, an analysis of the implications must be conducted where policies, technology functionality, and deviations are documented and adhered to. We think the IoT will play a big part in business and create "Smarter Cities".

1.4 WHAT ARE THE IOT'S CHALLENGES?

Over the past 20 years, there have been various hype cycles around IoT. To thrive, it needs coordinated technical advancements across industries, from advancing power and lowering chip and sensor prices to wireless communication that provides adequate power, speed, and flexibility (Al-Qaseemi, 2016).

Until now, innovators and enthusiasts have been the early adopters; Al-Qaseemi (2016) claims that moving to the mainstream will be the issue moving ahead.

The IoT presents a number of challenges:

1. **Security:** The connected nature of the IoT makes it vulnerable to cyberattacks, and protecting against these attacks can be a significant challenge. Ensuring the security of IoT devices and systems is essential to prevent the theft of sensitive data and maintain the integrity of connected systems.
2. **Privacy:** The vast amounts of data generated by the IoT raise concerns about privacy, as these data can be used to track and monitor individuals. Ensuring the privacy of individuals is a major challenge for the IoT and requires the implementation of strong privacy protections and the responsible use of data.
3. **Interoperability:** The IoT involves the integration of a wide range of devices and systems from different vendors, which can make it challenging to ensure interoperability between different devices and systems. Ensuring interoperability is essential for the effective functioning of the IoT and requires the development of standards and protocols to enable different devices and systems to communicate with each other.
4. **Complexity:** The IoT involves the integration of a wide range of complex technologies, which can make it challenging for organizations to implement and manage. Ensuring the reliability and stability of IoT systems requires a thorough understanding of the underlying technologies and the ability to manage and maintain these systems effectively.

5. **Regulation:** The rapid development of the IoT has led to a lack of clear regulatory frameworks, which can make it challenging for organizations to navigate the legal landscape and ensure compliance with relevant laws and regulations. Ensuring the development of clear and consistent regulatory frameworks is essential to support the growth of the IoT and to protect the interests of stakeholders.

1.4.1 SECURITY AND PRIVACY

The steering system of a Jeep was hacked, a casino's internet-connected aquarium thermostat was used to access the casino's information, and a baby monitor was used to threaten kidnapping.

Additionally, cyber researchers have illustrated potential vulnerabilities including pacemaker manipulation. Any business that integrates a good or service into an IoT system runs the danger of Internet Protocol (IP) and trade secret theft and exfiltration, and important physical national infrastructures like utilities, gas, electricity, and water can also be threatened by cyber threats (Pal, 2020).

As more customers use the internet, privacy is a related issue of worry. There will be further difficulties when AI and IoT combine, such as balancing the requirement to retain data privacy and confidentiality with the desire to personalize the digital experience. This is where the question of who owns data—the individual or the collector—becomes a key point of contention for the IoT (Khang et al., 2022).

The IoT provides businesses additional authority to gather data from people's homes, consolidate it, and resell it as a new product at a profit. What rights do/should you have to the data being gathered on you or the profits gained from it? To address the privacy and security challenges of the Internet of Things (IoT), the following recommendations can be considered:

1. **Implement strong security measures:** To ensure the security of IoT devices and systems, it is important to implement strong security measures such as encryption, secure authentication protocols, and regular software updates (Hussain et al., 2022).
2. **Follow best practices for data management:** To protect the privacy of individuals, it is important to follow best practices for data management such as collecting only the data that is necessary, obtaining consent before collecting personal data, and securely storing and protecting collected data.
3. **Use privacy-enhancing technologies:** To enhance privacy, it is recommended to use technologies such as anonymization, pseudonymization, and data minimization to protect personal data.
4. **Develop clear and concise privacy policies:** To ensure transparency and accountability, it is important to develop clear and concise privacy policies that outline the data that are being collected, how they are being used, and the rights of individuals regarding their data.
5. **Educate stakeholders:** To ensure that the privacy and security risks of the IoT are understood and addressed, it is important to educate stakeholders about the risks and best practices for managing these risks.

1.4.2 REGULATIONS

One of the challenges of the IoT is the lack of clear regulatory frameworks. The rapid development of the IoT has led to a patchwork of laws and regulations that can be difficult for organizations to navigate (Chatterjee, 2018). This lack of clear regulation can make it challenging for organizations to ensure compliance and create uncertainty for stakeholders.

Additionally, the IoT involves the integration of a wide range of technologies and sectors, which can make it difficult to develop regulatory frameworks that are appropriate for all stakeholders. Ensuring the development of clear and consistent regulatory frameworks is essential to support the growth of the IoT and protect the interests of stakeholders. There are a number of recommendations for addressing the regulatory challenges posed by the IoT:

1. **Develop clear and consistent regulatory frameworks:** To ensure the effective regulation of the IoT, it is essential to develop clear and consistent regulatory frameworks that address the unique characteristics of the IoT and the risks and challenges it presents. This could include the development of sector-specific regulations or the creation of broad-based frameworks that apply to the IoT as a whole.
2. **Establish guidelines for the responsible use of data:** As the IoT generates vast amounts of data, it is important to establish guidelines for the responsible use of these data to ensure that it is used in a way that respects the privacy and rights of individuals. This could include the development of codes of conduct or best practices for the use of IoT data, as well as the establishment of clear rules for the collection, storage, and use of these data.
3. **Promote the development of security standards:** To ensure the security of IoT systems, it is essential to promote the development of security standards that address the unique risks and challenges posed by the IoT. This could include the development of standards for the secure design, development, and operation of IoT systems, as well as the establishment of guidelines for the testing and certification of IoT devices and systems.
4. **Encourage the development of interoperability standards:** To ensure the interoperability of IoT systems, it is important to encourage the development of standards and protocols that enable different devices and systems to communicate with each other. This could include the development of industry-specific standards or the establishment of broader, cross-sector standards that apply to the IoT as a whole.
5. **Foster dialogue between regulators and stakeholders:** To address the regulatory challenges posed by the IoT, it is important to foster dialogue between regulators and stakeholders, including industry, academia, and civil society. This can help ensure that the regulatory framework for the IoT reflects the needs and concerns of all stakeholders and promotes the responsible and sustainable development of the IoT.

1.4.3 DIGITAL INFRASTRUCTURE

According to Stoyanova (2020), one of the main challenges of the IoT is the development of a robust digital infrastructure to support the connectivity and data exchange required by the IoT. Some specific challenges related to the development of this infrastructure include the following:

1. **Network coverage:** Ensuring adequate network coverage to support the widespread deployment of IoT devices can be a challenge, particularly in rural or remote areas.
2. **Bandwidth:** The vast amounts of data generated by the IoT can place a strain on networks, requiring the development of high-capacity networks to support the data transfer needs of the IoT.
3. **Latency:** The real-time nature of many IoT applications requires low-latency networks to enable timely data transfer and response.
4. **Interoperability:** Ensuring the interoperability of different networks and devices is essential for the effective functioning of the IoT and requires the development of standards and protocols to enable different devices and systems to communicate with each other.

To address gaps in digital infrastructure for the IoT, some recommendations include the following:

1. **Invest in infrastructure:** Governments and private companies should invest in the development of high-capacity networks and other infrastructure to support the data transfer needs of the IoT (Hajimahmud et al., 2022).
2. **Promote the deployment of new technologies:** The deployment of new technologies such as 5G and low-power wide-area networks (LPWANs) can help improve network coverage and capacity and support the growth of the IoT.
3. **Develop standards and protocols:** Establishing standards and protocols to ensure the interoperability of different devices and systems is essential for the effective functioning of the IoT. Governments and industry organizations should work together to develop these standards and protocols.
4. **Encourage the adoption of IoT technologies:** Governments and industry organizations can encourage the adoption of IoT technologies by providing incentives and support to businesses and organizations looking to implement the IoT (Khang et al., 2023b).
5. **Invest in research and development:** Investing in research and development to identify new technologies and approaches to improve the coverage, capacity, and interoperability of digital infrastructure is essential to support the growth of the IoT (Jaiswal et al., 2023).

1.5 CONCLUSION

In conclusion, the IoT has the potential to transform the way businesses operate, providing new opportunities for innovation and growth. By leveraging the power of the IoT, businesses can improve efficiency, reduce costs, and enhance the quality and safety of their products and services (Khang et al., 2023c).

Some specific ways in which the IoT can be used in business and production include monitoring and maintaining equipment and machinery, optimizing operations through the collection and analysis of data, improving product quality through the monitoring and control of the manufacturing process, enhancing customer service through the use of IoT-enabled products and services, and creating new products and services through the use of data collected by IoT devices (Khanh and Khang, 2021).

To fully realize the potential of the IoT, businesses must address challenges such as security, privacy, interoperability, complexity, and regulation (Khang et al., 2023d).

REFERENCES

Al-Qaseemi SA. "IoT Architecture Challenges and Issues: Lack of Standardization," *2016 Future Technologies Conference*, 731–738, 2016. https://ieeexplore.ieee.org/abstract/document/7821686/

Ammar Rayes SS. *Internet of Things from Hype to Reality*. Springer, 2022. https://dspace.agu.edu.vn/handle/agu_library/13494

Chatterjee S. "Regulation and Governance of the Internet of Things in India," *Digital Policy, Regulation and Governance*, 399–412, 2018. www.emerald.com/insight/content/doi/10.1108/DPRG-04-2018-0017/full/html

Hahanov V, Khang A, Litvinova E, Chumachenko S, Hajimahmud VA, Alyar AV. "The Key Assistant of Smart City—Sensors and Tools," *AI-Centric Smart City Ecosystems: Technologies, Design and Implementation* (1st Ed.). CRC Press, 2022. https://doi.org/10.1201/9781003252542-17

Hajimahmud VA, Khang A, Hahanov V, Litvinova E, Chumachenko S, Alyar AV. "Autonomous Robots for Smart City: Closer to Augmented Humanity," *AI-Centric Smart City Ecosystems: Technologies, Design and Implementation* (1st Ed.). CRC Press, 2022. https://doi.org/10.1201/9781003252542-7

Hossein Motlagh NM. "Internet of Things (IoT) and the Energy Sector," *Energies*, 13(2), 2020. www.mdpi.com/621434

Hussain SH, Sivakumar TB, Khang A (Eds.). "Cryptocurrency Methodologies and Techniques," *The Data-Driven Blockchain Ecosystem: Fundamentals, Applications, and Emerging Technologies* (1st Ed., pp. 149–164). CRC Press, 2022. https://doi.org/10.1201/9781003269281-2

Jaiswal N, Misra A, Misra PK, Khang A (Eds.). "Role of the Internet of Things (IoT) Technologies in Business and Production," *AI-Aided IoT Technologies and Applications in the Smart Business and Production*. CRC Press, 2023. https://doi.org/10.1201/9781003392224-1

Jayashree M. et al. (Eds.). "Vehicle and Passenger Identification in Public Transportation to Fortify Smart City Indices," *Smart Cities: IoT Technologies, Big Data Solutions, Cloud Platforms, and Cybersecurity Techniques*. CRC Press, 2024. https://doi.org/10.1201/9781003376064-13

Khan IH. "Internet of Things (IoT) in Adoption of Industry 4.0," *Journal of Industrial Integration and Management*, 7(4), 2022. www.worldscientific.com/doi/abs/10.1142/S2424862221500068

Khang A. "Material4Studies," *Material of Computer Science, Artificial Intelligence, Data Science, IoT, Blockchain, Cloud, Metaverse, Cybersecurity for Studies*, 2021. www.researchgate.net/publication/370156102_Material4Studies

Khang A, Gupta SK, Hajimahmud VA, Babasaheb J, Morris G. *AI-Centric Modelling and Analytics: Concepts, Designs, Technologies, and Applications* (1st Ed.). CRC Press, 2023d. https://doi.org/10.1201/9781003400110

Khang A, Gupta SK, Rani S, Karras DA. *Smart Cities: IoT Technologies, Big Data Solutions, Cloud Platforms, and Cybersecurity Techniques* (1st Ed.). CRC Press, 2023b. https://doi. org/10.1201/9781003376064.

Khang A, Hrybiuk O, Abdullayev V, Shukla AK. *Computer Vision and AI-integrated IoT Technologies in Medical Ecosystem* (1st Ed.). CRC Press, 2024. https://doi. org/10.1201/9781003429609

Khang A, Ragimova NA, Hajimahmud VA, Alyar AV. "Advanced Technologies and Data Management in the Smart Healthcare System," *AI-Centric Smart City Ecosystems: Technologies, Design and Implementation* (1st Ed.). CRC Press, 2022. https://doi. org/10.1201/9781003252542–16

Khang A, Rana G, Tailor RK, Hajimahmud VA. *Data-Centric AI Solutions and Emerging Technologies in the Healthcare Ecosystem* (1st Ed.). CRC Press, 2023a. https://doi. org/10.1201/9781032398570

Khang A, Rani S, Gujrati R, Uygun H, Gupta SK. *Designing Workforce Management Systems for Industry 4.0: Data-Centric and AI-Enabled Approaches* (1st Ed.). CRC Press, 2023c. https://doi.org/10.1201/9781003357070

Khanh HH, Khang A. "The Role of Artificial Intelligence in Blockchain Applications," *Reinventing Manufacturing and Business Processes Through Artificial Intelligence* (pp. 20–40). CRC Press, 2021. https://doi.org/10.1201/9781003145011-2

Mahore V, Aggarwal P, Andola N, Venkatesan S. "Secure and Privacy Focused Electronic Health Record Management System Using Permissioned Blockchain," *In 2019 IEEE Conference on Information and Communication Technology* (pp. 1–6). IEEE, December 2019. https://ieeexplore.ieee.org/abstract/document/9066204/

Pal SD. *IoT: Security and Privacy Paradigm*. CRC Press, 2020. www.sciencedirect.com/ science/article/pii/S1877050920305251

Sakthivel M, Gupta SK, Karras DA, Khang A, Dixit CK, Haralayya B. "Solving Vehicle Routing Problem for Intelligent Systems Using Delaunay Triangulation," *2022 International Conference on Knowledge Engineering and Communication Systems (ICKES)*, 2022. https://ieeexplore.ieee.org/abstract/document/10060807/

Shah V, Khang A. "Internet of Medical Things (IoMT) Driving the Digital Transformation of the Healthcare Sector," *Data-Centric AI Solutions and Emerging Technologies in the Healthcare Ecosystem* (1st Ed., p. 1). CRC Press, 2023. https://doi. org/10.1201/9781003356189-2

Shah V, Vidhi T, Khang A. "Electronic Health Records Security and Privacy Enhancement Using Blockchain Technology," *Data-Centric AI Solutions and Emerging Technologies in the Healthcare Ecosystem* (1st Ed., p. 1). CRC Press, 2023. https://doi. org/10.1201/9781003356189-1

Somayya Madakam RR. "Internet of Things (IoT): A Literature Review," *Mumbai: Journal of Computer and Communications*, 2015. www.scirp.org/html/56616_56616.htm

Stoyanova MN. "A Survey on the Internet of Things (IoT) Forensics: Challenges, Approaches, and Open Issues," *IEEE Communications Surveys & Tutorials*, 1191–1221, 2020. https://ieeexplore.ieee.org/abstract/document/8950109/

Sundmaeker HG. "Vision and Challenges for Realising the Internet of Things," *Cluster of European Research Projects on the Internet of Things*, 34–362010. http://docbox.etsi. org/zArchive/TISPAN/Open/IoT/CERP-IOT_Clusterbook_2009.pdf

Tailor RK, Pareek R, Khang A (Eds.). "Robot Process Automation in Blockchain," *The Data-Driven Blockchain Ecosystem: Fundamentals, Applications, and Emerging Technologies* (1st Ed., pp. 149–164). CRC Press, 2022. https://doi.org/10.1201/9781003269281-8

Zhifeng Diao FS. "Application of Internet of Things in Smart Factories Under the Background of Industry 4.0 and 5G Communication Technology," *Mathematical Problems in Engineering*, 8, 2022. www.hindawi.com/journals/mpe/2022/4417620/

2 Application of the Internet of Things within Business Process Management

Lokpriya Gaikwad M, Sandip Kanase, and Vinayak Patil

2.1 INTRODUCTION

The identification, recognizability, and ongoing tracking of products in supply chains have long been troublesome, in view of the heterogeneity of platforms and advancements utilized by different entertainers of the chain. The appearance of the Internet of Things (IoT) and distributed computing brings another methodology, empowering to gather, move, store, and offer.

Existing platforms neglected to take care of the issue in a few central issues: gathering information straightforwardly from sensors consolidated on merchandise for continuous handling and warnings; characterizing a typical strategy and correspondence convention for all partners, overseeing the interoperability between heterogeneous internet technology (IT) foundations for teammates and multi-tenure of heterogeneous heritage applications, making data accessible from cell phones with the goal of that information being processed from a distance and updates made, and dealing with the numerous communications between store network collaborators.

To solve the previously mentioned deficits, we propose planning a cooperative cloud-based stage to help with the information sharing, integration, and handling necessities for logistics goods following information on the logistics stream for better participation and interoperability between inventory network partners.

The added value of this architecture is mainly the integration of different layers of the IoT: the sensor layer, the data transmission layer, the storage layer in the cloud, and, finally, setting the data collected available to users.

And hence, to facilitate information sharing on logistic flows for traceability, collaboration, and interoperability between different actors in the supply chain, these requirements are the key challenges for enterprises in the field of flows management, collaborative supply chains, and future business intelligence.

DOI: 10.1201/9781003392224-2

2.2 UTILIZED TECHNOLOGIES AND MODELING APPROACHES

2.2.1 IoT

The IoT is a development in personal computer (PC) innovation and correspondence that expects to associate articles together by means of the web. Objects mean all that encompasses us and can convey or not.

1. The progression of data and occasions produced by the interconnection of these articles are utilized to work with the following: the board, control, and coordination. Logistics objects and streams are a substantial model (Rani et al., 2021).
2. The mix of heterogeneous advancements and concerns are some of the principal difficulties in accomplishing and benefiting from this new worldview.

2.2.2 Radio-Frequency Identification

Radio-frequency identification (RFID) is made out of a couple of readers/tags. The peruser sends a radio wave; the label thus sends a distinguishing proof casing.

When the chip is controlled, names and labels impart either a TTF (Tag Talk First) convention or an ITF (Interrogator Talk First) convention.

- In a TTF design, the tag sends the first data contained in the chip to the examiner.
- In the ITF mode, the cross-examiner sends a solicitation to the tag, and it meets later.

There are three kinds of tags: passive, active, and semiactive tags. The previous energy is given by the attractive field incited by their readers at the hour of identification. Active labels are taken care of by batteries and can send information without sales from a reader.

The semiactive tag uses a hybrid mechanism; self-powered, it is activated at the request of the tag reader, allowing for a lower power consumption than active tags. The reading distance of RFID chips ranges from a few centimeters to a few meters (10 m) and can go beyond (200 m) with long-range communication technologies.

- Data and information sharing for a collaborative supply chain
- Information and data sharing for a cooperative supply network

In customary production networks and businesses, the executive frameworks lacked in numerous ways since providers did not have worldwide permeability regarding clients' requests and market interest. Request-driven supply organizations are an IT approach empowering coordinated business-to-business efforts and interoperability.

Demand-driven supply networks (DDSNs) suggest information sharing between organization supply chains. By applying this methodology, rather than answering confined clients' requests separately, it would be better if providers would rearrange themselves and work together by sharing more information to more likely answer all market requests (Swaroop, 2012).

DDSNs utilize the draw procedure; for example, the store network is driven by clients' interests by responding, expecting, working together, and coordinating (Davis, 2013). For instance, the stock level is about 8% and could go up to 30%, as indicated by the situation, while reliable data sharing could bring down out-of-stock rates and further develop the demand chain management (Gruen et al., 2002).

2.2.3 GLOBAL SYSTEM OF MOBILE COMMUNICATIONS FRAMEWORK

Two methodologies exist together in business process modeling: the process-driven approach and the artifact-driven approach. In the process-driven approach, the business cycle is addressed by the different treatments or errands that are performed during the business interaction life cycle.

The most encouraging methodology is curio-driven, which consolidates the portrayal of the life cycle of the business substance with a data model that catches information connecting with elements of a similar sort.

The Global System for Mobile Communication (GSM) system is driven by three basic ideas: an achievement that addresses the business's important objective, a gatekeeper that is a condition to set off a phase, and a set of treatments. This stage addresses various ways of accomplishing the business objective (De Masellis, 2013).

2.3 FRAMEWORK DESCRIPTION: COMPONENTS AND TECHNOLOGIES

The worldwide design of the arrangement presents the principal parts of the platform and related advances as Figure 2.1. Without a doubt, we will zero in on chief undertakings, for example, RFID identification, geo-positioning, tracking and tracing, correspondence, transmission, and information sharing of pallets, and gathering logistics objects and moving them in compartments by geo-localized trucks.

2.3.1 IDENTIFICATION AND TRACKING UTILIZING RFID

EPCIS represents the Electronic Product Code Information System, a standard system that empowers partners to exchange and share data (what, where, when, why) about the actual developments and the status of items throughout the supply network.

As represented in Figure 2.2, pallets are outfitted with RFID chips containing item information like the Electronic Product Code (EPC), weight, and a description of the pallet content, item hazards, incongruent items, and so forth.

The main stage of the model named "RFID tag reading and transfer" is activated by two combined events or guards ("Pallet ready for loading" and "Transportation vehicle arrived") as shown in Figure 2.3.

When this stage is activated, the internal stage "tag information reading" is activated automatically with the same guards. The tasks "read pallet RFID tag information" and "send pallet information through General Packet Radio Service (GPRS)" are executed.

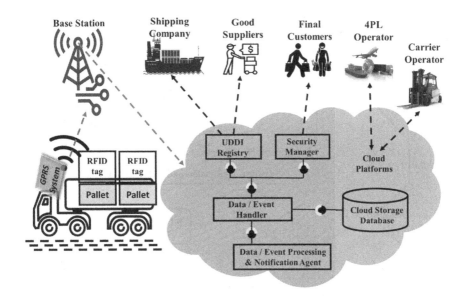

FIGURE 2.1 Global architecture of the platform.

Source: Khang (2021).

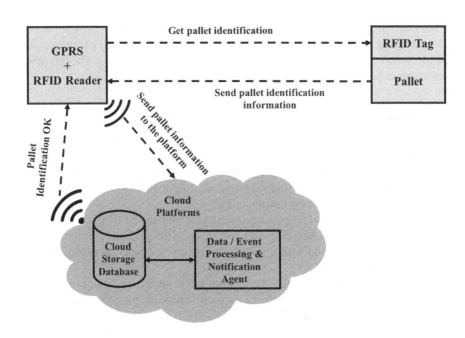

FIGURE 2.2 Transmission using GPRS.

Source: Khang (2021).

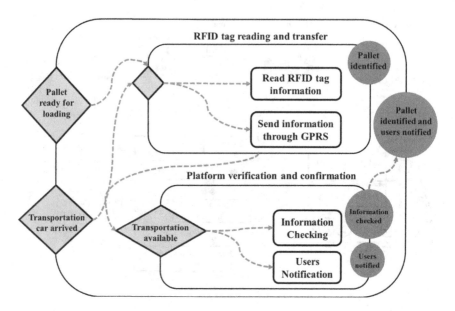

FIGURE 2.3   Global System of Mobile Components model for pallet identification.

Source: Khang (2021).

The internal stage "Platform verification and confirmation" is responsible for two tasks, verifying the read information and notifying related users of the results. After collected, the information must be stored somewhere for various uses. The following section shows how we store and process the collected data.

2.3.2   DATA STORAGE AND REAL-TIME EVENT PROCESSING WITH CLOUD STAGE

The design we proposed for the cloud stage is event-driven, an augmentation of administration-situated engineering (SOA) by event management and processing. The platform is composed of five blocks as we can find in Figure 2.4; the data/event overseer is liable for occasions and messages dealing with coming from heterogeneous end gadgets (RFID, client tablets, or PC).

The cloud storage data set for putting away data sent by the RFID tag or connected with clients. The occasion handling and notice specialist permits to cycle of every single received message and makes notices to applicable clients by producing new messages (Geetha et al., 2024).

A registry administration is utilized to store stage outside buyer administrations to be informed. Finally, a security part assists with encoding messages and oversees access privileges of shared data (Hahanov et al., 2022).

2.3.2.1   Data/Event Handler

The information/event overseer comprises two sections as shown in Figure 2.5: the solicitation controller and the information handling.

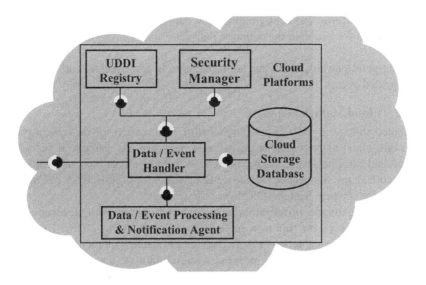

FIGURE 2.4 Cloud platform architecture data/event handler component.

Source: Khang (2021).

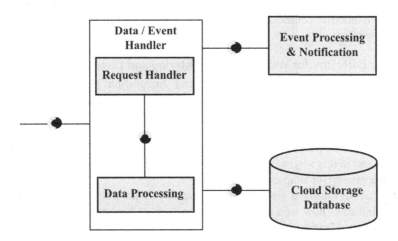

FIGURE 2.5 Illustration of data/event handler components.

Source: Khang (2021).

The solicitation overseer handles events (RFID reader or client solicitation) and moves the message to the Information handling unit that is answerable for handling.

The principal assignments of this unit follow:

1. Decide the solicitation type.
2. Extricate information in the envelope.

3. Confirm information configurations and consistency.
4. Record information in the common data set.

From that point onward, the notice administration is consequently called to illuminate expected clients.

2.3.2.2   Cloud Storage Database

Distributed storage is utilized here to store IoT information for its versatility, and accessibility, and to be available anyplace in any case. As referenced before, the conventional stage has its information spread in a conveyed data set, accessible from various colleagues' Sistema de Gestão de Banco de Dados (SGBDs) or in English is Database Management System (DBMS).

How much-gathered information, security, and convention heterogeneity are the main difficulties for this new methodology (Shah et al., 2023). To perform undertakings, the stage needs to store data about business elements, related business occasions, and clients. We propose a social data set mapping to store data on pretty much all previously mentioned things. Google Cloud structured query language (SQL) stage empowers us to utilize a social data set pattern (Bhambri et al., 2022).

Google will deal with the replication, and the way the executives guarantee execution and accessibility. We know that it isn't the correct method for resolving the issue of high volumetric; consequently, we intended to expand this work with a NoSQL information base.

2.3.2.3   Directory for Web Service Registering and Discovery

The Universal Description Discovery and Integration (UDDI) library is a help catalog in light of XML and uncommonly intended for web administration disclosure. UDDI registry can situate the organization with the ideal web administration.

We utilized a UDDI library on the grounds that our engineering is administration-situated, and in this way, we needed the web benefits that communicated to be registered in a catalog for simple revelations.

This part of the stage gives functionalities to web administration recording. The registry indicates how web administrations and other heritage applications or business cycles could register to be consequently informed when the business substance's setting changes (position, temperature, etc.).

We use for this reason the distribute/buy-in design permitting players to buy into a warning web administration and consequently naturally get cautions when the setting of the business element changes.

Distribute/Buy in design contrasts from other message trade designs on the grounds that only one membership permits an endorser to get at least one occasion notice without sending a request to the service producer (Rotem-Gal-Oz, 2012).

2.3.2.4   Event Processing and Notification

At the point when an event is taken care of (reading an RFID tag, remark on a pallet, updating the GPS facilitates, change of temperature, etc.), this module illuminates

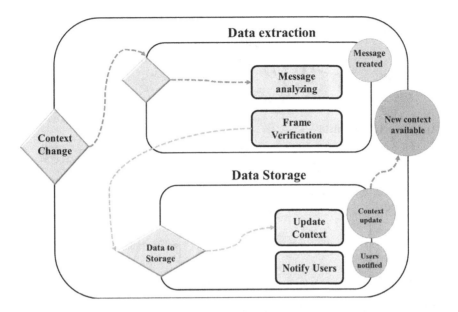

FIGURE 2.6 Event processing and notification process.

Source: Khang 2021).

related clients and partners. These reports are available from the cloud and permit product data access from any gadget (tablet, cell phone, notepad PC).

In this manner, entertainers in the store network who team up to screen the ware can make significant choices progressively, particularly in the event of issues (blocked container, pallet not adjusting to industry guidelines, etc.).

2.3.3 COMMUNICATION AND DATA TRANSMISSION

The communication network is used here for data transmission needs. Many communication technologies have been developed in the literature (Dar et al., 2010). In what follows, we center on the data transmission utilizing two correspondence advancements GPRS and Sigfox network.

A relative investigation of these innovations is introduced.

2.3.3.1 Transmission Using a GPRS Network

As represented in Figure 2.7, we notice the two-way correspondence between the Arduino board with a GPRS correspondence module (GPRS safeguard) and the cloud stage through the GPRS network. Data read from the RFID card is sent to the Arduino board.

The graph in Figure 2.8 shows the casing design. The casing comprises three sections: the first gives data on the EPC information. The Serial Shipping Container Code (SSCC) EPC conspire is utilized to exceptionally recognize a total calculated unit like a bed or delivery holder in coordinated operations.

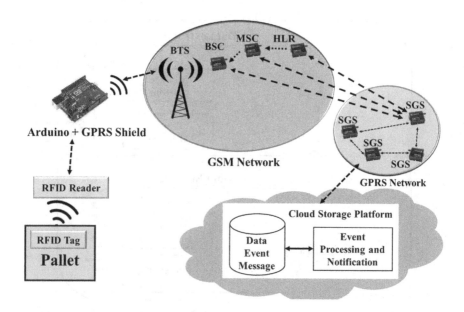

FIGURE 2.7   Data transmission using GPRS.

Source: Khang (2021).

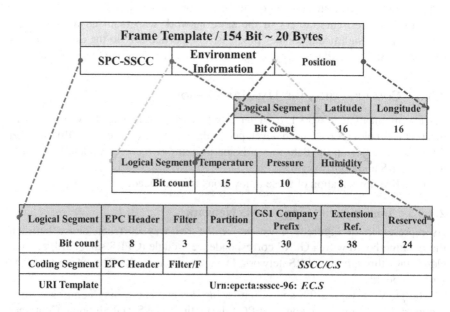

FIGURE: 2.8   Frame template.

Source: Khang (2021).

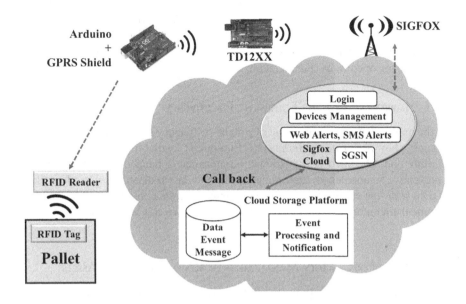

FIGURE 2.9   Data transmission using a Sigfox network.

Source: Khang (2021).

2.3.3.2   Transmission Using a Sigfox Cellular Network

Another way for data transmission is the use of a Sigfox cellular network. In Figure 2.9, we present a Sigfox-based architecture. The Telecom Design (TD) development boards are the only ones able to communicate directly with the SigFox network and the TD cloud platform (sensor web services). It is therefore not possible to provide another card (Arduino, Galileo, Raspberry Py, etc.) directly with the SigFox network currently.

2.3.3.3   Using GPS for Containers Tracking and Positioning

Tracking the position (GPS coordinates) of a commodity, a good, a pallet, or a container is very important, especially in case of loss or theft. Global Navigation Satellite Systems (GNSS) allows the tracking of goods in real time and can guarantee the quality of service (QoS) for transport logistics operators.

Incorporating a GPS sensor thus allows notifying the owner and his collaborators about the good position from the start of loading until delivery to the end customer. For this purpose, we use the signed-degrees format (DDD. dddd) for latitude and longitude coordinates. According to this format, latitude ranges from −90 to 90, and longitude ranges from −180 to 180.

2.4   IMPACT OF IOT SOLUTIONS ON OPERATIONAL EFFICIENCY IN BUSINESS PROCESS MANAGEMENT

The manufacturing industries are undergoing the Fourth Industrial Revolution or Industry 4.0. Technology advancements in manufacturing promoted by sensors, drives, actuators, and robotics (Khang et al., 2024a) have enabled a revolution in areas such as supply chain management, transportation, communication, energy, production, and so on are the few cases in which IoT solutions help in improving manufacturing operations (Khang et al., 2023a).

2.4.1 IMPROVING OVERALL EQUIPMENT EFFECTIVENESS

It is the most common measurement of a plant's efficiency and is a part of every plant manager's key performance indicators. Equipment availability gets impacted due to failure and scheduled or breakdown maintenance (Khang et al., 2023b). Instead of preventive maintenance, requirements can now be predicted by directly collecting data from the equipment, historical maintenance data, and inputs from original equipment manufacturers (OEMs). As a result, maintenance costs are drastically reduced.

2.4.2 ASSET PERFORMANCE MANAGEMENT OF REMOTE OR MOBILE ASSETS

As the asset availability, uptime, and utilization of an asset improve, the operational cost is reduced. The IoT makes this possible, provides a successful result for asset tracking, and collects inputs around operational subsystems (Subhashini and Khang, 2024).

2.4.3 OPTIMIZE SUPPLY CHAIN LOGISTICS AND WAREHOUSE OPERATIONS

A major expense for the consumer goods industry is the logistics and warehouse operations where any cost savings or productivity gains are directly reflected in the bottom line.

An IoT solution integrated with a transport management system provides vehicle visibility and improves planning and utilization that optimizes vehicle usage as shown in Figure 2.10. Early maintenance prediction improves uptime and availability of the vehicles, leading to the early movement of goods and on-time delivery (Shah et al., 2023).

2.5 CONCLUSION

The IoT in business process management is essential from the back end to the front end, creating smart products and campaigning for customers (Khang et al., 2022). IoT support can create value, strategy, innovation, design, and customer service security from the company side. Meanwhile, from the customer side, IoT is very influential on product purchase intentions (Khang et al., 2024b).

The still-limited research related to the IoT in business process management provides excellent opportunities for subsequent research, especially exploring the connection of the IoT with business model innovations, frameworks, development, dark-side behaviors (egoism, Machiavellianism, narcissism, psychopathy, sadism, and spitefulness are all traits that stand for the malevolent dark sides of human personality), demand services, customer preferences, marketers, new-age technology adoption, concepts, networks, and big data (Khang et al., 2023c).

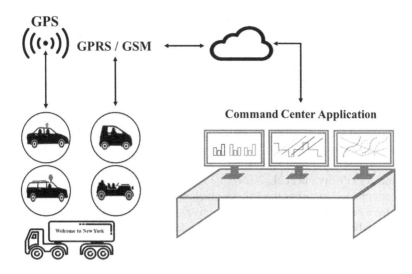

FIGURE 2.10 Role of the Internet of Things in transport management.

Source: Khang (2021).

Today, several megatrends are relevant for business process management within modern factories: globalization, progressing technological evolutions, the dynamism of product life cycles, and resource shortages.

Likewise, other relevant key factors seem to be the acceleration of innovation cycles and the increasing customer demand for individualized mass productions with the highest quality expectations. Within those industrial contexts, IoT projects and applications are being developed in manufacturing, supply chain, supervision and servicing (Khang et al., 2024b).

In addition to the revitalization of the technologies, a major question that arises in all those projects concerns the value and the benefit such an application can bring to the user and the entrepreneur (Hajimahmud et al., 2023).

IoT-based applications bring acceptance and wide use. However, they rely on several features, which are heavily application area–dependent. Therefore, the IoT adoption drives future research toward the many challenges in the international production systems and market demand and supply by creating value in the long term (Jaiswal et al., 2023).

So many examples nowadays clearly show that the main mechanism to create value from IoT technology is to generate actual and refined information from the real world, thus optimizing the technological and business process management based on it (Khang et al., 2023).

Managing data and handling them, as well as extracting relevant information and correlating IoT data with other factory information and processes, will define the achievement of business process-oriented results for IoT industrial applications (Shah et al., 2024).

The purpose of this chapter was to use IoT and cloud computing technologies associated with GNSS (GPRS/GPS) for real-time geo-positioning and tracing of goods. At first, we presented the adopted architecture and explained how the platform works. By introducing these technologies, the emphasis was made on the information-sharing aspect for interoperability and collaboration between involved actors in a logistic flow.

In addition, an artifact-centric-based approach is used to represent the behavior of each system component and model its interaction with other components. The next step is to implement and measure the real impact of this architecture in flows management in terms of optimization, coordination, and QoS. Simple establishment, normalization, setup, and adjusting are vital for saving IoT frameworks' functionality and subsequently offer an incentive for the business cycle of the executives inside each industry (Pritiprada et al., 2024).

Esteem created by using the IoT for mechanical renewal is probably going to be crucial for a business cycle from the executive's perspective. Influencing the utilization of IoT advancements in the business, on a logically higher scale, will be needed before very long.

REFERENCES

Bhambri P, Rani S, Gupta G, Khang A. *Cloud and Fog Computing Platforms for Internet of Things*. CRC Press, 2022. https://doi.org/10.1201/9781003213888

Dar K, Bakhouya M, Gaber J, Wack M, Lorenz P. "Wireless Communication Technologies for ITS Applications," *IEEE Communications Magazine*, 48(5), 156–162; *Physical Review*, 47, 777–780, 2010. https://ieeexplore.ieee.org/abstract/document/5458377/

Davis RA. *Demand-Driven Inventory Optimization and Replenishment: Creating a More Efficient Supply Chain*. Wiley, 2013. www.sas.com/storefront/aux/en/spscddior/66127_excerpt.pdf. Last visit May 2015.

De Masellis R. *The Guard-Stage-Milestone (GSM) for Modeling Artifact-Centric Workflows, Seminars in Software and Services*, 2013. https://dl.acm.org/doi/abs/10.1145/2002259.2002270

Geetha C, Neduncheliyan S, Khang A (Eds.). "Dual Access Control for Cloud Based Data Storage and Sharing," *Smart Cities: IoT Technologies, Big Data Solutions, Cloud Platforms, and Cybersecurity Techniques*. CRC Press, 2024. https://doi.org/10.1201/9781003376064-17

Gruen TW, Corsten D, Bharadwaj S. *Retail Out of Stocks: A Worldwide Examination of Causes, Rates, and Consumer Responses*. Grocery Manufacturers of America, 2002. www.emerald.com/insight/content/doi/10.1108/09590550310507731/full/html?fullSc=1

Hahanov V, Khang A, Abbas GL, Hajimahmud VA. "Cyber-Physical-Social System and İncident Management," *AI-Centric Smart City Ecosystems: Technologies, Design and Implementation* (1st Ed.). CRC Press, 2022. https://doi.org/10.1201/9781003252542-2

Hajimahmud VA. et al. (Eds.) "The Role of Data in Business and Production," *AI-Aided IoT Technologies and Applications in the Smart Business and Production*. CRC Press, 2023. https://doi.org/10.1201/9781003392224-2

Jaiswal N, Misra A, Misra PK, Khang A (Eds.). "Role of the Internet of Things (IoT) Technologies in Business and Production," *AI-Aided IoT Technologies and Applications in the Smart Business and Production*. CRC Press, 2023. https://doi.org/10.1201/9781003392224-1

Khang A. "Material4Studies," *Material of Computer Science, Artificial Intelligence, Data Science, IoT, Blockchain, Cloud, Metaverse, Cybersecurity for Studies*, 2021. www.researchgate.net/publication/370156102_Material4Studies

Khang A, Abdullayev V, Hahanov V, Shah V. *Advanced IoT Technologies and Applications in the Industry 4.0 Digital Economy* (1st Ed.). CRC Press, 2024b. https://doi.org/10.1201/9781003434269

Khang A, Gupta SK, Rani S, Karras DA (Eds.). *Smart Cities: IoT Technologies, Big Data Solutions, Cloud Platforms, and Cybersecurity Techniques*. CRC Press, 2023c. https://doi.org/10.1201/9781003376064

Khang A, Misra A, Abdullayev V, Eugenia L. *Machine Vision and Industrial Robotics in Manufacturing: Approaches, Technologies, and Applications* (1st Ed.). CRC Press, 2024a. https://doi.org/10.1201/9781003438137

Khang A, Rani S, Gujrati R, Uygun H, Gupta SK (Eds.). *Designing Workforce Management Systems for Industry 4.0: Data-Centric and AI-Enabled Approaches*. CRC Press, 2023a. https://doi.org/10.1201/9781003357070

Khang A, Rani S, Sivaraman AK. *AI-Centric Smart City Ecosystems: Technologies, Design and Implementation* (1st Ed.). CRC Press, 2022. https://doi.org/10.1201/9781003252542

Khang A, Shah V, Rani S. *AI-Based Technologies and Applications in the Era of the Metaverse* (1st Ed.). IGI Global Press, 2023b. https://doi.org/10.4018/9781668488515

Nait-Sidi-Moh A, Bakhouya M, Gaber J, Wack M. *Geopositioning and Mobility* (p. 272). Wiley-ISTE, Networks and Telecommunications Series, 2013. ISBN: 978-1-84821-567-2. www.sciencedirect.com/science/article/pii/S1877050915017329

Pritiprada P, Satpathy I, Patnaik BCM, Patnaik A, Khang A (Eds.). "Role of the Internet of Things (IoT) in Enhancing the Effectiveness of the Self-Help Groups (SHG) in Smart City," *Smart Cities: IoT Technologies, Big Data Solutions, Cloud Platforms, and Cybersecurity Techniques*. CRC Press, 2024. https://doi.org/10.1201/9781003376064-14

Rani S, Bhambri P, Kataria A, Khang A, Sivaraman AK. *Big Data, Cloud Computing and IoT: Tools and Applications* (1st Ed.). Chapman and Hall/CRC Press, 2023. https://doi.org/10.1201/9781003298335

Rani S, Chauhan M, Kataria A, Khang A (Eds.). "IoT Equipped Intelligent Distributed Framework for Smart Healthcare Systems," *Networking and Internet Architecture*. CRC Press, 2021. https://doi.org/10.48550/arXiv.2110.04997

Rotem-Gal-Oz A. *SOA Patterns*. Manning Publications Co., September 2012. https://dl.acm.org/doi/abs/10.1145/3147704.3147734

Shah V, Jani S, Khang A (Eds.). "Automotive IoT: Accelerating the Automobile Industry's Long-Term Sustainability in Smart City Development Strategy," *Smart Cities: IoT Technologies, Big Data Solutions, Cloud Platforms, and Cybersecurity Techniques*. CRC Press, 2024. https://doi.org/10.1201/9781003376064-9

Shah V, Khang A. "Internet of Medical Things (IoMT) Driving the Digital Transformation of the Healthcare Sector," *Data-Centric AI Solutions and Emerging Technologies in the Healthcare Ecosystem* (1st Ed., p. 1), CRC Press, 2023. https://doi.org/10.1201/9781003356189-2

Shah V, Vidhi T, Khang A. "Electronic Health Records Security and Privacy Enhancement using Blockchain Technology," *Data-Centric AI Solutions and Emerging Technologies in the Healthcare Ecosystem* (1st Ed., p. 1). CRC Press, 2023. https://doi.org/10.1201/9781003356189-1

Subhashini R, Khang A (Eds.). "The Role of Internet of Things (IoT) in Smart City Framework," *Smart Cities: IoT Technologies, Big Data Solutions, Cloud Platforms, and Cybersecurity Techniques*. CRC Press, 2024. https://doi.org/10.1201/9781003376064-3

Swaroop A. "Designing a Demand Driven Supply Network," *LBS Journal of Management & Research*, 10(2), 35–40, 2012. www.indianjournals.com/ijor.aspx?target=ijor:lbsjmr&volume=10&issue=2&article=005

3 Internet of Things–Generated Data in Business and Production

Alex Khang, Vugar Abdullayev, Eugenia Litvinova, Svetlana Chumachenko, Triwiyanto, and Vusala Abuzarova

3.1 INTRODUCTION

Data now play a crucial role in both business and industrial procedures. Businesses and manufacturers are using data to inform decisions and achieve a competitive advantage as a result of the advent of digital technology and the expanding availability of data (Khang et al., 2023b).

In today's fast-paced and data-driven corporate climate, the capacity to gather, manage, and analyze from has emerged as a critical success factor (Hajimahmud et al., 2023). Businesses and manufacturers may improve decision-making, lower costs, and optimize operations by using data properly. But there are hazards involved with data security and privacy, and the process of gathering and maintaining data can be difficult, especially data generated from the Internet of Things (IoT) and Internet of Everything (IoE) ecosystems (Rani et al., 2023).

We discuss the use of data in business and production in this session, emphasizing the advantages of data-driven decision-making, the difficulties in gathering and handling massive volumes of data, and the significance of data security and privacy (Rani et al., 2023). We also give instances from the real world of how manufacturers and companies are using data to enhance their processes and acquire a competitive edge, as shown in Figure 3.1.

3.1.1 DEFINITION OF DATA

A collection of unprocessed facts, statistics, or other types of information that can be measured, observed, or recorded is referred to as data as shown in Figure 3.2.

Data are essential in business and industrial environments for gaining insight into important elements that affect operations. Companies can learn more about customer behaviors, product performance, and other important aspects that have an impact on their operations by gathering and analyzing data (Wikipedia, 2013).

Because it may be used to guide decisions, data are important in business and manufacturing. For instance, data about customer behavior and preferences can be utilized to inform marketing tactics, product development, and customer support

DOI: 10.1201/9781003392224-3

FIGURE 3.1 Gathering and handling massive volumes of data.

FIGURE 3.2 Tools for processing and visualizing the data.

(Glass and Callahan, 2014). Similar information can be utilized to pinpoint problem areas and streamline processes, including information on production outputs, machine performance, and other elements. Financial information, such as revenue, costs, and profitability, can help investors and forecasters make more informed financial decisions (MajestEYE, 2023).

Data are necessary for monitoring and managing processes in production. Large volumes of data are produced by machine sensors and other monitoring devices, and these data can be used to track the functioning of the equipment, spot problems or abnormalities, and find areas for improvement (Vaughan, 2023).

For instance, a production facility may utilize sensor data to track the operation of its machines and spot patterns of failure (Khang et al., 2022a). This enables the business to solve problems before they become serious and result in downtime.

The use of data in predictive maintenance enables businesses to plan maintenance or repairs before equipment breaks down. This strategy lowers maintenance costs, decreases unscheduled downtime, and increases machine longevity. In some instances, businesses build a digital twin of their production processes using information from several sources (Davenport and Harris, 2007).

A digital twin is a virtual representation of the actual manufacturing process that may be used to simulate events and spot optimization opportunities (Provost and Fawcett, 2013). There are numerous instances of data being used in business and production settings. Here are a few typical instances:

1. **Information about the customer:** Information about consumer behaviors, preferences, and demographics is gathered in order to design new goods that cater to the needs of the market.
2. **Inventory levels:** Information on sales and inventory levels can be used to improve inventory control and make sure that goods are available when customers wish to buy them (Briney, 2015).
3. **Production outputs:** Information on production outputs, including the quantity, pace, and quality of the items produced, can be utilized to spot areas for process optimization and efficiency growth.
4. **Financial data:** Financial information can be used to guide financial predictions and investment decisions. Financial information includes income, expenses, and profitability.
5. **Machine performance:** It is possible to improve production procedures and cut maintenance costs by using data on machine performance, which includes measures like downtime, cycle time, and energy usage.
6. **Data about suppliers:** Performance, lead times, and quality can be utilized to streamline procedures and cut costs in the supply chain.
7. **Social media data:** Information gathered from social media sites can be utilized to gauge customer mood, spot trends, and develop niche marketing strategies.

These are only a few types of the numerous data types that are employed in commercial and industrial environments. Data are a crucial instrument for decision-making and process improvement, and modern businesses' performance depends heavily on their capacity for efficient data collection and analysis.

The process of employing statistical and computational approaches to draw conclusions and information from data is known as data analysis. Data analysis is used in commercial and manufacturing contexts to turn raw data into information that can guide decision-making. Here are a few instances of data analysis and decision-making using data (EMC Education Services, 2015):

1. **Analyzing** historical data in descriptive analytics helps us understand what happened in the past. Finding patterns and trends in data, such as changing consumer purchasing behaviors or production output over time, is possible

FIGURE 3.3 Benefits of data in several industries.

with this sort of analysis. Dashboards and reports that offer insights into previous performance and assist in decision-making can be produced using descriptive analytics.

2. **Analytics** for making predictions about the future: Analytics for making predictions about the future uses past data to build models. For instance, a business may utilize predictive analytics to project upcoming revenues based on previous sales data and other variables. Resource allocation and strategic planning can both benefit from predictive analytics.

3. **Prescriptive analytics** determines the most effective course of action to accomplish a specific goal using data. Prescriptive analytics, for instance, can be used to improve production processes by determining the most effective order of jobs or the ideal time for equipment maintenance. Operational choices can be supported by prescriptive analytics, which also increases effectiveness.

Data visualization is the practice of presenting data in a visual format to facilitate understanding and interpretation (Rani et al., 2021). Decision-makers can be given insights by using charts, graphs, and other visualizations to spot patterns and trends in data as shown in Figure 3.3.

3.1.2 The Benefits of Data in Business and Production

Companies that efficiently use data in both business and production stand to gain a lot from doing so (Stitch, 2023).

Some of the main advantages of using data are as follows:

1. **Better decision-making:** Data analysis offers useful information and insights that decision-makers can use to guide their decision-making processes. Businesses may minimize their risk of failure and avoid expensive mistakes by using data to inform their decisions.

2. **Enhanced efficiency:** Inefficient business and production processes can be found using data. Companies can find places where processes can be enhanced, lowering waste and raising efficiency, by evaluating data (Venkataramanan and Shriram, 2016).

3. **Better customer service:** Marketing efforts and product offerings can be made more individualized and targeted by using data on consumer behavior and preferences. Businesses may improve customer service and boost customer satisfaction by better understanding their customers.

4. **Cost savings:** Businesses can cut expenses associated with squandered resources and ineffective operations by using data to optimize business and production processes. Data can also be used to guide investment choices and financial predictions, lowering the likelihood of financial losses.

5. **Enhanced product quality:** Businesses can find opportunities to streamline processes and reduce product faults by examining production data. This may lead to higher-quality products and lower costs for returns and defective goods.

6. **Advantage over competitors:** Businesses that effectively use data have a distinct advantage over those that do not. Companies may keep ahead of industry trends and react rapidly to changes in the business environment by using data to make informed decisions (Khang et al., 2023c).

Here are some particular instances of how data has been applied to attain these advantages in real-world scenarios.

3.1.2.1 Better Decision-Making

Netflix makes decisions on which series to develop and how to advertise them based on data as shown in Figure 3.4. The popularity of *House of Cards* could be predicted thanks to data on user viewing patterns; thus, Netflix decided to invest in its development. *House of Cards* became one of Netflix's most popular series as a result of this choice (Davies, 2013).

3.1.2.2 Enhanced Efficiency

Amazon uses data to streamline its supply chain and cut waste, as shown in Figure 3.5. Amazon is able to pinpoint areas where it may streamline operations and cut costs

FIGURE 3.4 Useful information and insights that decision-makers can use to guide their decision-making processes.

Amazon

FIGURE 3.5   Inefficient business and production processes.

Starbucks

FIGURE 3.6   Using data on consumer behavior and preferences.

Walmart

FIGURE 3.7   Using data to optimize business and production processes.

by evaluating data on inventory levels, purchase history, and shipping schedules. In order to reduce the need for expensive storage space, Amazon, for instance, utilizes data to forecast which products will sell during peak seasons and adjusts its inventory levels appropriately (Mrkonjić, 2023).

3.1.2.3   Better Customer Service
Starbucks creates tailored offers for its customers using data, as shown in Figure 3.6. Starbucks is able to develop targeted offers and rewards that are customized to each individual customer by evaluating data on customer behavior and preferences. Due to this, Starbucks' customers are happier and more loyal (Sharaf, 2023).

3.1.2.4   Cost Savings
Walmart uses data to optimize its supply chain and reduce costs, as shown in Figure 3.7. By analyzing data on inventory levels and shipping times, Walmart is able to identify areas where it can reduce waste and inefficiencies. For example, Walmart uses data to determine the most efficient routes for its trucks to take, reducing fuel costs and transportation time (Thomas, 2023).

Tesla

FIGURE 3.8   Higher-quality products and lower costs for returns and defective goods.

Amazon

FIGURE 3.9   Changes in the business environment by using data to make informed decisions.

3.1.2.5   Enhanced Product Quality

Tesla uses data to raise the standard and security of its cars as shown in Figure 3.8. Tesla can find places where its design might be improved to lower the risk of accidents by evaluating data on car performance and crashes. For instance, Tesla uses data to enhance its autopilot system and lower the chance of collisions brought on by human mistakes (Tesla, 2023).

3.1.2.6   Competitive Advantage

In the e-commerce sector, Amazon leverages data to acquire a competitive advantage as shown in Figure 3.9. Amazon is able to develop targeted marketing campaigns and individualized recommendations that increase sales by evaluating data on customer behavior and interests. As a result, Amazon has grown to be among the most prosperous e-commerce businesses worldwide (Mrkonjić, 2023).

Here are some pertinent figures and studies that demonstrate the advantages of data in business and production (Jeffery, 2010; AWS, 2023):

1. **Better decision-making:** The global market for marketing analytics is currently around $3.2 billion and is predicted to grow to $6.4 billion by 2026, according to Business Wire. The data analysis has allowed companies to identify trends on social networks as well as generate consumer insights and improve strategies to reach consumers according to their concerns, affinities, and habits (Herrera, 2022).
2. **Enhanced efficiency:** A better way is to build an appreciation of data-driven decision-making gradually. This means introducing it as part of a

commitment to continuous improvement—an ongoing, staged process rather than an abrupt, one-off step change (Sher, 2022).

3. **Better customer service:** By using advanced data analytics, you can understand your consumer base, deliver on or exceed expectations, and proactively identify opportunities for improvement. Your business can establish a connection with customers that has the ability to stand the test of time (Phocas Software, 2023).

4. **Cost savings:** Companies with efficient supply chains reach 5–15% lower supply chain costs, 20–50% less inventory holdings, and up to 3X cash-to-cash cycle speeds. The breadth of data unmasks hidden players within the supply chain, connecting them to create a cohesive perspective. This thereby circumvents the problem that 76% of business leaders face (Mi, 2022).

5. **Enhanced product quality**: The efficiency of your production operation and value of your administrative activities can be directly influenced by the quality of your information and how well it is shared and understood the important of data (Frahm, 2021).

6. **Competitive advantage**: According to a study by the MIT Sloan Management Review, businesses that employ data analytics to obtain a competitive edge are twice as likely to perform financially in the top quartile of their respective industries (Crossan et al., 2022).

3.2 DATA COLLECTION AND MANAGEMENT

In order to ensure that data is successfully gathered, processed, and used to guide decision-making, the process of data collection and management in commercial and industrial contexts requires a number of processes.

Some of the crucial processes in this procedure follow:

1. **Define the data needs:** Identify the types of data required to achieve business objectives as the first step in the data collecting and management process. This includes deciding on the most important inquiries that must be addressed, the data sources that will be used, and the procedures for gathering and analyzing the data.

2. **Collect the data:** After the data requirements have been established, the data must be gathered. Numerous techniques, including surveys, interviews, observation, and the monitoring of automated systems, can be used to accomplish this.

3. **Process the data:** Data processing is necessary to make sure that the data are correct, full, and dependable after it has been acquired. This includes cleaning the data, getting rid of duplicates, and fixing any discrepancies.

4. **Store the data:** After processing, the data must be kept in a safe place that is also easily accessible. Databases, data warehouses, or cloud-based storage options can all be used.

5. **Analyze the data:** After the data has been saved, it must be examined in order to draw conclusions and guide decision-making. Several methods, including data visualization, statistical analysis, and machine learning algorithms, can be used to do this.

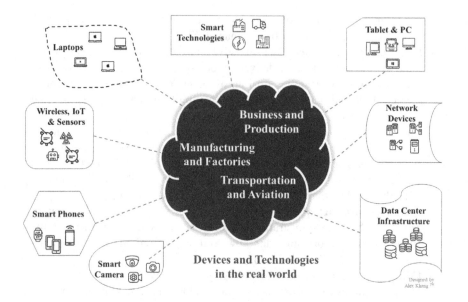

FIGURE 3.10   Several devices and technologies in the real world.

Source: Khang (2021).

6. **Communicate the findings:** Last but not least, relevant stakeholders must be informed of the findings of the data analysis in a concise and useful manner. In order to do this, reports and dashboards that offer a succinct summary of the conclusions and suggestions for action must be created.

For businesses and production processes to obtain a competitive advantage and make wise decisions, effective data gathering and administration are crucial, as shown in Figure 3.10. Businesses may make sure they are gathering accurate and trustworthy data, processing it efficiently, and using it to guide decision-making by taking the actions outlined earlier.

To reach their goals and objectives, businesses and production processes depend on accurate and trustworthy data gathering and administration (Rani et al., 2021). Here are a few explanations:

1. **Informed decision-making:** Making informed judgments is possible thanks to accurate and trustworthy data, which offer knowledge and insights. Decision-makers may choose poorly if they base their decisions on inadequate or inaccurate data in the absence of accurate data.
2. **Improved performance:** Businesses and production processes can discover areas for improvement and optimize their processes to boost productivity, cut costs, and enhance overall performance with the use of reliable data.
3. **Competitive advantage:** Having access to precise and trustworthy data can provide firms an edge over their rivals by giving them knowledge of market trends, consumer behavior, and rivalry activities. Using this knowledge,

business owners may create plans that keep them one step ahead of the competition.

4. **Risk management:** Reliable data may assist firms in identifying and reducing risks like supply chain interruptions, problems with regulatory compliance, and financial instability.

5. **Compliance**: Accurate and trustworthy data collection and administration are necessary for numerous businesses to meet legal and regulatory obligations. Failure to comply can result in penalties, legal action, and reputational damage.

To gather and handle data in commercial and production contexts, a variety of tools and strategies are employed (Stedman, 2023). Here are a few examples:

1. **Sensors:** In a manufacturing environment, sensors can be used to gather information about the machinery, temperature, pressure, and other elements that affect the way things are made. Utilizing this information will improve output, cut down on waste, and guard against equipment breakdowns.

2. **Databases:** Structured databases are used to store and manage massive volumes of data. This can include information about customers, inventories, sales, and more. Using sophisticated software tools, databases may be accessed and examined.

3. **Data analysis software:** To analyze and visualize data, software applications like Excel, Tableau, and Power BI are utilized. With the use of these technologies, organizations may find patterns, trends, and insights in their data and base decisions on them.

4. **IoT devices:** IoT devices, such as smart sensors and cameras, can be used to gather information on a variety of factors, including temperature, humidity, and motion. This information can be utilized to streamline production procedures, enhance quality assurance, and cut down on energy usage.

5. **Cloud computing:** Scalable, secure, and economical solutions to store and manage data are made available to enterprises by cloud computing services like Amazon Web Services (AWS) and Microsoft Azure. Businesses may have access to robust data analysis and machine learning tools thanks to cloud computing (Simplilearn, 2023).

3.3   DATA ANALYTICS AND DATA VISUALIZATION

In business and production contexts, data analytics and visualization are crucial because they offer a mechanism to transform unstructured data into insightful information that can guide decisions (Simranssonu19, 2023) as shown in Figure 3.11.

The following list of factors illustrates the significance of data analytics and visualization:

1. **Identify trends and patterns:** Businesses can spot trends and patterns that would be hard to find otherwise by examining data. This can aid firms in predicting future outcomes with greater accuracy and spotting development prospects (Khang et al., 2023d).

FIGURE 3.11   Data analytics and data visualization.

2. **Optimize processes:** Data analytics can be used to find inefficient regions in supply chains, production processes, or other company activities. This enables companies to make data-driven decisions that increase productivity, lower costs, and boost profitability.

3. **Measure success:** Success can be measured by using data analytics, which gives corporate strategies and activities a mechanism to be evaluated. Businesses can determine whether their efforts are yielding fruitful results by monitoring key performance indicators and modifying their strategy as necessary.

4. **Improve decision-making:** Businesses may communicate complex data in an understandable way by using data visualization tools. This makes it possible for decision-makers to recognize patterns and trends rapidly and take action based on the information supplied (Khang et al., 2023e).

5. **Stay competitive:** Companies that do not use data analytics and visualization risk slipping behind their rivals in today's data-driven business environment. Businesses can acquire a competitive edge and make wiser decisions that promote growth and success by utilizing the power of data.

Here are some particular instances of how data analytics and visualization can be used in business and production environments to obtain insights and make wise decisions:

1. **Sales forecasting:** By analyzing historical sales data, data analytics can be utilized to predict future sales trends. Businesses may decide wisely on inventory management, production planning, and marketing campaigns by visualizing these data in charts and graphs.

2. **Quality control:** Data analytics can be used for quality control to track production output and find errors or problems with the product's quality. Production managers can immediately spot issues by displaying this data in real-time dashboards and taking corrective action.

3. **Supply chain optimization:** Data analytics can be used to monitor inventory levels, delivery schedules, and other important supply chain indicators. Businesses may pinpoint bottlenecks and streamline their supply chains for increased effectiveness and cost savings by visualizing these data in heat maps and other visualizations.

4. **Customer behavior analysis:** Data analytics can be used to monitor consumer behavior, including online shopping patterns and website usage. Businesses can learn more about consumer preferences and make wise decisions about marketing and product development by visualizing this data in customer journey maps and other visualizations.

5. **Predictive maintenance:** By analyzing equipment performance data, data analytics can be utilized to foretell when maintenance is necessary. Production managers may schedule maintenance in advance, cutting downtime and raising productivity, by visualizing this data in predictive maintenance dashboards.

These are just a few instances of how data analytics and visualization may be used in business and production environments to obtain insights and make wise decisions. Businesses can open new doors for success and growth by utilizing the power of data.

In commercial and production contexts, a variety of tools and approaches for data analytics and visualization are frequently employed. Here are a few illustrations:

1. **Descriptive analytics:** Analyzing historical data in descriptive analytics helps us understand what has happened in the past. Charts, tables, and graphs can be utilized to visualize this type of information, which is frequently used for dash-boarding and reporting.

2. **Diagnostic analytics:** Analysis of data is used in diagnostic analytics to determine why something occurred in the past. This kind of analysis, which may be depicted using cause-and-effect diagrams and other visuals, is frequently used to pinpoint the sources of issues.

3. **Predictive analytics:** Predictive analytics is used for making predictions about the future Predictive models and other visualizations can be used to make this type of analysis, which is frequently used for forecasting and predictive maintenance, more understandable.

4. **Prescriptive analytics:** Prescriptive analytics entails data analysis to ascertain the optimal action to take in a specific circumstance. Simulators and other visual aids can be utilized to illustrate this type of analysis, which is frequently used for decision-making and optimization.

In addition to these kinds of analytics, data visualization can be done using a variety of tools and methods. These include the following:

1. **Tools for business intelligence (BI):** BI tools are computer programs that analyze and display data. They frequently have elements like dashboards, reports, and charts.

2. **Data visualization libraries:** Libraries comprising pre-built visualization components for use in the creation of customized visualizations are known

as data visualization libraries. D3.js, Highcharts, and Chart.js are a few examples.

3. **Geographic information systems (GISs):** GIS applications are made to analyze and display spatial data. Applications like mapping, environmental monitoring, and urban planning can all be done with them.

4. **Machine learning algorithms:** Machine learning algorithms can be used for data analysis and visualization, especially in applications like fraud detection and predictive maintenance.

These are just a few examples of the various kinds of data analytics and visualization tools and techniques that are frequently utilized in business and production contexts (Kirk, 2016).

The type and amount of data being analyzed, the analysis's objectives, and the resources at hand all play a role in selecting the best tools and methodologies for a given application (Healy, 2018).

3.4 DATA SECURITY AND DATA PRIVACY

Any company or production environment where sensitive or secret data are gathered, kept, and analyzed must prioritize data security and privacy (Khang et al., 2022b).

Businesses frequently hold sensitive and private information about their clients, staff, and business operations; therefore, it is critical to safeguard this information from theft, misuse, and unauthorized access, as shown in Figure 3.12.

Compliance with rules and regulations that control the processing of sensitive data, such as the General Data Protection Regulation in the European Union or the California Consumer Privacy Act in the United States, is one of the primary motivations for guaranteeing data security and privacy. These laws oblige businesses to

FIGURE 3.12 Illustration of data privacy security.

take steps to safeguard client information, stop data breaches, and alert them if their data are compromised.

Data privacy and security are important for establishing confidence with clients, suppliers, and business partners in addition to complying with the law (Khanh and Khang, 2021). Customers and partners, who are getting more worried about the security and privacy of their data, are more inclined to choose and stay with businesses that can show a commitment to protecting sensitive data.

Additionally, data breaches can have detrimental effects on a company's finances and reputation. For instance, a data breach may result in the loss of private information, such as financial and personal information, as well as monetary losses through theft, fraud, or fines from the authorities, it can also harm a business's reputation, lose customer confidence, and trigger legal action or lawsuits.

Companies utilize a variety of technologies and procedures, such as encryption, access restrictions, firewalls, and data loss prevention (DLP) software, to guarantee data security and privacy. Additionally, businesses develop policies and processes for managing and storing sensitive data and train staff on best practices for data protection. For organizations and production environments, critical data breaches and unauthorized access can have dire repercussions. These dangers consist of the following:

1. **Financial loss:** Theft or fraudulent activity caused by data breaches may result in financial losses. For instance, hackers may obtain credit card numbers or personal information, which they may subsequently utilize for illegal actions like identity theft.
2. **Loss of reputation:** Data breaches can damage the reputation of a company or production setting, leading to loss of customer trust and, eventually, loss of revenue. Customers might stop doing business with a company after a data breach, particularly if the incident resulted in the loss of sensitive personal data.
3. **Legal and regulatory repercussions:** Businesses that experience data breaches may be subject to penalties, legal action, and other legal and regulatory repercussions. In addition to regulatory sanctions for failing to protect consumer data, businesses may be held accountable for the losses brought on by the data breach.
4. **Operations disruption:** Data breaches can lead to considerable downtime and productivity losses in business operations. To investigate and lessen the effects of the data breach, businesses might need to temporarily stop operations.
5. **Theft of intellectual property:** In industrial settings, unauthorized access to data can lead to the theft of proprietary information, such as trade secrets, formulas, and other sensitive data. Financial loss, harm to the company's reputation, and a loss of competitive advantage might result from this.

Businesses and production environments must employ efficient data security measures, such as strict access controls, encryption, and data loss prevention techniques, to reduce these risks.

Additionally, they must keep abreast of the most recent security risks and adopt a proactive security approach by routinely monitoring and analyzing their systems for vulnerabilities (Andress, 2014).

Having a response strategy in place can also aid businesses in minimizing the harm brought on by a data breach and facilitating a speedy recovery. Here are some instances of good data security and privacy practices in commercial and production settings (DataPrivacyManager, 2021).

1. **Encryption:** One of the finest methods for securing sensitive data is encryption. Data must be transformed into an unintelligible format that can only be cracked with a unique key. Data in transit, such as emails or data being moved between computers, are frequently encrypted.
2. **Secure data storage:** Secure data storage is necessary to guard against illegal access. This can entail putting data on secure storage media, like encrypted hard drives, or in a trustworthy provider's secure cloud.
3. **Access controls:** Since they restrict who has access to sensitive data, access controls are a crucial component of data security. As part of this, strong password guidelines, multiple-factor authentication, and role-based access controls must be implemented (Pınarbaşı, 2023).
4. **Regular backups:** Backups should be made on a regular basis to ensure that information can be restored in the case of a security breach or system failure. To ensure that backups can be successfully restored, they should be securely kept and tested on a regular basis (Khang, 2023a).
5. **Security training:** Employees should receive security training on best practices for data protection and privacy, including how to spot phishing scams and steer clear of risky actions like downloading untrusted software or clicking on dubious links (Murphy, 2018).

3.5 CONCLUSION

Data are essential to giving insights that help businesses make wise decisions in today's fast-paced business and production environments. Data are a crucial asset for companies looking to stay profitable and competitive, from recognizing market trends to simplifying production processes.

Businesses may streamline operations, cut costs, and boost profits through the gathering, administration, and analysis of data. Decision-makers can gain useful insights from vast amounts of data using data analytics and visualization technologies, which gives them a clear picture of performance and identifies areas that need improvement (Khang et al., 2024).

However, to guarantee the quality, dependability, and security of sensitive data, data collection and management must be done with caution. Companies must adopt best practices for data security and privacy since data breaches and illegal access can have serious repercussions for both organizations and customers.

The advantages of data in business and manufacturing environments are obvious notwithstanding these difficulties. Businesses may remain flexible and responsive to market changes by using data-driven insights, which enhances efficiency and effectiveness across a range of domains (Khang et al., 2023b).

Businesses that prioritize data collecting, administration, and analysis will have a distinct advantage over those who do not see its worth as the value of data continues to rise. Companies may promote growth and success in an environment of growing commercial competition by embracing the power of data and putting best practices for its usage and protection into place (Khang, 2023a).

REFERENCES

Andress J. "The Basics of Information Security: Understanding the Fundamentals of InfoSec in Theory and Practice," *Syngress*, 2014. www.sciencedirect.com/book/9781597496537/the-basics-of-information-security

AWS. *AWS Supply Chain Features*, 2023. https://aws.amazon.com/ru/aws- supply-chain/features/

Briney K. *Data Management for Researchers: Organize, Maintain and Share Your Data for Research Success*. Pelagic Publishing, 2015. https://pelagicpublishing.com/products/data-management-for-researchers-briney

Crossan M, Furlong W (Bill), Austin RD. *Make Leader Character Your Competitive Edge*, 2022. https://sloanreview.mit.edu/article/make-leader-character-your-competitive-edge/

DataPrivacyManager. *Data Privacy vs. Data Security, January 10, 2021 in Blog, Data Privacy*, 2023. https://dataprivacymanager.net/security-vs-privacy/

Davenport TH, Harris JE. *Competing on Analytics: The New Science of Winning*. Harvard Business Review Press, 2007. https://hbr.org/2006/01/competing-on-analytics

Davies A. *How Big Data Helped Netflix Series House of Cards Become a Blockbuster?* 2013. https://sofy.tv/blog/big-data-helped-netflix-series-house-cards-become-blockbuster/

EMC Education Services. *Data Science and Big Data Analytics: Discovering, Analyzing, Visualizing and Presenting Data*. Wiley, 2015. www.wiley.com/en-us/Data+Science+and+Big+Data+Analytics%3A+Discovering%2C+Analyzing%2C+Visualizing+and+Presenting+Data-p-9781118876138

Frahm B. *Article Shop Management. Improving Product Quality with Data Analytics. Defining Data Flow Encourages Efficiency throughout Metal Stamping Shops*, 2021. https://www.thefabricator.com/thefabricator/article/shopmanagement/improving-product-quality-with-data-analytics

Glass R, Callahan S. *The Big Data-Driven Business: How to Use Big Data to Win Customers, Beat Competitors, and Boost Profits*. Wiley, 2014. www.wiley.com/en-ag/The+Big+Data+Driven+Business:+How+to+Use+Big+Data+to+Win+Customers,+Beat+Competitors,+and+Boost+Profits-p-9781118889800

Hajimahmud VA. et al. (Eds.). "The Role of Data in Business and Production," *AI-Aided IoT Technologies and Applications in the Smart Business and Production*. CRC Press, 2023. https://doi.org/10.1201/9781003392224-2

Healy K. *Data Visualization: A Practical Introduction*. Princeton University Press, 2018. https://press.princeton.edu/books/hardcover/9780691181615/data-visualization

Herrera P. *Decision Makers Must Prioritize Data Analytics*, 2022. https://www.forbes.com/sites/forbesbusinesscouncil/2022/02/25/decision-makers-must-prioritize-data-analytics-in-2022/?sh=735c7ef74cce

Jeffery M. *Data-Driven Marketing: The 15 Metrics Everyone in Marketing Should Know*. Wiley, 2010. www.wiley.com/en-aw/Data+Driven+Marketing:+The+15+Metrics+Everyone+in+Marketing+Should+Know-p-9780470504543

Khang A. "AlexKhangMaterial4Studies," *Material of Computer Science, Artificial Intelligence, Data Science, IoT, Blockchain, Cloud, Metaverse, Cybersecurity for Studies*, 2021. www.researchgate.net/publication/370156102_AlexKhangMaterial4Studies

Khang A (Ed.). *AI-Oriented Competency Framework for Talent Management in the Digital Economy: Models, Technologies, Applications, and Implementation*. CRC Press, 2023a. https://doi.org/10.1201/9781003440901

Khang A. *Advanced Technologies and AI-Equipped IoT Applications in High-Tech Agriculture* (1st Ed.). IGI Global Press, 2023b. https://doi.org/10.4018/9781668492314

Khang A, Abdullayev V, Hahanov V, Shah V. *Advanced IoT Technologies and Applications in the Industry 4.0 Digital Economy* (1st Ed.). CRC Press, 2024. https://doi.org/10.1201/9781003434269

Khang A, Gupta SK, Shah V, Misra A (Eds.). *AI-aided IoT Technologies and Applications in the Smart Business and Production.* CRC Press, 2023d. https://doi.org/10.1201/9781003392224

Khang A, Hahanov V, Abbas GL, Hajimahmud VA. "Cyber-Physical-Social System and Incident Management," *AI-Centric Smart City Ecosystems: Technologies, Design and Implementation* (1st Ed.). CRC Press, 2022b. https://doi.org/10.1201/9781003252542-2

Khang A, Hahanov V, Litvinova E, Chumachenko S, Triwiyanto T, Hajimahmud VA, Nazila Ali R, Alyar AV, Anh PTN. "The Analytics of Hospitality of Hospitals in Healthcare Ecosystem," *Data-Centric AI Solutions and Emerging Technologies in the Healthcare Ecosystem* (p. 4). CRC Press, 2023b. https://doi.org/10.1201/9781003356189-4

Khang A, Ragimova NA, Hajimahmud VA, Alyar AV. "Advanced Technologies and Data Management in the Smart Healthcare System," *AI-Centric Smart City Ecosystems: Technologies, Design and Implementation* (1st Ed.). CRC Press, 2022a. https://doi.org/10.1201/9781003252542-16

Khang A, Rani S, Gujrati R, Uygun H, Gupta SK (Eds.). *Designing Workforce Management Systems for Industry 4.0: Data-Centric and AI-Enabled Approaches.* CRC Press, 2023c. https://doi.org/10.1201/9781003357070

Khang A, Shah V, Rani S. *AI-Based Technologies and Applications in the Era of the Metaverse* (1st Ed.). IGI Global Press, 2023e. https://doi.org/10.4018/9781668488515

Khanh HH, Khang A. "The Role of Artificial Intelligence in Blockchain Applications," *Reinventing Manufacturing and Business Processes Through Artificial Intelligence* (pp. 20–40). CRC Press, 2021. https://doi.org/10.1201/9781003145011-2

Kirk A. *Data Visualisation: A Handbook for Data Driven Design.* Sage Publications Ltd, 2016. https://uk.sagepub.com/en-gb/eur/data-visualisation/book266150

MajestEYE. *Why Is Data Important for Your Business?* 2023. www.majesteye.com/why-is-data-important-for-your-business/

Mi K, *Big Data Analytics: The Future of Supply Chain Management.* Columbia GRC, 2022. https://insights.grcglobalgroup.com/big-data-analytics-the-future-of-supply-chain-management/

Mrkonjić E. *How Amazon Uses Big Data,* 2023. https://seedscientific.com/how-amazon-uses-big-data/

Murphy D. *Data Protection and Privacy: The Basics.* Bloomsbury Professional, 2018. www.bloomsbury.com/us/data-protection-and-privacy-volume-13-9781509941759/

Phocas Software. *3 Ways Advanced Data Analytics Can Improve Customer Service,* 2023. https://www.phocassoftware.com/resources/blog/3-ways-business-intelligence-can-improve-customer-service

Pınarbaşı AT. *Data Privacy vs. Data Security vs. Data Protection: In-Depth Look,* 2023. https://termly.io/resources/articles/data-privacy-vs-data-security-vs-data-protection/

Provost F, Fawcett T. *Data Science for Business: What You Need to Know About Data Mining and Data-Analytic Thinking.* O'Reilly Media, 2013. www.oreilly.com/library/view/data-science-for/9781449374273/

Rani S, Bhambri P, Kataria A, Khang A, Sivaraman AK. *Big Data, Cloud Computing and IoT: Tools and Applications* (1st Ed.). Chapman and Hall/CRC Press, 2023. https://doi.org/10.1201/9781003298335

Rani S, Chauhan M, Kataria A, Khang A (Eds.). "IoT Equipped Intelligent Distributed Framework for Smart Healthcare Systems," *Networking and Internet Architecture.* CRC Press, 2021. https://doi.org/10.48550/arXiv.2110.04997

Sharaf A. *How Starbuck Uses Big Data AI Its Business*, 2023. www.linkedin.com/pulse/how-starbuck-uses-big-data-ai-its-business-anshu-sharaf-/?trk=articles_directory

Sher R. *How Midsized Companies Must Drive Efficiency with Data*, 2022. https://www.forbes.com/sites/robertsher/2022/05/04/how-midsized-companies-must-drive-efficiency-with-data/?sh=611f6d06ad3e

Simplilearn. *What Is Data Collection: Methods, Types, Tools, and Techniques*, 2023. www.simplilearn.com/what-is-data-collection-article

Simranssonu19. *Difference Between Data Visualization and Data Analytics*, 2023. www.geeksforgeeks.org/difference-between-data-visualization-and-data-analytics/

Stedman C. *Data Collection*, 2023. www.techtarget.com/searchcio/definition/data-collection

Stitch. *5 Benefits of Data Analytics for Your Business*, 2023. www.stitchdata.com/resources/benefits-of-data-analytics/

Tesla. *Vehicle Safety and Security Features*, 2023. www.tesla.com/support/vehicle-safety-security-features

Thomas L. *Walmart Changed the Way It Buys Shopping Bags and Saved $60 Million—and That's Just One Way It Cut Costs*, 2023. www.cnbc.com/2020/02/18/walmart-saves-millions-of-dollars-each-year-by-making-these-small-changes.html

Vaughan J. *Definition of Data*, 2023. www.techtarget.com/searchdatamanagement/definition/data

Venkataramanan N, Shriram A. *Data Privacy: Principles and Practice*. Chapman & Hall/CRC Press, 2016. https://dl.acm.org/doi/book/10.5555/3073998

Wikipedia. *Wikipedia Data Definition*, 2013. https://en.wikipedia.org/wiki/Data

4 Data-Oriented Internet of Things in Intelligent Analytics and Visualization

Jainam Shah, Aakansha Saxena, and Akash Gupta

4.1 INTRODUCTION

Data-oriented artificial intelligence (AI) is an AI architecture that has been around for a while. It is based on the idea of data-driven learning. Data-oriented AI is a type of machine learning (ML) that uses data to drive the process of ML. Let's dive more deeply into the connection between IoT and AI (Patel et al., 2017).

The Internet of Things (IoT) is a network of physical objects that are connected to the internet. They can be anything from a fridge, a car, and even a pacemaker. This network can be used for various purposes such as home automation, transportation, and healthcare (Khang et al., 2022a).

AI is going to be the next big thing in the IoT. AI will help with decision-making, analysis, and prediction for IoT devices. It will help with the self-learning and self-healing of devices as well. AI and IoT integrated with visualization is the next generation of technology experience (Rana et al., 2021). The IoT has made it possible for devices to communicate with each other and share data in real time. This means that any device with an internet connection can be used to collect and transmit data (Khanh and Khang, 2021).

4.1.1 On Using the Intelligent Edge for IoT Analytics

This chapter employs the concept of cloud computing to present a flexible framework for IoT data analytics. To come up with solutions for adaptive IoT business intelligence, the authors identify key individuals and their functions (Rani et al., 2021).

The proposed methodology can be used to productively build reliable IoT applications that need to make a choice between cloud and advantage computing based on changing application requirements (Bhambri et al., 2022). The scenarios of smart cities, security monitoring, and advanced manufacturing, where the level of user encounter is important, are some examples of possible use cases for this technology (Rani et al., 2023).

DOI: 10.1201/9781003392224-4

4.1.2 AI AND BIG DATA

This chapter examines some of the fundamental concerns and applications of AI for big data as part of the brand-new department of IEEE AI-based systems called AI Development in Industry (AI has been used in several different ways to facilitate capturing and structuring big data, and it has been used to analyze big data for key insights).

Even though collecting data has always been important to business, contemporary digital tools have made it easier than ever. Because data sets are expanding at an exponential rate, it is almost impossible for anyone or a company to use the data they are collecting effectively. Understanding big data and AI is crucial for this reason (Geetha et al., 2024).

Applications with AI capabilities can process any data set fast, whether it was obtained in real time or was taken from a database (Khang et al., 2022b). Businesses are utilizing AI solutions to increase efficiency, personalize experiences, facilitate decision-making, and save costs. Data and AI are regularly added to analytics and automation to help firms transform their business processes.

Big data and AI complement each other effectively. Big data analytics uses AI to better data analysis, and AI requires a massive amount of data to learn and improve decision-making processes (Hajimahmud et al., 2023a).

With this convergence, it will be easier to apply advanced analytics tools like augmented or predictive analytics and to uncover insightful information more quickly from your huge data stores. By giving your users the user-friendly tools and dependable technology they need to extract high-value insights from data using big data AI-powered analytics, you can promote data literacy throughout your organization and enjoy the benefits of becoming a fully data-driven organization (Khang et al., 2023b).

4.2 LITERATURE SURVEY

The Role of AI, Machine Learning, and Big Data in Digital Twinning: A Systematic Literature Review, Challenges, and Opportunities (Hajimahmud et al., 2023b).

4.2.1 THE ROLE OF AI

Due to its wide range of industrial applications, digital twinning has appeared as one of the 10 leading technological trends of the past decade. Digital twinning's significance and research possibilities are further reinforced by the incorporation of big data analytics and AI-ML techniques, which present new opportunities and interesting challenges (Khang et al., 2022b).

Various scientific models have been developed and put into effect in relation to this developing subject to date. However, there is no complete study of digital twinning that directs academia and business toward upcoming breakthroughs, especially one that focuses on the function of AI-based and big data (Rani et al., 2022).

4.2.2 ETHICS AND PRIVACY IN AI AND BIG DATA

Big data and burgeoning AI technologies, as well as the implementations they enable, are attracting a lot of media and policy attention. Confidentiality, anonymity, and other ethical implications are given a lot of attention.

In this chapter, we contend that what is considered necessary here and now is a way to deeply grasp these issues and formulate ways to address them that involve all relevant parties, including civil society, to make sure that the advantages of these technologies outweigh their drawbacks. To ensure that the technologies are culturally acceptable, beneficial, and sustainable, we contend that the idea of responsible innovation and study could indeed offer the necessary framework.

4.2.3 PRIVACY AND SECURITY OF BIG DATA IN AI SYSTEMS

Big data's enormous volume, diversity, and velocity have aided ML and AI systems in their development. But a sizable portion of the data used to train AI systems contains sensitive data. Therefore, any weakness could have a disastrous effect on security and privacy concerns (Khang et al., 2022d).

However, the increased demand for high-quality AI from both businesses and governments necessitates the use of big data in the systems. Numerous studies have highlighted the risks posed by big data on various platforms and the protective measures that can be taken. In this chapter, we gave a general overview of the security risks and privacy violations that big data poses as a major driving force in the AI/ML workflow as Table 4.1.

4.3 THE COMPLETE GUIDE TO THE IOT IN AI AND VISUALIZATION

The IoT is the process of connecting physical objects to the internet, and AI is a very important part of it. IoT devices are not just limited to sensors and actuators. There are many other devices that can be connected to the internet, such as smart TVs, refrigerators, security cameras, and others.

This chapter provides an overview of IoT in AI and visualization that includes a brief introduction to IoT and its impact on AI development. It will also provide some examples of how AI can be used in conjunction with IoT for various purposes.

4.3.1 THE POTENTIAL OF THE IOT IN AI FOR A MORE INTELLIGENT FUTURE

Intelligent objects are embedded with AI software that is used to track and meet your requests, be it simple requests like smart bulb lights or complex ones such as identifying a person on the other side of an ocean.

The IoT goes hand in hand with AI because these devices need to have some intelligence at play so that they can react accordingly depending on the number and type of data sent or stored by them. Henceforth, the IoT can be used in AI for a more intelligent future because they take on any type of task and make information accessible in near real time by storing data locally. In the field of AI, there are many different applications and ways it is used.

TABLE 4.1
Literature Review

Sr. No	Title	Reported Outcomes	Challenges
1	On Using the Intelligent Edge for IoT Analytics (Patel et al., 2017)	The use of intelligent edges for IoT analytics can result in improved data processing speed, reduced latency, increased accuracy, better privacy and security, increased efficiency, and reduced costs compared to traditional cloud-based IoT analytics solutions. It can also enable real-time decision-making and enable new use cases for IoT data analysis, such as edge computing and autonomous systems.	The main challenges in using the intelligent edge for IoT analytics are limited computational resources, network connectivity and reliability, security and privacy concerns, a lack of standardization, and difficulty in managing and maintaining edge devices at scale.
2	Artificial Intelligence and Big Data (O'Leary, 2013)	AI models can analyze and process large amounts of data in real time, resulting in improved accuracy and efficiency. AI models can uncover new insights and predictions that would be difficult or impossible to detect using traditional data analysis techniques. AI can be used to personalize the customer experience and improve customer satisfaction.	Ensuring the quality and cleanliness of big data is essential for accurate AI models. AI models can perpetuate existing biases, so it is important to ensure the data used to train the models is diverse and unbiased. AI models can be difficult to understand and interpret, making it challenging to ensure that decisions made by AI are transparent and fair.
3	The Role of AI, Machine Learning, and Big Data in Digital Twinning: A Systematic Literature Review, Challenges, and Opportunities (Rathore et al., 2021)	AI and machine learning can improve the accuracy of digital twin simulations and predictions. AI and machine learning can automate processes and tasks, resulting in increased efficiency and freeing up time for other valuable activities. Digital twin models can be used to support better decision-making by providing real-time data and simulations.	Integration and interoperability of digital twin systems with AI and ML technologies can be challenging. Ensuring the quality and availability of data is essential for accurate digital twin models.

(*Continued*)

TABLE 4.1 *(Continued)*
Literature Review

Sr. No	Title	Reported Outcomes	Challenges
4	Privacy and Security of Big Data in AI Systems: A Research and Standards Perspective (Dilmaghani et al., 2019)	Addressing privacy and security concerns can increase trust in AI systems and encourage wider adoption. Ensuring the privacy and security of sensitive information can protect individuals and organizations from potential harm. Compliance with privacy regulations can reduce the risk of legal or financial penalties and improve public perception of AI systems.	Ensuring that sensitive information remains confidential and protected from unauthorized access or misuse is a major challenge. Ensuring that big data is protected from cyber threats, such as hacking or malware, is a critical concern. It is important to comply with privacy regulations, such as the General Data Protection Regulation, to ensure that data is handled in an ethical and legal manner.
5	Ethics and Privacy in AI and Big Data: Implementing Responsible Research and Innovation (Stahl and Wright, 2018)	Ensuring the privacy and security of sensitive information can protect individuals and organizations from potential harm. Ensuring the ethics and privacy of big data can improve data quality by reducing the risk of data breaches or tampering. Addressing ethics and privacy in AI and big data can lead to the development of AI systems that are more responsible and aligned with societal values.	AI models can perpetuate existing biases in the data used to train them, leading to unfair or unethical outcomes. Ensuring that sensitive information remains confidential and protected from unauthorized access or misuse is a major challenge. AI models can be complex and difficult to understand, making it challenging to ensure that decisions made by AI are transparent and fair.

4.3.2 Connectivity Is Everywhere

Connectivity is everywhere. It is a key part of the IoT, and it is also an integral part of how we live today. It is not just about connecting people to one another but also about connecting things to one another. Connectivity is the key to unlocking so much potential in our lives and in our businesses.

IoT is the next step in the evolution of AI. The potential for it is endless. It will not only change our lives but also the way we work, shop, and interact with each other. It will bring about a more intelligent future where we can automate many tasks and create new opportunities for everyone to learn and grow.

The IoT has opened up a whole new world of possibilities for AI by providing data that were previously unavailable to it. It has made AI smarter by feeding it more information than ever before, which in turn means that it can make better decisions based on this data. IoT has the potential to transform our future in many ways. It will be able to monitor and manage our environment, transportation systems, buildings, utilities, and more. It will provide a wide range of services, from metering and billing to monitoring and control.

The IoT is not just about connecting devices that are already in our environment but also about creating new types of sensors that can monitor parts of the world we have never seen before. This will allow us to understand the Earth in ways we have never been able to before. Two emerging technologies have shown promise for delivering on this vision: augmented reality (AR) and AI.

4.3.3 Combination of Two Technologies

You may wonder why there are so many benefits to combining these two technologies, the benefits of combining these two technologies are plentiful. They can be used to create content that is more authentic and engaging. This can be done by using the emotional intelligence of AI writers to write human-like content or by using the data analytics of AI to write marketing messages that resonate with customers. They can be used to create content that is more authentic and compelling than ever before (Khang, 2023).

For example, the use of computer graphics has helped transform the way audiences experience media content. If a person wants to watch an action movie or play a game that is more realistic, they can sit in their living room and see it as if they were on screen. This gives them a more immersive experience and provides them with content that is not possible with traditional methods.

4.3.4 Benefits of IoT Devices

Now let's see the benefits of IoT devices when implementing AI into your business strategy, IoT devices are often overlooked when it comes to the implementation of AI in business strategies. But they can be a great tool for businesses to use and implement in their strategies. They can be used as data sources, automate processes, and help in the creation of a more personalized customer experience.

The growth of IoT devices is one of the most important trends in the AI industry. This is because there are certain advantages that come with these devices. For example, with these devices, we can have more data and insights than ever before. This means that we can make better decisions and predictions about our plans. To understand the importance of data analytics of IoT and the evolution of business intelligence, let's understand the importance of data analytics and how AI is changing the face of data analytics and business intelligence in the many industries in next sections.

4.3.5 IMPORTANCE OF DATA ANALYTICS IN THE IoT

First of all, data architecture and IoT data modeling are essential elements of every successful IoT implementation. Data are the lifeblood of the IoT, and how those data are organized and kept is crucial to its usefulness.

Data are collected, processed, and stored in a way that is effective, secure, and scalable thanks to a well-designed data architecture. To better understand how data will be used and analyzed, IoT data modeling entails developing a conceptual framework for how data will be organized and managed.

Making data flow diagrams, data entity relationship diagrams, and other visual representations of the data may be necessary for this. Overall, an effective IoT data architecture and data modeling methodology can assist organizations in making better use of their data, resulting in better business outcomes and decision-making.

Data analytics in the IoT is the process of collecting, storing, analyzing, and interpreting data from IoT devices to gain insights. The insights are then used to improve operational efficiency and make better business decisions. There are many use cases of data analytics in IoT. A few examples follow:

1. Data analytics can be used to predict the likelihood of a machine failure before it actually happens.
2. Data analytics can be used to get an insight into customer behavior and preferences, which helps businesses make better decisions about their products.
3. Data analytics can also be used for predictive maintenance by using predictive models that forecast when a machine is likely to fail next based on its usage history.

As we believe it gives a brief understanding of the importance of data analytics, let's dive deeper into the question we had in mind ahead in the next sections of this chapter.

4.3.6 EVOLUTION OF BUSINESS INTELLIGENCE

The emergence of IoT has completely changed the way companies operate. It has enabled them to collect real-time data from their products and customers, which, in turn, has led to a change in their business models.

With IoT data analytics, companies can now optimize their supply chains more efficiently by monitoring inventory levels and predicting demand for products before they run out. Then what's the role of AI in this chapter? Let's discuss this in more detail.

4.3.7 How AI Is Changing the Face of Data Analytics and Business Intelligence

As the world becomes more and more digitized, the need for data analytics and business intelligence has increased. The amount of data that can be analyzed is immense, and it is becoming increasingly difficult for humans to process this information.

This is where AI comes in to do the heavy lifting. AI algorithms are able to scan vast amounts of raw data and find patterns that would be impossible for humans to do so alone. This means that companies can use AI as a tool to make sense of their data, which, in turn, helps them make better decisions.

AI is changing the face of data analytics and business intelligence because it has introduced a new level of automation which is making the process more efficient. It has also introduced new insights into data that have never been seen before, which can help businesses make better decisions and increase profitability.

4.4 AI LEVERAGE

How does AI leverage the benefits of the IoT to grow your business? AI can leverage the benefits of the IoT to grow your business in a variety of ways. It can help you find new customers by analyzing data from IoT devices and then matching them with similar profiles in your customer database.

AI can also help optimize service delivery by monitoring IoT-connected assets such as machinery or equipment for maintenance issues before they become critical. Now let's derive some conclusions from the earlier discussion.

4.4.1 AI-Powered IoT Devices

AI-powered IoT devices and how they will enhance our lives, IoT analytics that is data-driven or data-oriented use cutting-edge algorithms and machine learning approaches to glean insights from the vast amount of data that IoT devices created, which help organizations to analyze and get the best view from their business.

Organizations may make data-driven choices, streamline operations, and even forecast the future by analyzing data in real time. Data gathering, data preprocessing, data analysis, and data visualization are just a few of the steps that make up data-driven IoT analytics. Data pipeline diagrams, which show how data moves through the various phases of analytics, frequently include these steps. Overall, data-driven IoT analytics can give businesses a competitive edge by revealing previously undiscovered insights in their data.

Today, IoT is a buzzword used in many industries. It is the future of our lives and it will make us more connected than ever before. IoT devices are everywhere, and they have become a part of our lives. They are in our homes, offices, cars, on the streets, and even in the skies. And it is not just about connecting to Wi-Fi or Bluetooth anymore—IoT devices can connect to other devices as well as collect data from them and process it. IoT devices will enhance our lives by making it easier for us to do things that were once difficult or impossible.

4.4.2   IoT-Powered Smart Cities

While taking on the role of AI in IoT-powered smart cities, experts should include a definition of what an IoT-powered smart city is and how AI can help make it smarter.

An IoT-powered smart city has sensors to collect data from various sources, such as traffic, air quality, noise levels, and water quality. These sensors are then analyzed through AI algorithms to understand the patterns in the data and make predictions about what might happen next (Hahanov et al., 2022).

AI can help make these cities smarter by improving our understanding of the world around us through data analysis, making predictions about what might happen next, and optimizing systems for better outcomes.

4.5   VISUALIZATION WITH AI

A cutting-edge method for making sense of the enormous amounts of data generated by IoT devices is AI-based IoT data visualization. Organizations may produce effective data visualizations that aid users in quickly comprehending complex data sets by utilizing AI and ML technologies.

AI-based IoT data visualizations entail real-time data analysis, pattern recognition, and trend detection, as well as automatically producing visualizations that highlight the most important data. These visualizations can be created as dashboards, heat maps, three-dimensional (3D) models, or interactive maps, among other formats. Organizations may better comprehend their data by utilizing AI-based IoT data visualization, which will help them make better decisions and produce better business results.

4.5.1   How to Select the Right AI Visualization Tool for Your Business

Visualization tools are widely used in the market. Their main purpose is to make it easier for people to understand and interpret data. But you can choose from many different types of visualization tools.

This chapter will help you learn how to select the right one for your business needs. There are many ways of selecting a visualization tool depending on what you want to do with it and who is going to be using it.

The following list provides some guidelines on how to select the right AI visualization tool for your business needs:

- Decide what type of information you want to visualize.
- Determine who will be using the visualization tools.
- Know what type of data you have available.
- Consider what type of device the user has access to.
- Know if there is a need for interactivity or animation in your design.
- Understand if there is a need for support for other languages or cultures.

4.5.2   Need for an AI Visualization Tool

An AI visualization tool is software that automatically generates data visualizations. It is a very useful tool for people who want to improve their knowledge of data

TABLE 4.2
Market Share of AI Visualization Tools in 2021–2022

AI visualization tool and software	Market share (%)
Power BI	36
Tableau	20
Qlik Sense	11
SAP Analytics Cloud	11
IBM Cognos Analytics	7
Looker	6
MicroStrategy Analytics	5
Sisense	3
Oracle Analytics	1

analysis and statistics. It can also be used by people who are skilled in data science to create visualizations that can be shared with other professionals.

Some people are not that great at visualization. Their brains might not be wired in a way that they can easily grasp the meaning of data visualizations. In some cases, they might even struggle with understanding the purpose of the visualization. This is where AI visualization tools come in handy as Table 4.2.

These tools help you simplify complex data into something easier to understand and comprehend. They also allow you to create interactive data visualizations which will help you get a better understanding of what your data are trying to tell you.

AI visualization projects are becoming more and more popular these days. This can be seen in both business and personal fields of life. AI visualizations can be used to make predictions about future events, solve problems, and make data easier to understand for humans. So, readers, as you all know, the conclusion is the last part of a report, essay, article, or another type of writing. It is an important part of the writing process because it summarizes what has been said and gives a general idea about the writer's opinion on the topic as Figure 4.1.

4.6 IMPLEMENTATION

4.6.1 IoT Data Modeling and Data Architecture

Data architecture and IoT data modeling are essential elements of every successful IoT implementation. In order to better understand how data will be utilized and analyzed, data modeling entails developing a conceptual framework for how data will be organized and managed.

Understanding the needs for data collecting and storage is crucial before implementing a successful IoT data modeling and data architecture solution.

Create a data processing pipeline that comprises data cleansing, filtering, and transformation after the requirements for data collection and storage have been determined. This pipeline should be created to guarantee data consistency and accuracy while lowering the possibility of mistakes or abnormalities in the analysis. Another

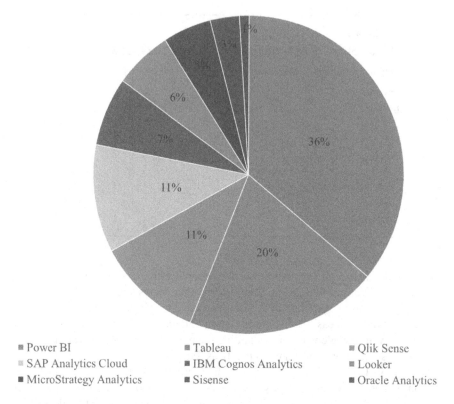

FIGURE 4.1   Artificial intelligence visualization tool and software in the market.

Source: Khang (2021).

important factor to consider while implementing IoT data modeling and data architecture is data analysis. ML and AI are also used to get data insights. Organizations can improve operations, forecast results, and spot possible problems or anomalies with the insights provided by data analysis.

Finally, integrating IoT data modeling and data architecture requires a strong focus on data visualization. Insights and patterns must be presented in a form that is simple to comprehend and interpret. Dashboards, heat maps, and other visualizations that emphasize the most crucial data points and patterns can be made using visualization tools. To maximize the value of IoT data, it is essential to apply an effective IoT data modeling and data architecture methodology.

4.6.2   DATA-DRIVEN OR DATA-ORIENTED IOT ANALYTICS

Inspired by data or focused on data in the context of IoT, analytics refers to the use of data to inform decision-making and enhance business outcomes. This strategy entails gathering, processing, and analyzing data produced by IoT devices to learn more and spot trends that may be utilized to streamline procedures, enhance goods, and improve consumer experiences.

It is crucial to first define the primary data sources and the categories of data that need to be gathered in order to successfully implement a data-driven or data-oriented IoT analytics strategy. This might comprise, among other things, sensor data, environmental data, and user-generated data.

- The establishment of a data processing pipeline that comprises data cleansing, filtering, and transformation is crucial after the data sources have been identified.
- Utilizing IoT analytics is crucial to enhancing business outcomes and making informed decisions based on data from IoT devices.
- Successful implementation of a data-driven or data-oriented IoT analytics strategy requires identifying primary data sources and categories of data, such as sensor data, environmental data, and user-generated data.
- Efficient data management is essential to effectively analyze data generated by IoT devices, and a data processing pipeline including data cleansing, filtering, and transformation can help achieve this.
- Effective data analysis is key to optimizing processes, identifying issues, and detecting potential anomalies for better business outcomes.
- Data visualization tools such as dashboards and heat maps enable businesses to easily understand complex data sets and make informed decisions.

For example, a hospital can utilize data-driven IoT analytics to monitor patients' health and vital signs, optimize resource allocation, and ultimately improve patient outcomes. Data visualization tools can help medical professionals make informed decisions based on real-time data, and maintaining IoT data security is essential for protecting patient privacy and confidentiality.

4.6.3 IoT-Based Visualization

AI-based IoT data visualization is changing the game for organizations handling massive amounts of data generated by IoT devices (Khang et al., 2024).

Thanks to advanced AI techniques, this technology quickly extracts valuable insights from complex data sets, enabling faster and more informed decisions. With its ability to automatically identify patterns and anomalies that even human analysts may miss, AI-based data visualization helps organizations gain a competitive edge (Khang et al., 2023c).

Plus, it scales to accommodate ever-increasing amounts of data, without slowing down traditional visualization techniques. In addition, it enhances security by detecting potential breaches and taking swift action (Luke et al., 2024).

To implement this powerful technology, organizations must establish a robust data processing pipeline and determine which machine learning algorithms and visualization techniques best suit their needs. With its impressive capabilities, AI-based IoT data visualization is revolutionizing how organizations analyze and visualize IoT data (Subhashini and Khang, 2024).

4.7 CONCLUSION

Can we predict the future of AI? AI is a double-edged sword. On one hand, it can help us achieve our goals more efficiently; on the other hand, it can take away our jobs. It is not easy to predict the future of AI, but we need to be aware of the risks and impacts.

AI will free up our time from doing boring work and allow us to pursue more meaningful things. But the flip side of that coin is the threat of AI taking away our jobs. The key for us as humans is to find a way to make AI work for us rather than against us.

How will AI change the personal and professional lives of people in the next 10 years? In the next 10 years, AI will be able to do many of the tasks that we consider too difficult for machines to do. For example, AI can handle legal and medical research.

AI will also affect the way we work. It will take over more jobs that are repetitive and mundane. It may also create new jobs in areas like data analysis and designing algorithms. In the next 10 years, AI will change how we live our lives as well as how we work.

Industries that have traditionally been the backbone of the economy are now being disrupted by AI. The manufacturing industry is also not immune to this trend. AI has been used in manufacturing for quite a while now and it is only going to get more advanced in the years to come. From design and production, to supply chain management, AI will be able to do it all. The future of manufacturing looks bright with AI on board (Shah et al., 2024).

Do you like building, learning, and exploring in games? The future of gaming will be very interesting with AI. The gaming industry is a major pillar of the economy. It provides jobs and generates revenue that supports everything from charities to schools. This industry will likely get more creative with AI's assistance, resulting in ever-more engaging content for players.

REFERENCES

Bhambri P, Rani S, Gupta G, Khang A. *Cloud and Fog Computing Platforms for Internet of Things*. CRC Press, 2022. https://doi.org/10.1201/9781003213888

Dilmaghani S. et al. "Privacy and Security of Big Data in AI Systems: A Research and Standards Perspective," *2019 IEEE International Conference on Big Data (Big Data)*, 5737–5743. IEEE Xplore, 2019, https://doi.org/10.1109/BigData47090.2019.9006283

Geetha C, Neduncheliyan S, Khang A (Eds.). "Dual Access Control for Cloud Based Data Storage and Sharing," *Smart Cities: IoT Technologies, Big Data Solutions, Cloud Platforms, and Cybersecurity Techniques*. CRC Press, 2024. https://doi.org/10.1201/9781003376064-17

Hahanov V, Khang A, Litvinova E, Chumachenko S, Hajimahmud VA, Alyar AV. "The Key Assistant of Smart City—Sensors and Tools," *AI-Centric Smart City Ecosystems: Technologies, Design and Implementation* (1st Ed.). CRC Press, 2022. https://doi.org/10.1201/9781003252542-17

Hajimahmud VA, Khang A, Gupta SK, Babasaheb J, Morris G. *AI-Centric Modelling and Analytics: Concepts, Designs, Technologies, and Applications* (1st Ed.). CRC Press, 2023a. https://doi.org/10.1201/9781003400110

Hajimahmud VA. et al. (Eds.). "The Role of Data in Business and Production," *AI-Aided IoT Technologies and Applications in the Smart Business and Production.* CRC Press, 2023b. https://doi.org/10.1201/9781003392224-2

Khang A. "Material4Studies," *Material of Computer Science, Artificial Intelligence, Data Science, IoT, Blockchain, Cloud, Metaverse, Cybersecurity for Studies,* 2021. www. researchgate.net/publication/370156102_Material4Studies. Last visit 2023.

Khang A (Ed.). *AI-Oriented Competency Framework for Talent Management in the Digital Economy: Models, Technologies, Applications, and Implementation.* CRC Press, 2023. https://doi.org/10.1201/9781003440901

Khang A, Abdullayev VA, Hahanov V, Shah V. *Advanced IoT Technologies and Applications in the Industry 4.0 Digital Economy* (1st Ed.). CRC Press, 2024. https://doi.org/10.1201/9781003434269

Khang A, Gupta SK, Rani S, Karras DA (Eds.). *Smart Cities: IoT Technologies, Big Data Solutions, Cloud Platforms, and Cybersecurity Techniques.* CRC Press, 2023b. https://doi.org/10.1201/9781003376064

Khang A, Hahanov V, Abbas GL, Hajimahmud VA. "Cyber-Physical-Social System and İncident Management," *AI-Centric Smart City Ecosystems: Technologies, Design and Implementation* (1st Ed.). CRC Press, 2022b. https://doi.org/10.1201/9781003252542-2

Khang A, Ragimova NA, Hajimahmud VA, Alyar AV. "Advanced Technologies and Data Management in the Smart Healthcare System," *AI-Centric Smart City Ecosystems: Technologies, Design and Implementation* (1st Ed.). CRC Press, 2022d. https://doi.org/10.1201/9781003252542-16

Khang A, Rani S, Sivaraman AK. *AI-Centric Smart City Ecosystems: Technologies, Design and Implementation* (1st Ed.). CRC Press, 2022a. https://doi.org/10.1201/9781003252542

Khang A, Shah V, Rani S. *AI-Based Technologies and Applications in the Era of the Metaverse* (1st Ed.). IGI Global Press, 2023c. https://doi.org/10.4018/9781668488515

Khanh HH, Khang A. "The Role of Artificial Intelligence in Blockchain Applications," *Reinventing Manufacturing and Business Processes through Artificial Intelligence* (pp. 20–40). CRC Press, 2021. https://doi.org/10.1201/9781003145011-2

Luke J, Khang A, Chandrasekar V, Pravin AR, Sriram K (Eds.). "Smart City Concepts, Models, Technologies and Applications," *Smart Cities: IoT Technologies, Big Data Solutions, Cloud Platforms, and Cybersecurity Techniques* (1st Ed.). CRC Press, 2024. https://doi.org/10.1201/9781003376064-1

O'Leary DE. "Artificial Intelligence and Big Data," *IEEE Intelligent Systems,* 28(2), 96–99, March 2013. IEEE Xplore. https://doi.org/10.1109/MIS.2013.39

Patel P. et al. "On Using the Intelligent Edge for IoT Analytics," *IEEE Intelligent Systems,* 32(5), 64–69, September 2017. IEEE Xplore. https://doi.org/10.1109/MIS.2017.3711653

Rana G, Khang A, Sharma R, Goel AK, Dubey AK (Eds.). *Reinventing Manufacturing and Business Processes through Artificial Intelligence.* CRC Press, 2021. https://doi.org/10.1201/9781003145011

Rani S, Bhambri P, Kataria A, Khang A. "Smart City Ecosystem: Concept, Sustainability, Design Principles and Technologies," *AI-Centric Smart City Ecosystems: Technologies, Design and Implementation* (1st Ed.). CRC Press, 2022. https://doi.org/10.1201/9781003252542-1

Rani S, Bhambri P, Kataria A, Khang A, Sivaraman AK. *Big Data, Cloud Computing and IoT: Tools and Applications* (1st Ed.). Chapman and Hall/CRC Press, 2023. https://doi.org/10.1201/9781003298335

Rani S, Chauhan M, Kataria A, Khang A (Eds.). "IoT Equipped Intelligent Distributed Framework for Smart Healthcare Systems," *Networking and Internet Architecture.* CRC Press, 2021. https://doi.org/10.48550/arXiv.2110.04997

Rathore MM. et al. "The Role of AI, Machine Learning, and Big Data in Digital Twinning: A Systematic Literature Review, Challenges, and Opportunities," *IEEE Access*, 9, 32030–32052, 2021. https://doi.org/10.1109/ACCESS.2021.3060863

Shah V, Jani S, Khang A (Eds.). "Automotive IoT: Accelerating the Automobile Industry's Long-Term Sustainability in Smart City Development Strategy," *Smart Cities: IoT Technologies, Big Data Solutions, Cloud Platforms, and Cybersecurity Techniques*. CRC Press, 2024. https://doi.org/10.1201/9781003376064-9

Stahl BC, Wright D. "Ethics and Privacy in AI and Big Data: Implementing Responsible Research and Innovation," *IEEE Security & Privacy*, 16(3), 26, May 2018. IEEE Xplore. https://doi.org/10.1109/MSP.2018.2701164

Subhashini R, Khang A (Eds.). "The Role of Internet of Things (IoT) in Smart City Framework," *Smart Cities: IoT Technologies, Big Data Solutions, Cloud Platforms, and Cybersecurity Techniques*. CRC Press, 2024. https://doi.org/10.1201/9781003376064-3

5 Applications of Wireless, 5G, 6G, and Internet of Things Technologies

Pankaj Kumar Dubey, Bindeshwar Singh, Varun Kumar, and Aryan Kumar Dubey

5.1 INTRODUCTION

The very first wireless telephone conversation took place about 1880 when Bell (1980) and Tainter (1880) invented the photophone, a device that communicated sound through a beam of light.

Marconi (1922) started working on a wireless telegraph network in 1909 utilizing radio waves, which had already been recognized since Heinrich Hertz demonstrated their presence in 1888 but were disregarded as a communication style at the stage since they appeared to be a short-range event. In his studies, Chandra Bose et al. (1858) obtained an exceptionally high frequency of up to 60 GHz, which was the first time millimeter connectivity had been studied.

The 1990s saw the start of the wireless famous insurrection as a result of transformative mobility and a radical shift from wired to wireless advancement, as well as the widespread use of commercial wireless devices like cell phones, mobile telephony, fax machines, wireless computers, data networks, the wireless internet, and laptop and portable computers with communication links.

Computers, digital communication devices, network components, and several other devices are connected by radio waves and radio frequency (RF) in mobile broadband. It is used in locations wherein wired equipment cannot be set up or whether the cost of doing so is prohibitive.

The four basic types of wireless connections—wireless local area networks (WLANs), wireless metropolitan area networks (MANs), wireless personal area networks (PANs), and wireless wide area networks (WANs)—each serve a specific function.

5.1.1 WLANs

WLAN techniques make it feasible to link to the internet indoors or in a small outdoor space. Initially used in homes and companies, WLAN innovation is frequently found in stores and restaurants.

5.1.2 Wireless MANs

In addition to giving accessibility outside a workplace or personal network, wireless MANs have already been set up in urban areas. Although these networks have a larger coverage area than networks in offices or homes, the fundamentals are equivalent.

American Physical Society (APs) can be spotted throughout the covered area on telephone poles or the exteriors of residences. Even though they are tethered to the internet, APs broadcast a wireless signal throughout the area.

5.1.3 Wireless PANs

Wireless PAN Connectivity, which uses protocols like Bluetooth and Zigbee, frequently only covers a relatively short area—usually no more than 100 meters in most circumstances.

Bluetooth enables hands-free phone calls, earbud hookups, and data transfer between smart devices (Khang et al., 2022a).

5.1.4 Wireless WANs

To enable accessibility from outside coverage areas of a wireless LAN or MAN, wireless WANs use mobile connectivity. These linkages allow users to make phone calls to other people.

WANs manage both voice and data exchange using the same method. Customers can also access the internet to discover a website and client applications. Versions include 3G, 4G, 5G, and 6G as noted in Table 5.1.

Merits: Rapid information transmission is a result of enhanced data connections, and installing wireless networks can be simpler and less expensive, particularly in historically significant buildings. By using wireless networking, you might introduce new goods or services, additionally, office-based wireless staff could communicate without using separate computers and carry on with their task when out of the office.

TABLE 5.1

Differences between the Various Wireless Networking Topologies

Types	Wireless LAN	Wireless MAN	Wireless PAN	Wireless WAN
Parameters				
Diversity of networks	Local area network	Metropolitan area network	Personal area network	Wide area network
Range	Within building	Outside office	Commonly 100 meters	Beyond the reach of WLAN and WMAN
Connectivity	Cellular	IEEE 802.16 WiMax	Bluetooth, infrared, and Zigbee	LTE, 4G, 5G, and 6G

Source: The comparison table was designed by Pankaj Kumar Dubey, Bindeshwar Singh, Varun Kumar, and Aryan Kumar Dubey.

Demerits: Users must pay careful attention to privacy because wireless transmission is much more susceptible to attacks from unauthorized users. Obtaining constant coverage in particular structures can be challenging, resulting in "black spots" when the signal is unavailable, in comparison to "wired" networks, wireless transmission can be sluggish and a little less effective, and wireless networks also often offer lesser speeds.

This chapter covers the Internet of Things (IoT), 5G, and 6G, with recent trends, comparisons, conclusions, and future scope discussed.

This chapter is also organized into five sections: Introduction, Comparison, Results and Discussion in which recent trends discuss, Challenges and Opportunities, and Conclusion.

5.2 IOT

The IoT relates to the total system of interconnected gadgets but also the technology to allow interaction between products as well as with the cloud.

The phrase "Internet of Things" describes real physical objects with sensors, computing power, and software, as well as other innovations that may connect to other sensors and devices over the internet or other network infrastructure and transfer files with them as shown in Figure 5.1.

The IoT primarily consists of seven levels:

- **Sensors:** The core components of any IoT device are sensors and other data-gathering gear. They act as a bridge, connecting the real and artificial realms (Hahanov et al., 2022).
- **Sensors to the gateway network:** This is the first network architecture of every IoT environment. It is responsible for sending information from the sensors on the first level to the third level (gateways).

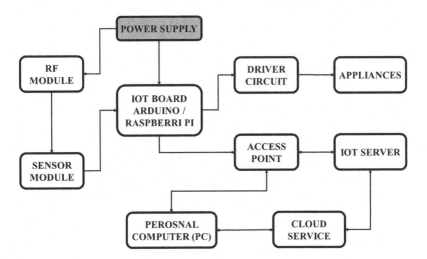

FIGURE 5.1 Block diagram of the Internet of Things.

Source: Khang (2021).

- **Gateways:** Data aggregators called gateways gather information from various sensors and transmit it to a backend service. They serve as a connection between both the local sensor area and the internet and are essentially routers or modems.
- **Gateways to the internet network:** These services enable the transfer of data from the gateway to the internet/backend services, just like the sensor to the gateway network does.
- **Automation of information and data:** The raw data gathered from the first four layers are transformed into useful information at this layer (Tailor et al., 2022).
- **Internet to the user network:** This is the top-level network layer for every IoT device. This term refers to the unprocessed data stored in the cloud network as value-added data and displays it on the user's screen.
- **Additional worth of technical data:** This final layer represents the front end of the overall IoT device. The obtained data and value-added information are displayed on users' screens.

The IoT platform, which serves as the app's brain, is where programs can be written and loaded. According to the goal, the IoT board could be either an Arduino or a Raspberry Pi (Srinivasan et al., 2019).

- **Advantages:** Enhanced employee productivity and decreased human labor, improved administration of operations, greater utilization of the available resources and equipment, enhanced work security, cost-effective operations, and enhanced customer support and maintenance (Hajimahmud et al., 2022)
- **Disadvantages:** IoT as individuals are accepting and utilizing these smarter devices in their activities, which results in them abandoning their security, rising unemployment, excessive reliance on technology, and individuals relinquishing control of a situation as the IoT takes over

There are four main types of IoT technology and networks: cellular, LAN/PAN, low-power WAN, and mesh networks.

5.3 5G

The fifth generation of cellular telecommunications is known as 5G. Customers may expect better download speeds and fewer transmission delays thanks to the latest tech. It also offers an increase in capacity for a service that is more effective.

The versatility of 5G is being built to handle technologies and activities that might not even currently exist. A full-length movie, for instance, can take approximately 6 minutes to download using 4G. A movie can be downloaded via 5G in as low as 15 seconds.

Technically, yes, 5G services are anticipated to enable rates of up to 300 megabytes per second (Mbps) or more, while current 4G speeds range from about 12–36 Mbps as shown in Figure 5.2.

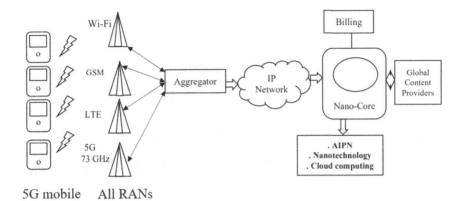

5G mobile   All RANs

FIGURE 5.2   Network architecture for 5G.

Source: Arunachalam et al. (2018).

- **Merits:** Quick transfer rates, lower congestion, which boosts the capacity for off-site computing, a greater number of smart devices, and the ability to build virtual networks (network slicing), which provides more customized access to particular demands, are the main advantages of 5G (Rani et al., 2022).
- **Demerits:** The obtrusions will impede, interfere with, or consume high-frequency broadcasts. Rural residents may not always profit from the connection, since the mobile battery is overheated or rapidly exhausted, even though 5G may bring about actual connectivity for primarily metropolitan areas.

With 5G technology, download rates can reach up to 1.9 Gbps in some circumstances. The upload capabilities, which are rarely greater than 100 Mbps, are less impressive than initially claimed and detract from an area's overall aesthetic appeal.

5.4   6G

Sixth-generation wireless has surpassed innovations in 5G cell phones. According to its capacity to function at higher frequencies, 6G services will have much greater bandwidth and congestion than 5G services.

The 1-microsecond latency of communications is among the goals of the 6G network. Some significant ways that 6G varies from 5G include additional mobility, better use of the radio frequencies, and flexible exposure to technology links as depicted in Figure 5.3.

Merits: With a potential of around 10×105 mobile links per km^2 and very high data throughput (Tb/s), 6G is intended to handle more mobile connectivity than 5G (sub-ms).

As a result, 6G wireless can be used for a variety of areas. It utilizes visible lights that take advantage of LEDs' illumination and high data transmission capabilities,

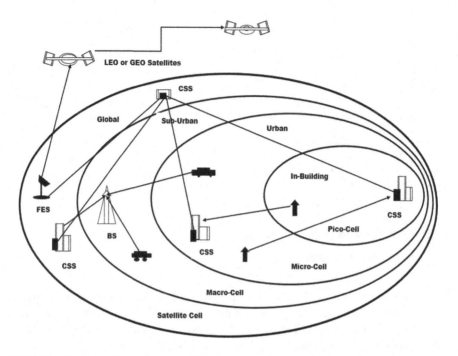

FIGURE 5.3   Network architecture of 6G.

and 6G will simulate additional devices, including the PHY layer and MAC layer. Dedicated hardware iterations are considered necessary for PHY/MAC solutions.

Demerits: The 6G networks' improved speeds and dependability will consume more energy, which could result in higher energy prices for consumers, difficulty in usage, a threat to privacy, a detrimental effect on the environment, and compatibility problems.

5.5   COMPARISON OF DIFFERENT TECHNOLOGY GENERATIONS

This section analyses different generations based on their updating and downloading rates, among other performance metrics as laid out in Table 5.2.

5.6   RESULTS AND DISCUSSION

In this section, recent trends in the IoT, along with 5G and 6G, are discussed through 31 research articles in tabular form. To assess recent research findings, a study of articles in the literature was done (Khang et al., 2022a). In this analysis, each journal's description includes details on its subject, purpose, technology, constraints, performance measures, methodologies, future scope, and rate of citation (ROC) as shown in Tables 5.3, 5.4, and 5.5.

This study developed multidimensional data transfer protocols (MDDCPs) at the IoT's practical level that are based on 5G communication (COMMS; Mao, 2020).

TABLE 5.2

An Evaluation of Various Technological Eras

Generation Parameters	1G	2G	3G	4G	5G	6G
Years	1980	1990	2000	2010	2020	2030
Max DL Speed	50 Kbps	475 Kbps	41.9 Mbps	2.99 Gbps	19.9 Gbps	1Tbps
Avg DL Speed	10 Kbps	49.9 Kbps	7.89 Mbps	99.99 Mbps	299.8 Mbps	0.99 Gbps
Max UL Speed	50 Kbps	469 Kbps	12 Mbps	1.499 Gbps	9.9 Gbps	9.999 Gbps
Avg UL Speed	5 Kbps	49.99 Kbps	1.99 Mbps	49.9 Mbps	99.99 Mbps	0.99 Gbps
E2E Latency (ms)	3000	599.9	119.9	29.9	9.9	0.999
Reliability	90%	98.9%	99.89%	99.989%	99.9989%	99.99989%
Correlation Density (devices/km^2)	–	–	–	10^5	10^6	10^7
Mobility (km/h)	50	149.9	300	350	500	1000
Spectral Efficiency gain	1 time	2 times 1G	5 times 2G	3 times 3G	3 times 4G	10 times 5G
A Rise in Energy Conservation	1time	3 times 1G	5 times 2G	5 times 3G	10 times 4G	100 times 5G
Satellite Integration	No	No	No	No	Partially	Yes
Coverage	29%	35%	50%	60%	70%	99%
Receiver Sensitivity (dBm)	–35	–50	–75	–100	–120	<–130
Time buffer	No actual time	No real time	No actual time	No actual time	No actual time	Actual time
Artificial Intelligence	Absent	–	–	Incomplete	Half	Completely
Extended Fantasy	Absent	No	Absent	Partial	Incompletely	Present
Haptic Interaction	–	Absent	No	Half	Partial	Fully
Technology	AMPS, NMT, TACS	TDMA, CDMA	GPRS	WMax, LTE	LTE advance, OMA, NOMA.	Advanced edge computing, spatial mapping

Source: The comparison table is designed by Pankaj Kumar Dubey, Bindeshwar Singh, Varun Kumar, and Aryan Kumar Dubey.

Note: DL = downloading rate; UL = uploading rate; E2E = end-to-end latency; AMPS = Advanced Mobile Phone System; NMT = Nordic Mobile Telephone; TACS = Total Access Communication System; TDMA = Time Division Multiple Access; CDMA = Code-Division Multiple Access; GPRS = General Packet Radio Service; LTE = long-term evolution; OMA = orthogonal multiple access; NOMA = non-orthogonal multiple access.

TABLE 5.3

MDDCP for the 5G COMMS–Based Practical Layer of the IoT Description Includes Details on Its Subject, Purpose, Technology, Constraints, Performance Measures, Methodologies, Future Scope, and ROC

No.	Title	Objective	Component	Limitation	Performance parameter	Method Utility	Future Scope	ROC
1	MDDCP for the 5G COMMS–based practical layer of the IoT.	Enhance cloud computing and security efficiency.	IoT, and 5G.	Theoretical analysis.	100 times more channel knowledge is essential.	High-band broad detecting technique and MDDCP.	Assess ideas in the future.	1
2	An overview of 5G's potential uses and IoT possibilities.	Enhance transfer capability.	IoT, and 5G.	Low uploading speed.	5G user experience rate can reach 100Mbps to 1Gbps.	OFDM technology.	6G development.	5
3	Toward 6G IoT and the convergence with RoF system.	Enhance convergence rate.	IoT, and 6G.	Lack of infrastructure.	–	Artificial Intelligence method.	6G setup requirements in activity.	12
4	6G Visions: Mobile Ultra-Broadband, Super IoT, and AI.	Adaptability to a challenging and changing radio environment.	IoT, and 6G.	Expensive.	The potential for state-action combos was increased by 150%.	AI technique with IoT COMMS with satellite assistance.	Hardware designs need to be done.	35
5	In the 6G-enabled large IoT area, neural design searches for stable systems.	Improve search efficiency.	IoT, and 6G.	Less accurate method.	Accuracy increased from 8.76 to 15.45.	Neural architecture search and AI method.	Nature-based techniques should be used.	5
6	EdgeGO is a framework for large-scale IoT platforms to exchange mobile services for 6G edge computing.	An increase in resource efficiency.	IoT, and 6G.	Radiation minimization is absent.	Reduce the 6G edge computing installation costs by 25.58%.	6G edge computing, and EdgeGO.	6G advancement.	4

7	The promotion of intelligent vessel traffic services through data-driven trajectory quality enhancement in 6G-enabled marine IoT systems	Promote vessel traffic services.	IoT, and 6G.	Various maritime traffic data are missing.	–	Density-based clustering method and BLSTM-based Supervised learning techniques.	Additional kinds of maritime traffic data, such as those collected by cameras, radar, and BeiDou location.	35
8	In 6G IoT contexts, a multi-objective sanitization paradigm with privacy protection.	By deleting the trade, you can conceal private or delicate data.	IoT, and 6G.	Mainly useful in tiny IoT domains.	Metric hyper-volume and second metric convergence are 360 and 0.86%.	ACO.	Future developments for this approach could include an emphasis on federated learning and edge computing.	47
9	URLLC and eMBB in 5G Industrial IoT: A Survey.	Performance enhancement.	IoT, and 5G.	Less uploading speed.	–	URLLC, eMBB, cloud, and edge computing.	6G.	3
10	URLLC key TECH and standardization for 6G power IoT.	Improve resource utilization and network performance.	IoT, and 6G.	More radiation.	Energy efficiency improved by 2 times while spectrum improves by 2–3 times.	URLLC.	Radiation reduction.	6

(Continued)

TABLE 5.3 *(Continued)*

MDDCP for the 5G COMMS–Based Practical Layer of the IoT Description Includes Details on Its Subject, Purpose, Technology, Constraints, Performance Measures, Methodologies, Future Scope, and ROC

No.	Title	Objective	Component	Limitation	Performance parameter	Method Utility	Future Scope	ROC
11	Trust aspects of IoT in the context of 5G and beyond.	Optimizing network utilization.	IoT, and 5G.	Always about the low-level part.	Uploading speed by 1.5 times than conventional 5G.	URLLC, BC, and IBE.	When looking at fog/edge computing, the initial trust offered by BC is the most important factor.	1
12	A review of opportunities and challenges for the integration of 5G, IoT, and AI in the next generation of wireless services.	Integration of coordination controlling.	IoT, and 5G.	Challenges are not defined.	Downloading speed increases by 1.3 times of conventional one.	AI.	We looked at the feasibility of implementation through modeling as well as the confluence of AI, IoT, and 5G for next-generation SG.	6

Source: The comparison table was designed by Pankaj Kumar Dubey, Bindeshwar Singh, Varun Kumar, and Aryan Kumar Dubey.

Note: MDDCP = multidimensional data transfer protocol; COMMS = communication; IoT = Internet of Things; ROC = rate of citation; OFDM = orthogonal frequency division multiplexing; RoF = radio over fiber; AI = artificial intelligence; BLSTM = bidirectional long short-term memory; ACO = ant colony heuristics; URLLC = ultra-reliable low-latency communication; eMBB = enhanced mobile broadband; TECH = technology; BC = blockchain; IBE = identity-based encryption; SG = smart grid.

TABLE 5.4

A detailed examination of AI and ML methods for the IoT in the 6G period includes details on its subject, purpose, technology, constraints, performance measures, methodologies, future scope, and ROC

No.	Title	Objective	Component	Limitation	Performance parameter	Method Utility	Future Scope	ROC
13	A detailed examination of AI and ML methods for the IoT in the 6G period.	Promoting automated network operations.	IoT, and 6G.	Less accurate.	Energy efficiency increased by 10 times.	AI/ML Algorithms.	Research opportunities in AI/ML-based COMMS systems for future IoT.	3
14	Coordinated resource management managed by AI for 6G-enabled massive IoT task completion.	Decreases the typical decision delay time by around 7%.	IoT, and 6G.	Parallel operation is less accurate.	An 8 percent increase in the average resource hit rate.	Cooperative asset scheduling approach powered by AI.	Interdependent effects of many jobs being carried out in tandem.	6
15	Events use, needs, technology, and future research on the 6G frontiers are surveyed.	Enhanced uploading speed.	6G.	Expensive.	Increase efficiency by 2 times.	AI, MCS, GA, and PSO.	Propose a roadmap for future research directions toward 6G.	60
16	SDT deployment at the margin for IoT networks over 5G.	Preserving polynomial time complexity.	IoT, and 5G.	Sustainable efforts are missing.	The suggested placement policy guarantees SDT latency that is up to 1.4 times less than that of the established proximity-based.	Edge computing, the quadratic assignment issue, etc.	ML techniques for (re)allocating mobility- and time-dependent SDTs.	1

(Continued)

TABLE 5.4 (Continued)
A detailed examination of AI and ML methods for the IoT in the 6G period includes details on its subject, purpose, technology, constraints, performance measures, methodologies, future scope, and ROC

No.	Title	Objective	Component	Limitation	Performance parameter	Method Utility	Future Scope	ROC
17	CSUNs for 6G wide-area IoT.	Improve network efficiency and observe channel patterns.	IoT, and 6G.	Expensive.	Increase efficiency by 10 times.	CSUNs, techniques for allocating resources between domain and spontaneous spectrum sharing.	Practical analysis.	30
18	Implementation challenges and opportunities beyond 5 G and 6G COMMS.	Performance parameter enhancement.	5G, and 6G.	–	Satellite-based operation is 2 times more efficient than others.	MIMO, and SWIPT.	–	21
19	Huge data collecting using the DRLR technique in IoT systems that rely on 6G.	Decrease staffing costs while increasing coverage ratio.	IoT, and 6G.	More chances to reduce cost at more ratio.	Increases gathering rates by 19.015% and the coverage ratio of collecting data by 14.961%.	DRLR, and ML.	Reducing the expense of collecting large amounts of data on the 6G-based infrastructure.	7
20	Multiple-tenant 5G IoT systems with adaptive network slicing.	Excellent privacy, flexibility, and durability while being able to achieve the highest QoS standards.	IoT, and 5G.	Theoretical approach.	A network slice is created in the data plane in about 5.7 ms on average.	SDN, and QoS.	Recommended remedy in 5G and beyond demonstrations in the real world.	6

| 21 | A coalition of 6G and BC in AR/VR space: challenges and future directions (Khang et al., 2023). | Improves the real-time payment service. | 6G. | Integrated coordination improvement. | 43% more surplus incoming packets are stored at the edge computing. | BC. | Consider the potential impact of BC-based reliable 6G solutions on mIoT scenarios. | 5 |
| 22 | Evolution of non-terrestrial networks from 5G to 6G: A Survey. | Network deployment optimization, and channel estimation. | 5G and 6G. | Theoretical analysis. | – | MEC, ML, AI, and RL. | Practical analysis. | 5 |

Source: The comparison table is designed by Pankaj Kumar Dubey, Bindeshwar Singh, Varun Kumar, and Aryan Kumar Dubey.

Note: AI = artificial intelligence; ML = machine learning; IoT = Internet of Things; ROC = rate of citation; COMMS = communication; MCS = Monte Carlo simulation; PSO = particle swarm optimization; SDT = social digital twin; CSUN = cognitive satellite-UAV; MIMO = multiple-input multiple-output; SWIPT = simultaneous wireless information and power transfer; DRLR = dynamic random linear regression; SDN = software-defined networking; QoS = quality of service; BC = blockchain; AR = augmented reality; VR = virtual reality; mIoT = ; MEC = mobile edge computing; RL =.

TABLE 5.5

Comparing Wi-Fi 6 and 5G downlink performance for industrial IoT description includes details on its subject, purpose, technology, constraints, performance measures, methodologies, future scope, and ROC

No.	Title	Objective	Component	Limitation	Performance parameter	Method Utility	Future Scope	ROC
23	Comparing Wi-Fi 6 and 5G downlink performance for industrial IoT.	Compared to Wi-Fi 6, 5G NR-FDD performs better with URLLC and satisfies the sub-ms latency criterion at loads up to five times higher.	IoT, and 5G.	Small network analysis.	In comparison to WiFi6, 5G supports 2–4 times better spectral efficiency with a delay need of less than 1 ms.	AI, and ML.	6G.	4
24.	Low-loss supplementary deposited ultra-short copper paste connectors for 5G and IoT uses in 3D antenna integrated devices.	It was established that the copper paste has great flexibility to bring the place of solder in flip-chip assemblies.	IoT, and 5G.	Conductivity can be increased at a high level.	By using a nano-copper paste, the dB scale at 28 GHz was reduced by 53%.	Evaluation of nano-copper paste, MS, and AI.	Materials for silver plating.	1
25	IoT drone with ultra-low latency edge assistance and QoS awareness for RT stride assessment.	Connected Client Ratio and SDN Coverage Ratio of 0.99 and 0.95.	6G.	Expensive.	20 percent maximum resource utilization after hyper-parameter optimization.	Intelligent disease detection architecture relies on IoDT (Sai Kumar et al., 2023).	Creation of mission-critical, social, and real-time IoT solutions.	1

26	NTNs with MEC capabilities for time-sensitive IoT on 6 Gwide areas.	Boost the MEC-powered NTN's resource utilization efficiency.	IoT, and 6G.	Lacking real-world experience.	With MEC latency enhanced by 2 times.	MEC. ML. and NTN.	7G.	7
27	A detailed examination of the 6G communication's basic blockchain and IoT integration for intelligent urban (Khanh and Khang, 2021).	Enhanced application-related synchronization effectiveness.	IoT, and 6G.	Expensive.	The delay is 10–100 s, and dependability is 99.99999%.	AI, ML, QoS, GA.	By employing 6G, the case study is more genuine.	20
28	With assistance from space, the 6G era will see grant-free broad exposure for satellite-based IoT.	Preserving little signal overhead and enhancing resilience.	IoT, and 6G.	Waveform design is missing.	With N having a MAX value of 1000, the performance boost is almost 22%.	NoMaGFA, NOMA, and RA technologies.	Architecture for Satellite-based IoT with AI enhancements in the 6G period.	8
29	Mega reliable 5G/ B5G enabled IoT MU-MIMO detector relying on DL.	Boost the effectiveness of the system's recognition.	IoT, and 5G.	Less accurate.	MIN the error of client identification based on criteria and random client choice to around 0.15 percent and 0.14 percent, respectively.	DL, and MUMIMO.	Add some additional DL-based techniques to the system under consideration to improve reliability even more.	18

(*Continued*)

TABLE 5.5 *(Continued)*
Comparing Wi-Fi 6 and 5G downlink performance for industrial IoT description includes details on its subject, purpose, technology, constraints, performance measures, methodologies, future scope, and ROC

No.	Title	Objective	Component	Limitation	Performance parameter	Method Utility	Future Scope	ROC
30	Formalization of the metrics for the 5G-IoT environment's subject-system engagement session quality assessment.	The quantities of all distinctive characteristics of the metric grow as the IoT state's workload increases.	IoT, and 5G.	-	Between 5% and 18%, the likelihood of losing an incoming request climbed significantly.	Discrete Markov chain.	AI-based approach.	7
31	Effectual NOMA-enabled IoT network power distribution in the 6G age.	The NOMA scheme relies on SQP and is superior at supporting more IoT technologies.	IoT, and 6G.	Less accurate.	The suggested NOMA technique only reaches EE of 2.7 b/J/Hz when Smax = 2.	KKTOM, MCS, and SQP.	Analyze the influence of IoT device mobility.	18

Note: IoT = Internet of Things; ROC = rate of citation; NR-FDD = New Radio-Frequency division duplexing; AI = artificial intelligence: ML = machine learning; URLLC = ultra-reliable low-latency communication; QoS = quality of service; SDN = software-defined networking; IoDT = Interoperability Device Testing: NTN = non-terrestrial network; MEC = mobile edge computing; GA = Global Accelerator; NoMaGFA = non-orthogonal massive grant-free access; NOMA = non-orthogonal multiple access; RA = random access; MUMINO = multiuser multiple-input multiple-output; DL = Dedicated Line; SQP = sequential quadratic programming; EE = energy efficiency: KKTOM = KKT-based optimization technique.

This present study explains the history of 5G's evolution and several of its major technologies (TECH), including flexible frame layout and orthogonal frequency division multiplexing (OFDM) technologies (Gai et al., 2021).

This chapter evaluates the integration of the radio-over-fiber (RoF) infrastructure with the IoT's upcoming 6G ambition (Chen and Okada, 2021). We presented three key components of 6G aspirations in this chapter: mobile ultra-broadband, super IoT (Rani et al., 2021), and artificial intelligence (AI; Rana et al., 2021; Lin et al., 2019). Throughout this work, a durable methodology is offered in terms of deep learning models in IoT systems with 6G support despite micro disturbances (He et al., 2020).

In this study, we present EdgeGO, a cutting-edge 6G edge computing platform for expansive IoT networks (Cong et al., 2022). Smart vessel tracking facilities in maritime IoT implementations can be ensured by assembling the large volume of spatial file transitions from automatic identification systems (AISs) using density-based clustering methods and bidirectional long short-term memory (BLSTM)–based ensemble training strategies (Wen Liu et al., 2021).

To protect crucial and confidential information, a transactional elimination method named ant colony heuristics (ACO) is presented in this study. This approach was motivated by the requirement to safeguard 6G IoT networks (Jerry et al., 2021).

An in-depth analysis of B5G-assisted Industrial Internet of Things (IIoT) wireless connections is presented in this study, with an emphasis on services for enhanced mobile broadband and ultra-reliable low-latency communication (URLLC; Sharfeen Khan et al., 2022). We conducted a thorough examination of URLLC's development tools and standardization for 6G PIoT in this research (Yang et al., 2021).

In this study, we consider a two-level technique in which trust is simulated by combining a high-level component relying on blockchain and identity-based encryption to control unit IDs and a low-level structure—tied to temporal degradation and comparative entropy characteristics (Evandro et al., 2020).

This study aims to offer a thorough analysis of the next phase of smart grid research demand and technological backdrop and address a prospective SG powered by AI and utilizing IoT and 5G (Ebenezer et al., 2022).

The focus of this chapter is to explain AI/machine learning techniques that can be used to create IoT network activities and services that are safe, effective, and successful (Rezwanul et al., 2022). In addition to supporting the 6G-enabled huge IoT and the real-time order of assets by portable appliances, this chapter examined the dynamic resource provisioning decisions (Kai Lin et al., 2021). It offers a comprehensive examination of the most recent advancements toward 6G in this study (De Alwis et al., 2021), through various techniques artificial intelligence (AI), Monte Carlo Simulation (MCS), Genetic algorithms (GA), and Particle swarm optimization (PSO).

This study, establishment a model for the dynamic deployment of digital twins connected to actual IoT equipment to create associations between social digital twins (Chukhno et al., 2022). To increase the effectiveness of enormous access in large areas, we examine the multi-domain distribution of resources for the cognitive satellite-unmanned aerial vehicle (UAV) network (CSUNs), which includes a satellite and a swarm of UAVs (Chengxiao et al., 2020). Multiple-input multiple-output (MIMO) and simultaneous wireless information and power transfer are two of the fundamental challenges confronting the growth of wireless technology beyond 5G that we have discussed in this research report (Gustavsson et al., 2021).

In addition to using deep reinforcement learning (DRL)–based routing policies, this research provides an innovative ML for collecting information from numerous IoT sensors (Ting et al., 2021). In this study, a software-defined networking compliant system slicing framework for 5G IoT platforms is provided (Matencio Escolar et al., 2021).

The study delivers a brand-new methodical analysis that illustrates how BC and 6G networks have joined forces to enable augmented reality (AR) and virtual reality (VR) verticals (Bhattacharya et al., 2021). An in-depth examination of the development of non-terrestrial networks is provided, with a focus on their significance for 5G technology and, more crucially, how they will be essential for the development of the 6G ecosystem (Mahdi et al., 2022).

The goal of this study is to evaluate and optimize Wi-Fi 6 and 5G New Radio licensed and unlicensed wireless networks for critical IIoT scenarios that have high packet latency and frequency stability (Maldonado et al., 2021).

The creation of nano-copper interfaces for chip-last or flip-chip construction in packaging with incredibly low interconnect losses was the emphasis of this chapter's material synthesis and operational creation sections (Watanabe et al., 2019). In this study, we offer a 6SDN centered on 6G and an interoperability development testing (IoDT)-based intelligent illness identification infrastructure for permeating and extensive surveillance of a severe neurodegenerative ailment (Mukherjee et al., 2021).

In this chapter, we suggest a process-oriented model for developing time-divisional telecommunication and mobile edge computing platforms (Chengxiao et al., 2022). We also offer a detailed and methodical evaluation of the IoT-envisioned sustainability, smart societies relying on BC innovation, and 6G communications infrastructure in this study (Kumari et al., 2021).

Regarding asymmetrical transmissions in satellite-based IoT, a non-orthogonal massive grant-free access system that gains from both random access (RA) and non-orthogonal multiple access (NOMA) techniques is presented (Neng et al., 2022). With the 5G/B5G-enabled IoT, in which the scheme is functioning in conflicting surroundings associated over the time or frequency domain, we suggest an ultra-reliable multiuser MIMO (MUMIMO) analyzer relying on deep learning in this article (Ke et al., 2020).

As opposed to the current approaches, the investigated process is characterized by a discrete Markov chain with a simpler queue mechanism, where the controlled parameter is the overall resources used by all test runs (Kovtun et al., 2022).

In this research, we present a novel power allotment strategy for enhancing the energy and spectrum effectiveness of NOMA-enabled IoT devices using sequential quadratic programming, Karush–Kuhn–Tucker (KKT)-based optimization technique, and Monte Carlo simulation (Ullah Khan et al., 2020).

5.7 CHALLENGES AND OPPORTUNITIES

Opportunities and challenges have been defined in this part. The problems of IoT, 5G, and 6G have been covered in Table 5.6, while the prospects of IoT, 5G, and 6G have been examined in Table 5.7.

TABLE 5.6

Challenges of IoT, 5G, and 6G Technology

No.	IoT	5G	6G
1	Lack of encryption.	Compared to the low-frequency spectrum, millimeter waves have far more complicated hardware architecture criteria.	Transparency and personalization.
2	Inadequate upgrading and assessment.	MIMO technology requires complex algorithms.	Artificial intelligence network.
3	Battery life is a limitation.	Ultra-low Latency Service.	Terahertz communication effect.
4	Connectivity problem.	Confidentiality concerns.	Network security (Khang et al., 2022b).
5	Increased cost and time to market.	Ultra-reliability Network.	Diversity, adaptability, and the opportunity to self.

Source: The comparison table is designed by Pankaj Kumar Dubey, Bindeshwar Singh, Varun Kumar, and Aryan Kumar Dubey.

TABLE 5.7

Opportunities of IoT, 5G, and 6G technology.

IoT	5G	6G
Possibilities in corporate application development, cloud computing, and enabled platforms, connectivity, and devices.	Thanks to 5G's lower latency, faster speed, and reliability, uses for virtual reality (VR) and augmented reality (AR) are getting closer. These could be used in numerous industries, including entertainment, gaming, real estate, industrialization, retail, and tourism.	The THz-band operating framework, resource planning, pervasive AI, huge network digitization, intelligent network contexts, ambient backscatter communication, the internet of things, and a massive MIMO cellular system are just a few of the novel capabilities that the 6G network will propose, which will address the deficiencies of the 5G network.

Source: This table is designed by Pankaj Kumar Dubey, Bindeshwar Singh, Varun Kumar, and Aryan Kumar Dubey.

5.8 CONCLUSION

For learners to evaluate current 5G and upcoming 6G IoT, as well as comprehend their common standards, capabilities, requirements, and uses, this chapter seeks to condense a quantity of information and put it in table form.

This chapter presents a brief literature overview on the subject, outlining the definition, examples, advantages, disadvantages, current trends, problems, and prospects of IoT, 5G, and 6G technologies as well as parametric comparison (Khang et al., 2023c).

The investigation queries about directions and possible research possibilities in AI/ML-based COMMS systems for upcoming IoT, 5G, and 6G are featured at the end of this study. Indirectly illustrative of the research area, key performance indicators, and prospects for 6G in the coming is the utilization paradigm of 6G, which is enlarged from the 5G and classified into more particular applications (Khang, 2023).

The assessment on the topic of 5G network progression, terrestrial radio access networks, and aerial radio access networks are the main foci of this chapter. Massive IoT networks also require a lot of electricity to run, therefore recent advances in the study of achieving sustainability and energy harvesting have become crucial (Jaiswal et al., 2023).

Large IoT networks also require a lot of electricity to run, therefore recent advances in the study of energy sustainability and energy generation have become crucial.

Additionally, the advancement of 6G wireless technologies is heavily dependent on the current generation of advanced monolithic microwave integrated circuits, so previous and forthcoming research in the field of semiconductors should be very active (Hajimahmud et al., 2023).

REFERENCES

Arunachalam S, Kumar S, Kshatriya H, Patil M. "Analyzing 5G: Prospects of Future Technological Advancements in Mobile," *IOSR Journal of Engineering (IOSRJEN)* www.iosrjen.orgISSN(e): 2250-3021, ISSN (p): 2278-8719PP 06-11. Conference: ICIATE 2018. At Mumbai. www.researchgate.net/publication/324941597

Bell AG. *The Photo Phone*, 1980. https://games4esl.com/wp-content/uploads/Reading-Comprehension-Worksheet-Alexander-Graham-Bell.pdf

Bhattacharya P, Saraswat D, Dave A, Acharya M, Tanwar S, Sharma G, Davidson IE. *Coalition of 6G and Blockchain in AR/VR Space: Challenges and Future Directions*, 2021. ISSN: 2169-3536. https://doi.org/10.1109/ACCESS.2021.3136860

Chandra Bose J, Mukherjee DC, Sen D. "Photosynthesis Research," *A Tribute to Sir Jagadish Chandra Bose (1858–1937)*, Springer 1858. https://link.springer.com/article/10.1007/s11120-006-9084-6. Last visit 2007.

Chen N, Okada M. *Toward 6G Internet of Things and the Convergence with RoF System*, 2021. ISSN: 2327-4662. https://doi.org/10.1109/JIOT.2020.3047613

Chengxiao L, Feng W, Chen Y, Wang C-X, Ge N. *Cell-Free Satellite-UAV Networks for 6G Wide-Area Internet of Things*, 2020. ISSN: 0733-8716-1558-0008. https://doi.org/10.1109/JSAC.2020.3018837

Chengxiao L, Feng W, Tao X, Ge N. *MEC-Empowered Non-Terrestrial Network for 6G Wide-Area Time-Sensitive Internet of Things*, 2022. ISSN: 2095–8099. https://doi.org/10.1016/j.eng.2021.11.002

Chukhno O, Chukhno N, Araniti G, Campolo C, Iera A, Molinaro A. *Placement of Social Digital Twins at the Edge for Beyond 5G IoT Networks*, 2022. ISSN: 2327-4662-2372-2541. https://doi.org/10.1109/JIOT.2022.3190737

Cong R, Zhao Z, Min G, Feng C, Jiang Y. *EdgeGO: A Mobile Resource-sharing Framework for 6G Edge Computing in Massive IoT Systems*, 2022. ISSN: 2327-4662. https://doi.org/10.1109/JIOT.2021.3065357

De Alwis C, Kalla A, Pham Q-V, Kumar P, Dev K, Hwang W-J, Liyanage M. "Survey on 6G Frontiers: Trends, Applications, Requirements, Technologies, and Future Research," *IEEE Open Journal of the Communications Society*, 2021. ISSN: 2644-125X. https://doi.org/10.1109/OJCOMS.2021.3071496

Esenogho E, Djouani K, Kurien AM. *Integrating Artificial Intelligence Internet of Things and 5G for Next-Generation Smart Grid: A Survey of Trends Challenges and Prospect*, 2022. ISSN: 2169-3536. https://doi.org/10.1109/ACCESS.2022.3140595

Evandro LC, Silva RS, de Moraes LFM, Fortino G. *Trust Aspects of Internet of Things in the Context of 5G and Beyond*, 2020. ISBN: 978-17281-9541-4. https://doi.org/10.1109/CIoT50422.2020.9244297

Franjic, S. "Wireless Technology Is Easy to Use," *International Journal of Mathematical, Engineering, Biological and Applied Computing*, 2022, https://doi.org/10.31586/ijmebac.2022.294

Gai R, Du X, Ma S, Chen N, Gao S. *A Summary of 5G Applications and Prospects of 5G in the IoT*, 2021. ISBN: 978-1-6654-1540-8. https://doi.org/10.1109/icbaie52039.2021.9389985

Gustavsson U, Frenger P, Fager C, Eriksson T, Zirath H, Dielacher F, Studer C, Parssinen A, Correia R, Matos JN, Belo D, Carvalho NB. *Implementation Challenges and Opportunities in Beyond-5G and 6G Communication*, 2021. ISSN: 2692-8388. https://doi.org/10.1109/JMW.2020.3034648

Hahanov V, Khang A, Litvinova E, Chumachenko S, Hajimahmud VA, Alyar AV. "The Key Assistant of Smart City—Sensors and Tools," *AI-Centric Smart City Ecosystems: Technologies, Design and Implementation* (1st Ed.). CRC Press, 2022. https://doi.org/10.1201/9781003252542-17

Hajimahmud VA, Khang A, Hahanov V, Litvinova E, Chumachenko S, Alyar AV. "Autonomous Robots for Smart City: Closer to Augmented Humanity," *AI-Centric Smart City Ecosystems: Technologies, Design and Implementation* (1st Ed.). CRC Press, 2022. https://doi.org/10.1201/9781003252542-7

Hajimahmud VA. et al. (Eds.). "The Role of Data in Business and Production," *AI-Aided IoT Technologies and Applications in the Smart Business and Production*. CRC Press, 2023. https://doi.org/10.1201/9781003392224-2

He K, Wang Z, Li D, Zhu F, Fan L. *Ultra-Reliable MU-MIMO Detector Based on Deep Learning for 5G/B5G-Enabled IoT*, 2020. ISSN: 1874-4907. https://doi.org/10.1016/j.phycom.2020.101181

Heinrich Hertz-Theorist and Experimenter. *JD Kraus—IEEE Transactions on Microwave Theory*, 1888, https://ieeexplore.ieee.org/abstract/document/3601/

Jaiswal N, Misra A, Misra PK, Khang A (Eds.). "Role of the Internet of Things (IoT) Technologies in Business and Production," *AI-Aided IoT Technologies and Applications in the Smart Business and Production*. CRC Press, 2023. https://doi.org/10.1201/9781003392224-1

Jerry C-Wl, Srivastava G, Zhang Y, Djenouri Y, Aloqaily M. *Privacy-Preserving Multi-Objective Sanitization Model in 6G IoT Environments*, 2021. ISSN: 2327-4662-2372-2541. https://doi.org/10.1109/JIOT.2020.3032896

Kai L, Li Y, Zhang Q, Fortino G. *AI-Driven Collaborative Resource Allocation for Task Execution in 6G-Enabled Massive IoT*, 2021. ISSN: 2327-46622372–2541. https://doi.org/10.1109/JIOT.2021.3051031

Ke W, Xu P, Chen CM, Kumari S, Shojafar M, Alazab M. *Neural Architecture Search for Robust Networks in 6G enabled Massive IoT Domain*, 2020. ISSN: 2327-4662. https://doi.org/10.1109/JIOT.2020.3040281

Khang A. "Material4Studies," *Material of Computer Science, Artificial Intelligence, Data Science, IoT, Blockchain, Cloud, Metaverse, Cybersecurity for Studies*, 2021. www.researchgate.net/publication/370156102_Material4Studies

Khang A (Ed.). *AI-Oriented Competency Framework for Talent Management in the Digital Economy: Models, Technologies, Applications, and Implementation*. CRC Press, 2023. https://doi.org/10.1201/9781003440901

Khang A, Gupta SK, Shah V, Misra A (Eds.). *AI-Aided IoT Technologies and Applications in the Smart Business and Production*. CRC Press, 2023. https://doi.org/10.1201/9781003392224

Khang A, Hahanov V, Abbas GL, Hajimahmud VA. "Cyber-Physical-Social System and Incident Management," *AI-Centric Smart City Ecosystems: Technologies, Design and Implementation* (1st Ed.). CRC Press, 2022a. https://doi.org/10.1201/9781003252542-2

Khang A, Rani S, Sivaraman AK. *AI-Centric Smart City Ecosystems: Technologies, Design and Implementation* (1st Ed.). CRC Press, 2022b. https://doi.org/10.1201/9781003252542

Khang A, Shah V, Rani S. *AI-Based Technologies and Applications in the Era of the Metaverse* (1st Ed.). IGI Global Press, 2023c. https://doi.org/10.4018/9781668488515

Khanh HH, Khang A. "The Role of Artificial Intelligence in Blockchain Applications," *Reinventing Manufacturing and Business Processes through Artificial Intelligence* (pp. 20–40). CRC Press, 2021. https://doi.org/10.1201/9781003145011-2

Kovtun V, Izonin I, Gregus M. *Formalization of the Metric of Parameters for Quality Evaluation of the Subject-System Interaction Session in the 5G-IoT Ecosystem*, 2022. ISSN: 1110-0168. https://doi.org/10.1016/j.aej.2022.01.054

Kumari A, Gupta R, Tanwar S. "Amalgamation of Blockchain and IoT for Smart Cities Underlying 6G Communication: A Comprehensive Review," *Computer Communications*, 2021. ISBN 0140-3664. https://doi.org/10.1016/j.comcom.2021.03.005

Lin Z, Liang YC, Niyato D. "6G Visions: Mobile Ultra-broadband, Super Internet of Things, and Artificial Intelligence," *China Communications*, 2019, https://doi.org/10.23919/jcc.2019.08.001

Mahdi Azari M, Solanki S, Chatzinotas S, Kodheli O, Sallouha H, Colpaert A, Mendoza Montoya JF, Pollin S, Haqiqatnejad A, Mostaani A, Lagunas E, Ottersten B. *Evolution of Non-Terrestrial Networks From 5G to 6G: A Survey*, 2022. ISSN: 1553-877X-2373-745X. https://doi.org/10.1109/COMST.2022.3199901

Maldonado R, Karstensen A, Pocovi G, Esswie AA, Rosa C, Alane O, Kasslin M, Kolding T. *Comparing Wi-Fi 6 and 5G Downlink Performance for Industrial IoT*, 2021. ISSN: 2169-3536. https://doi.org/10.1109/ACCESS.2021.3085896

Mao Y. *Multi-Dimensional Data Communication Protocols for the Physical Level of the IoT Based on 5G Communication*, 2020. ISBN: 978-1-7281-4685-0. https://doi.org/10.1109/icict48043.2020.9112399

Marconi G. "Radio Telegraphy," *Journal of the American Institute of Electrical*, 1922. https://ieeexplore.ieee.org/abstract/document/6591020/

Matencio Escolar A, Alcaraz-Calero JM, Salva-Garcia P, Bernabe JB, Wang Q. "Adaptive Network Slicing in Multi-Tenant 5G IoT Networks," *IEEE Access*, 2021, https://doi.org/10.1109/ACCESS.2021.3051940

Mukherjee A, Mukherjee P, De D, Dey N. "QoS-Aware 6G-Enabled Ultra-Low Latency Edge-Assisted Internet of Drone Things for Real-Time Stride Analysis," *Computers and Electrical Engineering*, 2021. ISSN: 0045-7906. https://doi.org/10.1016/j.compeleceng.2021.107438

Neng Y, Yu J, Wang A, Zhang R. *Help from Space: Grant-Free Massive Access for Satellite-Based IoT in the 6G Era*, 2022. ISSN: 2352–8648. https://doi.org/10.1016/j.dcan.2021.07.008

Rana G, Khang A, Sharma R, Goel AK, Dubey AK (Eds.). *Reinventing Manufacturing and Business Processes Through Artificial Intelligence*. CRC Press, 2021. https://doi.org/10.1201/9781003145011

Rani S, Bhambri P, Kataria A, Khang A. "Smart City Ecosystem: Concept, Sustainability, Design Principles and Technologies," *AI-Centric Smart City Ecosystems: Technologies, Design and Implementation* (1st Ed.). CRC Press, 2022. https://doi.org/10.1201/9781003252542-1

Rani S, Chauhan M, Kataria A, Khang A (Eds.). "IoT Equipped Intelligent Distributed Framework for Smart Healthcare Systems," *Networking and Internet Architecture*. CRC Press, 2021. https://doi.org/10.48550/arXiv.2110.04997

Rezwanul MM, Matin MA, Sarigiannidis P, Goudosa SK. *Comprehensive Review on Artificial Intelligence/Machine Learning Algorithms for Empowering the Future IoT Toward 6G Era*, 2022. ISSN: 21693536. https://doi.org/10.1109/ACCESS.2022.3199689

Sai Kumar DV, Chaurasia R, Misra A, Misra PK, Khang A. "Heart Disease and Liver Disease Prediction using Machine Learning," *Data-Centric AI Solutions and Emerging Technologies in the Healthcare Ecosystem* (1st Ed., p. 4). CRC Press, 2023. https://doi.org/10.1201/9781003356189-13

Sharfeen Khan B, Jangsher S, Ahmed A, Al-Dweik A. "URLLC and eMBB in 5G Industrial IoT: A Survey," *IEEE Open Journal of the Communications Society*, 2022. ISSN: 2644-125X. https://doi.org/10.1109/OJCOMS.2022.3189013

Srinivasan CR, Bodduna R, Saikalyan P, Yadav ES. "A Review on the Different Types of Internet of Things (IoT)," *Journal of Advanced Research in Dynamical and Control Systems*, 11(1), 154–158 January 2019, Lab: S Mamatha's Lab. www.researchgate.net/publication/332153657

Tailor RK, Pareek R, Khang A (Eds.). "Robot Process Automation in Blockchain," *The Data-Driven Blockchain Ecosystem: Fundamentals, Applications, and Emerging Technologies* (1st Ed., pp. 149–164). CRC Press, 2022. https://doi.org/10.1201/9781003269281-8

Tainter CS. "Arabic Youth Mobile Usages in Egypt & Bahrain Mobile as 'Black Box'," *A Qualitative Approach*, 1880. https://core.ac.uk/download/pdf/234652868.pdf

Tilahun Mihret E, Haile G. "4G, 5G, 6G, 7G and Future Mobile Technologies February 2021," *American Journal of Computer Science and Technology*, 9(2), 75. www.researchgate.net/publication/349392966

Ting L, Liu W, Zeng Z, Xiong NN. *DRLR: A Deep Reinforcement Learning Based Recruitment Scheme for Massive Data Collections in 6G-Based IoT Networks*, 2021. ISSN: 2327-4662-2372-2541, https://doi.org/10.1109/JIOT.2021.3067904

Ullah Khan W, Jameel F, Jamshed MA, Pervaiz H, Khan S, Liu J. *Efficient Power Allocation for NOMA-Enabled IoT Networks in 6G Era*, 2020. ISSN 1874-4907. https://doi.org/10.1016/j.phycom.2020.101043

Watanabe AO, Wang Y, Ogura N, Markondeya Raj P, Smet V, Tentzeris MM, Tummala RR. *Low-Loss Additively-Deposited Ultra-Short Copper-Paste Interconnections in 3D Antenna-Integrated Packages for 5G and IoT Applications*. IEEE, 2019. ISBN: 978-1-7281-1500-9. https://doi.org/10.1109/ECTC.2019.00152

Wen Liu R, Nie J, Garg S, Xiong Z, Zhang Y, Shamim Hossain M. *Data-Driven Trajectory Quality Improvement for Promoting Intelligent Vessel Traffic Services in 6G-Enabled Maritime IoT Systems*, 2021. ISSN: 2327-4662-2372-254. https://doi.org/10.1109/JIOT.2020.3028743.

Yang X, Zhou Z, Huang B. *URLLC Key TECH and Standardization for 6G Power IoT*, 2021. ISSN: 2471-2825-2471-2833. https://doi.org/10.1109/MCOMSTD.001.2000042

6 Monitoring Vehicle Noise and Pollution with a Smart Internet of Things System

*Ravi Chandra B., Thota Teja, Thirupathaiah S.,
Syed Shashavali, and Patan Sohail Khan*

6.1 INTRODUCTION

Environmental issues are becoming increasingly severe. Asthma attacks can be brought on by ground-level ozone and particulate matter from automobiles, buses, and trucks. More than half of the carbon monoxide (CO) in the atmosphere may be attributed to transportation. The health of people could be endangered by this CO (EHMS/01l2012, 2014).

Air pollution increases the risk of cancer and contributes to chronic obstructive pulmonary disease (COPD). Pollution from vehicles (Andersen et al., 2011) and trucks can also harm public health in large metropolitan areas. One of the main sources of air pollution, accounting for 70% of all air pollution, is the release of harmful gases from moving cars. Monitoring pollution levels and identifying polluting vehicles are necessary for controlling air pollution.

Cities may benefit from using the Internet of Things (IoT) to track vehicle-related air pollution as well as collect and analyze information on the degree of pollution along various city routes. Vehicle pollution is becoming a big issue in today's age. The incidence of automotive pollution has surged in recent years due to the exponential growth in the number of automobiles.

The pollution created by vehicles disrupts the overall ecological balance that exists in nature because it is hazardous to not only humans but also the environment. Vehicular pollution is the outcome of massive urbanization and population growth over the last decade. The automobile industry is becoming the backbone of economic development in emerging countries such as India, yet pollution created by autos is also a risk (Postscapes, 2023).

Gas sensor technologies are still in development and have yet to attain their full capability and application potential. Some systems are extremely accurate, but they are also prohibitively expensive for large-scale adoption. However, low-cost solutions can be implemented using a sensor network, and the issue of false positives may be reduced by using data multiplicity.

DOI: 10.1201/9781003392224-6

A huge number of outputs from various sensors can be compared for a more accurate analysis. As a result, wireless sensor networks provide strong new methods for monitoring air quality. This experiment indicates a promising path for monitoring engine emissions, particularly carbon dioxide (CO_2) emissions. A gas sensor is used to detect motor vehicle pollution. This enables data tracking to be done cheaply and in real time (Khang et al., 2022).

The car's owner may easily determine the emission level ahead of time. The goal of this system is to reduce CO emissions in the atmosphere while also developing a tiny car pollutant-detecting device that could be installed on the vehicle itself. Sensors monitor the vehicle's smoke ratio, and the data are stored on the owner's phone, where it may be submitted to officials when they inquire about the vehicle's emissions report, such as when they visit a hill station.

Winters are no longer cold, and glaciers are melting faster than before due to changed weather patterns. It could be quite useful to have a CO gas detector on hand. Knowing the proper level of pollution in the environment might help us plan for future problems and follow their sources to avert those (Xiaofeng et al., 2015).

Air and noise pollution are primarily brought on by transportation. Air pollution is responsible (Euro.WHO, 2023) for about 100,000 premature adult deaths per year. Forty million people are exposed to air that exceeds World Health Organization (WHO) air quality guideline values for at least one pollutant in the 115 largest cities of the European Union.

The largest cause of climate change and the source of CO_2 emissions from fossil fuels is transportation. Due to a rise in transport volumes that exceeded improvements in the vehicle economy, transportation accounted for roughly 35% of all energy consumption in the 25 EU countries in 2004, leading to a 20% net increase in greenhouse gas emissions over the preceding decade.

Automobiles produce sulfur dioxide (SO_2), nitrogen oxide (NO_2), and CO. Using the Internet of Things, our suggested technique resolves this issue. The term *Internet of Things* (IoT) describes how physical objects, structures, cars, and other items are connected by electronics, sensors, software, actuators, and network connectivity. This enables each object to gather and share data. To detect the pollutant level in this chapter, semiconductor sensors were utilized.

6.2   LITERATURE SURVEY

6.2.1   Minimization of CO and CO_2 from Two-Wheeler Exhaust Gases

This chapter discusses a CO and CO_2 removal method for two-wheeler motorcycle exhaust emissions based on adsorption. The main objective of this chapter (Lee and Lee, 2001) was to develop an adsorption model that can actively reduce contaminants from vehicle exhaust emissions while being economical and environmentally benign.

In this instance, adsorption is carried out in a device that resembles an absorber and has a charcoal pad through which the exhaust gases pass (Khang et al., 2024a).

The cost of a product can be greatly decreased by using charcoal powder as an adsorbent. A fraction of emissions from two-wheelers can simply be added to the adsorption-based model. This device reduces CO_2 emissions significantly after adsorption from high-output exhaust gases. In this situation, the performance of charcoal for CO_2 adsorption is estimated to be 20%.

6.2.2 WIRELESS SENSOR NETWORK FOR REAL-TIME AIR POLLUTION MONITORING

This study outlines a methodology for monitoring ambient air quality in real time. A machine-to-machine link connects the network's numerous scattered monitoring stations, which interact wirelessly with a backend server (Kularatna et al., 2008).

Data collected in real time from the stations are transformed into information that consumers can access via web portals and mobile apps via the backend server. Gas and weather sensors are installed in every building (Hahanov et al., 2022).

Additionally, it has data-logging capabilities and Wi-Fi connectivity. Four solar energy stations are installed over a 1-kilometer region as part of the system's experimental phase. Performance analysis and testing are done 4 months after the information collection period. India is the second-largest nation in the world by population.

The nation's economy is growing quickly, and the transportation industry is essential to that growth. To detect and predict air pollution brought on by vehicles, this project makes use of an Arduino board, three gas sensors (MQ-2, MQ-7, MQ-135), a Global System for Mobile Communication (GSM) module, and a solenoid valve.

The project is primarily an IoT one. The sensors pick up the gaseous pollution emissions from the car's exhaust. Depending on the threshold value, each sensor's corresponding LED flashes to show how many hazardous gases it can detect. A warning message is sent to the owner, notifying them that the car has exceeded the BS IV safe emission standards.

After conceptualizing and analyzing it using a few IEEE papers and our ideas for preventing and controlling air pollution, we propose this method. An important aspect of this gadget dramatically lowers air pollution.

The system stops/chokes the fuel flow from the fuel injector via the solenoid valve when the emission rate exceeds the specified threshold value. The IoT is a developing field, and its technology aids in the automation of practically everything. As a result, we employ the benefits of this field to reduce a key environmental concern, air pollution. This approach has the potential to cause a paradigm change in the idea of preventing and reducing air pollution.

Ghewari et al. (2017) outline methods for monitoring roadside air pollution and tracking vehicles that pollute above a certain level. Because of increased industrialization and urbanization, a large amount of particulate matter and harmful gases are created. Emissions are poorly controlled, and catalytic converters are rarely used. A critical issue that has existed for a long time is the increased use of automobiles.

There was an air quality monitoring system (AQMS) that adhered to IEEE standards. This procedure made use of the GSM wireless communication module. Real-time monitoring of dangerous gases like CO_2, CO, NO_2, and SO_2 was done using IoT sensor arrays. To hold the gas values in the IoT, the graphical user interface (GUI) was developed (Ayele and Mehta, 2018).

A global air emission monitoring program has been started with a three-phase air pollution surveillance system. Gas sensors, an Arduino integrated development environment (IDE), and a Wi-Fi module are all included in this kit. Gas sensors are included in this IoT kit. Air quality data are collected by sensors placed in a town or region, which is then are communicated in real-time to a server where they may be observed in a particular area (Dhingra et al., 2019).

PIC (2015) has a low-cost geo-referenced air-pollution measuring system that is used as an early warning tool. The system is linked to a low-cost board with built-in Wi-Fi, allowing the data to be sent to the IoT cloud over Message Queuing Telemetry Transport (MQTT) protocol in real time, allowing the geo-referenced information to be printed on an open-access platform utilizing IoT.

6.3   THE IOT

Many definitions of the IoT succinctly outline the technology's primary attributes and qualities as well as what users might expect when connecting things to the internet. With the use of various data collection and networking technologies; distributed computing, such as cloud computing; and connecting physical and virtual domains, the IoT is a worldwide network architecture.

The IoT enables objects to communicate with one another, access information on the internet, store and retrieve data, and interact with users, resulting in smart, pervasive, and always-connected environments. Major technological advancements and developments are required to achieve such intelligence within computing environments (Hajimahmud et al., 2022).

The experts believe that the IoT will soon take on a new structure and that the number of ubiquitous devices will skyrocket. According to the IoT vision, individual everyday objects like cars, roads in transportation systems, pacemakers, wirelessly connected pill-shaped cameras in digestive tracks for healthcare applications, refrigerators, or other household items like cattle can be fitted with sensors that can track useful information about these objects.

Uniquely addressable objects and their virtual representations on a topology resembling a network are what the IoT is supposed to be composed of. Such objects might be able to connect to information about them or broadcast real-time sensor data about their condition or other important properties related to the object.

The uniquely addressable items are connected to the internet, and data about them can be transferred using the same protocol that links computers to the internet. Due to their ability to interact with one another through feeling and communication, the objects can understand complex environmental behaviors and frequently enable autonomic reactions to challenging situations without human intervention (Khang, 2023b).

Pervasive and ubiquitous computing is made possible by the simultaneous automatic production of data from the environment by a huge number of devices.

The future internet is thought to include the IoT. Research needs to concentrate on significant issues including identity, interoperability, privacy, and security for IoT technology to advance quickly (Khang et al., 2022). Future networks like 5G, big data, and cloud technologies must all be integrated with IoT (Jianli et al., 2011).

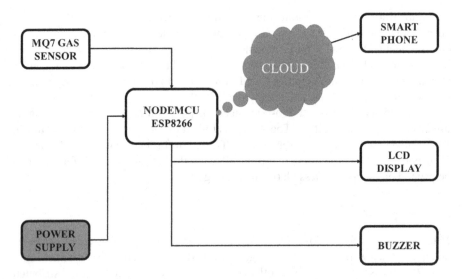

FIGURE 6.1   Block diagram of gas sensors and Node MCU ESP8266.

Source: Khang (2021).

6.4   PROPOSED METHODOLOGY

We show how the IoT may be utilized to successfully address the problem of car pollution in this section. It is necessary to continuously evaluate the quality of the air to ascertain pollution levels and the presence of particular dangerous chemicals as Figure 6.1.

6.4.1   Liquid Crystal Display

Displays that use liquid crystals, or liquid crystal displays (LCDs) use a technology that is a combination of the solid and liquid phases of materials. It is a 16×2 LCD electronic display module with 16 characters and 2 rows. A crystal liquid is used to form the viewable image on an LCD.

This method is used to display graphics on laptop computers and other electronic devices. It contains 16 pins, including 8 input pins and a 5-volt external input voltage pin. It works by blocking light rather than creating light. It has the particular advantage of consuming less energy than an LED or a cathode ray tube as shown in Figure 6.2.

6.4.2   Node MCU

The Node MCU-ESP8266 (Skraba, 2016) is a Wi-Fi-capable microcontroller. It is an open-source IoT platform. This little device connects microcontrollers to a Wi-Fi network and allows them to create simple Transmission Control Protocol/Internet Protocol (TCP/IP) connections using Hayes-style commands.

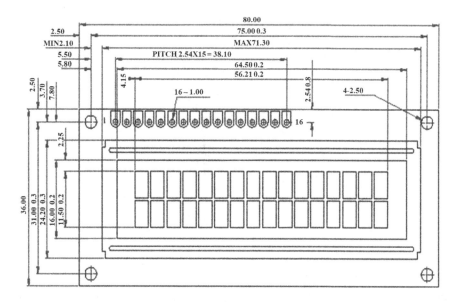

FIGURE 6.2   General pin diagram of liquid crystal display.

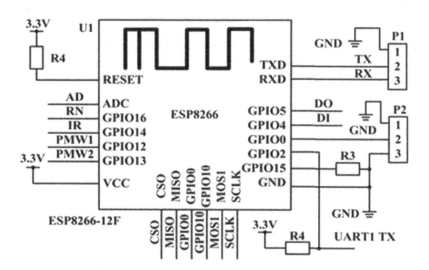

FIGURE 6.3   General Node MCU ESP8266.

The firmware is referred to as the Node MCU. This firmware makes use of the Lua scripting language. Its operating system and CPU are XTOS and ESP8266 as shown in Figure 6.3. It contains 128KB of memory and 4MB of storage. The controller's power is supplied via USB.

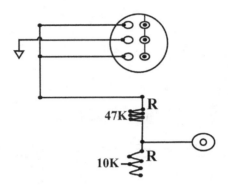

FIGURE 6.4   MQ7 Gas Sensor.

Source: Khang (2021).

6.4.3   MQ-7 GAS SENSORS

MQ-7 sensors (Stankovic, 2014) detect CO levels ranging from 20 to 2000 ppm. The sensitivity of the sensor can be altered using a potentiometer. There are four pins (power, ground, digital and analog output).

The production and CO gas density are inversely related. Analog output is how the sensor data are presented. The sensitive chemical SnO_2 is used in the MQ135 gas sensor.

When the concentration of a gas, which has lower conductivity in clear air, increases, so does its conductivity. It reacts violently to fumes, ammonia, benzene steam, and hydrogen sulfide. It is used to identify gases with concentrations ranging from 10 to 10,000 ppm.

6.4.4   THE IoT

The IoT is a network of physical objects with electronics, sensors, software, actuators, and the internet that connect to exchange data. The IoT enables objects to sense data and control them remotely as Figure 6.5.

6.4.5   BUZZER

A buzzer, also referred to as a beeper, is a mechanical, electromechanical, or piezoelectric audio signaling device (piezo for short). Common uses for buzzers and beepers include alarm clocks, timers, and confirming human input like a mouse click or a keyboard.

We are making use of a piezoelectric buzzer. This buzzer generates sound by reversing the piezoelectric effect. The fundamental concept is the application of electric power across a piezoelectric material to produce pressure fluctuation or strain as Figure 6.6.

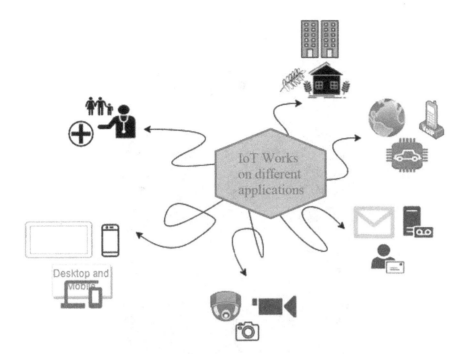

FIGURE 6.5 General Internet of Things.

FIGURE 6.6 Buzzers.

6.4.6 CLOUD-BLYNK

All this information is sent through the cloud to rescue systems, family members, and others as shown in Figure 6.7.

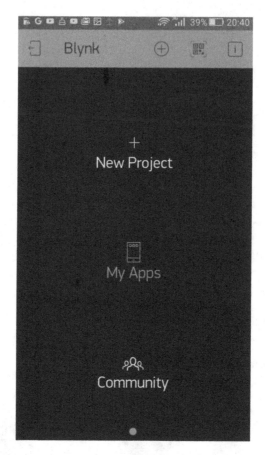

FIGURE 6.7 Blynk app on mobile.

6.5 SOLUTIONS ANALYSIS

Step 1: The Node MCU ESP8266 receives its output from the MQ7 gas sensor, which is installed in the vehicle's smoke outlet. The MQ7 gas sensor will measure the amount of CO emissions in parts per million.

Step 2: The Blynk app stores the sensor data on a smartphone. The CO value is saved and accessible at any time by date. Figure 6.8 shows the vehicle's carbon monoxide emissions over time, where the x-axis is the data and the y-axis is the Co and SnO_2 values.

Step 3: If the CO emission level of the car exceeds the required threshold, Blynk receives the data and alerts the specified user through email. The owner of the car receives a message.

Step 4: Blynk is linked to the system. Blynk is a cloud-based IoT analytics platform that allows you to gather, view, and analyze live data streams. Blynk

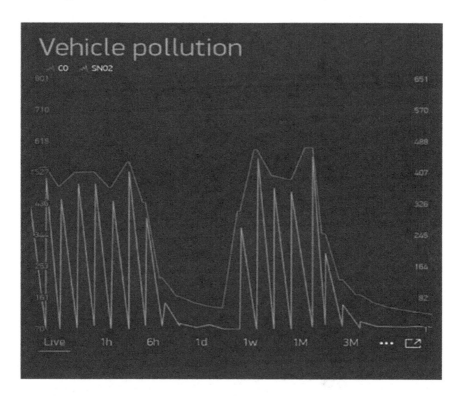

FIGURE 6.8   Blynk cloud data view in smart mobile.

provides real-time visualizations of knowledge posted to Blynk by your devices when there is no gas around as shown in Figure 6.9.

The sensor returns a value of 90, indicating that the emission level is safe up to 350 ppm and should not exceed 1000 ppm. When it exceeds the 1000 ppm limit, it causes headaches, lethargy, and stagnant, stuffy air as shown in Figure 6.10.

Step 5: Because of air pollution, we receive distinct graphical representations with various parameters.

Step 6: The output devices are an LCD and a buzzer. The LCDs gas data in ppm, and the buzzer sounds when the ppm exceeds a preset limit. Blynk is an IoT solution that is used to control Arduino, Raspberry Pi, and other similar devices through the internet as shown in Figure 6.11.

6.6   CONCLUSION

To protect our environment, it is necessary to lessen the environmental pollution brought on by the gas emissions from vehicle exhaust, including CO, hydrocarbons, and NO_2 (Jaiswal et al., 2023).

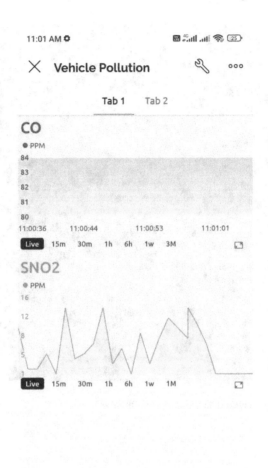

FIGURE 6.9   Vehicle pollution.

The suggested approach gives the most effective technique to keep track of the gases that a vehicle's exhaust releases into the atmosphere, extending the life of the engine and reducing environmental pollution (Bhambri et al., 2022).

The suggested remedy is low-cost and easy to keep up. In the future, GPS might be used to track data and figure out how many gases are generated from driving cars in a certain location (Khang et al., 2023b).

FIGURE 6.10   Results of different levels of ppm.

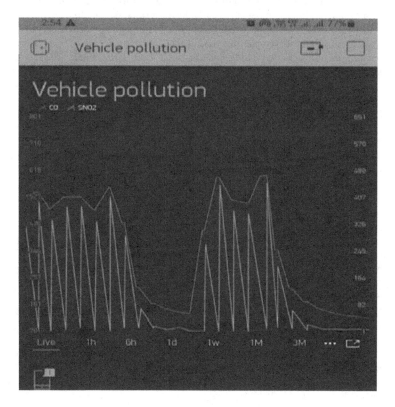

FIGURE 6.11   A digital dashboard on a smartphone that displays real-time air quality values for the local surroundings in this project created by Blynk.

REFERENCES

Andersen ZJ, Hvidberg M, Jensen SS, Ketzel M, Loft S, Sorensen M, Raaschou-Nielsen O. "Chronic Obstructive Pulmonary Disease and Long-Term Exposure to Traffic-Related Air Pollution: A Cohort Study," *American Journal of Respiratory and Critical Care Medicine*, 1, 2011. www.atsjournals.org/doi/abs/10.1164/rccm.201006-0937OC

Ayele TW, Mehta R. "Air Pollution Monitoring and Prediction Using IoT," *International Conference on Inventive Communication and Computational Technologies (ICICCT) 2018*, 1741–1745, 2018. https://doi.org/10.1109/ICICCT.2018.8473272

Bhambri P, Rani S, Gupta G, Khang A. *Cloud and Fog Computing Platforms for Internet of Things.* CRC Press, 2022. ISBN: 978-1-032-101507. https://doi.org/10.1201/9781003213888

Dhingra S, Madda RB, Gandomi AH, Patan R, Daneshmand M. "Internet of Things Mobile-Air Pollution Monitoring System (IoT-Mobair)," *IEEE Internet Things*, 6(3), 5577–5584, 2019. https://doi.org/10.1109/JIoT.648890710.1109/JIOT.2019.2903821

EHMS/0112012. *Epidemiological Study on Effect of Air Pollution on Human Health (Adults) in Delhi.* Environmental Health Management Series: EHMS/0112012, Central Pollution Control Board, Government of India, 2014. https://ieeexplore.ieee.org/abstract/document/6909157/

Euro.WHO. *Air Pollution and Climate Change*, 2023. www.euro.who.int/en/healthtopics/environment-andhealth/Transport-and-health/data-and-statistics/air-pollution-andclimate-change2

Ghewari R. S., *Action Recognition from Videos using Deep Neural Networks*, 2017, UNIVERSITY OF CALIFORNIA, SANDIEGO. https://escholarship.org/content/qt2mr798mn/qt2mr798mn_noSplash_9068e3bea835fde00a6fd904e0039c79.pdf

Hahanov V, Khang A, Litvinova E, Chumachenko S, Hajimahmud VA, Alyar AV. "The Key Assistant of Smart City—Sensors and Tools," *AI-Centric Smart City Ecosystems: Technologies, Design and Implementation* (1st Ed.). CRC Press, 2022. https://doi.org/10.1201/9781003252542-17

Hajimahmud VA, Khang A, Hahanov V, Litvinova E, Chumachenko S, Alyar AV. "Autonomous Robots for Smart City: Closer to Augmented Humanity," *AI-Centric Smart City Ecosystems: Technologies, Design and Implementation* (1st Ed.). CRC Press, 2022. https://doi.org/10.1201/9781003252542-7

Jaiswal N, Misra A, Misra PK, Khang A (Eds.). "Role of the Internet of Things (IoT) Technologies in Business and Production," *AI-aided IoT Technologies and Applications in the Smart Business and Production.* CRC Press, 2023. https://doi.org/10.1201/9781003392224-1

Jianli P, Paul S, Jain R. "A Survey of the Research on Future Internet Architectures," *IEEE Communications Magazine*, 49(7), 26–36, 2011. https://ieeexplore.ieee.org/abstract/document/5936152/

Khang A. "Material4Studies," *Material of Computer Science, Artificial Intelligence, Data Science, IoT, Blockchain, Cloud, Metaverse, Cybersecurity for Studies*, 2021. www.researchgate.net/publication/370156102_Material4Studies

Khang A (Ed.). *AI-Oriented Competency Framework for Talent Management in the Digital Economy: Models, Technologies, Applications, and Implementation.* CRC Press, 2023. https://doi.org/10.1201/9781003440901

Khang A, Gupta SK, Rani S, Karras DA. *Smart Cities: IoT Technologies, Big Data Solutions, Cloud Platforms, and Cybersecurity Techniques* (1st Ed.). CRC Press, 2023b. https://doi.org/10.1201/9781003376064

Khang A, Ragimova NA, Hajimahmud VA, Alyar AV. "Advanced Technologies and Data Management in the Smart Healthcare System," *AI-Centric Smart City Ecosystems: Technologies, Design and Implementation* (1st Ed.). CRC Press, 2022. https://doi.org/10.1201/9781003252542-16

Kularatna N, Senior Member IEEE, Sudantha BH. "An Environment Air Pollution Monitoring System Based on the IEEE1451 Standard for Low Cost Requirements" *IEEE Sensors Journal*, 8, 415–422, April 2008. https://ieeexplore.ieee.org/abstract/document/4459727/

Lee DD, Lee DS. "Environmental Gas Sensors," *IEEE Sensors Journal*, 1(3), 214–215, October 2001. www.researchgate.net/profile/Dae-Sik-Lee/publication/3430880_Environmental_Gas_Sensors/links/53eccca30cf2981ada10eaa5/Environmental-Gas-Sensors.pdf

PIC. 2015 International Conference on Technologies for Sustainable Development (ICTSD). *Date Added to IEEE Xplore: April 30, 2015 INSPEC Accession Number: 15092517.* IEEE Conference, February 4–6, 2015. https://doi.org/10.1109/ICTSD.2015.7095909

Postscapes. "IoT Standards and Protocols," *Postscapes*, 2023. www.postscapes.com/internet-of-things-protocols/

Shah V, Khang A. "Internet of Medical Things (IoMT) Driving the Digital Transformation of the Healthcare Sector," *Data-Centric AI Solutions and Emerging Technologies in the Healthcare Ecosystem* (1st Ed., p. 1), CRC Press, 2023. https://doi.org/10.1201/9781003356189-2

Skraba A. "Streaming Pulse Data to the Cloud with Bluetooth LE or NODEMCU ESP8266," *2016 5th Mediterranean Conference on Embedded Computing (MECO), Bar*, 428–431, 2016. www.china-total.com/Product/meter/gas-sensor/Gassensor.htm

Stankovic JA. "Research Directions for the Internet of Things," *IEEE Internet Things*, 1(1), 3–9, 2014. https://ieeexplore.ieee.org/abstract/document/6774858/

Xiaofeng M, Jianhua G, Peng C, Chaozhong W. "An Inexact Bus Departure Frequency Model for Traffic Pollution Control," *International Conference on Transportation Information and Safety (ICTIS), Wuhan, China*, June 25–28, 2015. https://ieeexplore.ieee.org/abstract/document/7232195/

7 Ubiquitous Sensor-Based Internet of Things Platform for Smart Farming

Neha Jain, Yogesh Awasthi, and Jain R. K.

7.1 INTRODUCTION

The coalescence of technology in agriculture makes a system self-efficient, and such a system developed to be called a smart agriculture (SA)–based system. To get the best improvement and benefits, SA adjusts regular cultivating strategies to particular states for each purpose of the yield by applying various advances: self-actuating miniaturized low-power-consuming devices, ubiquitous networks of arbitrary sensors, personal computer (PC) frameworks, and upgraded hardware (Gebbers and Adamchuk, 2010).

The correct use of this technique ensures acceptable outcomes. Such a system can be developed for any category of agribusiness range and can be applicable in any area, for a kitchen garden or a huge farm field or may be at the nursery in the greenhouse.

The whole has to be specifically divided into three categories, which take the form of smart agribusiness. The ground data are perceived through a mechanism called a sensor, which determines the classified information associated with the crop being cultivated, and hence, this section of an SA system is known as the determination stage or the primary stage of the system (Rana et al., 2021).

The associated information about the crop is fed to the middle stage, when the information is compared and an estimation is made with the ideal information that suits that crop, and hence, the system is calibrated in the secondary or the final stage. Equipped with propelled machinery, measures are taken up in this stage to maximize the efficiency of the whole system (Khanh and Khang, 2021).

In previous years, new patterns have developed in the farming segment. On account of improvements in the field of WSN just as scaling down of the perceivable units, SA has begun to rise.

Be that as it may, despite the fact that there have been significant advances in various innovative regions, not many of them are centered on the plan and execution of specific ease frameworks for farming situations.

Generally, data advancements are accessible, yet they have not been broadly presented in horticultural situations: costly frameworks and complex installing, controlling, and keeping up these establishments are the primary hindrances.

DOI: 10.1201/9781003392224-7

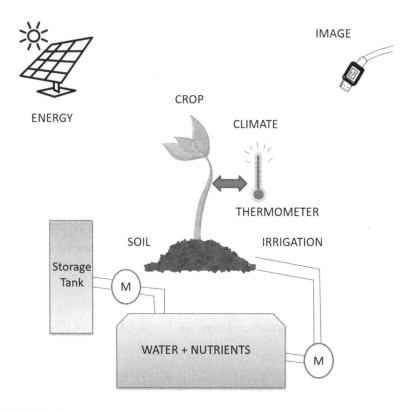

FIGURE 7.1 Procedural farming and necessities.

In such a manner, new stages especially coordinated to expel these boundaries are required. This covers an answer dependent on a heterogeneous and versatile stage, a ready-to-get procedure to store and screen information from crops, developing frameworks that utilize a portable pervasive methodology.

The methodology is developed with an arbitrary network of ubiquitous sensors working with the Internet of Things (IoT) to create an ideal model of agribusiness (Rani et al., 2021).

The development of such a system creates an ideal environment for the crop to nurture at a high pace, which is possible because of the integration of a technology-based automated system that maintains the ideal condition suitable for the crop, for example, temperature, barometrical moistness, humidity, sunlight, the pH of water being used for irrigation, soil conductivity, and climatic conditions, and the actuators are the executives which impact these variables as Figure 7.1.

7.2 DESIGN OF AN IOT-BASED SYSTEM

IoT alludes to implementable machine-to-machine (M2M) communications, which is an urgent segment of late development in the computerized market (Pritiprada et al., 2024).

The most appropriately characterized IoT can be a unique worldwide system framework with self-arranging capacities dependent on standard and interoperable communication protocols where physical and virtual "things" have personalities, physical qualities, and virtual characters and utilize canny interfaces, and are flawlessly coordinated into the information network, regularly impart information partner with clients and their surroundings (Kranenburg, 2008).

Be that as it may, different wordings have likewise been utilized for IoT as "IoT consolidates individuals, procedure, gadget and innovation with sensors and actuators." This general mix of IoT with person in regard to correspondences, joint effort and specialized investigation empowers to seek after continuous choice (Khang et al., 2022a).

7.2.1 IoT-Based System Blocks

An IoT framework includes various useful functional blocks to encourage different utilities to the framework, for example, the scrutinizer, the identifier, activation, communication, and the board (Sethi et al., 2017) as shown in Figure 7.2.

7.2.1.1 Device

An IoT framework depends on a scrutinizer that detects, incites, controls, and checks exercises.

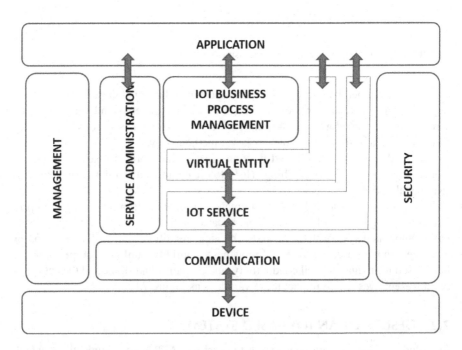

FIGURE 7.2 Generalized block diagram of an IoT-based system.

Source: Bauer et al. (2013).

IoT-based gadgets can trade information with other associated devices and application or gather information from different perceivable setups and procedures whether the information is either locally sent or brought together on servers or cloud-based applications that closes for preparing the information or plays out certain errands locally; different assignments inside the IoT framework depend on transient and space requirements (i.e., memory, handling abilities, communication latencies, and velocities, and cutoff times; Bhambri et al., 2022).

An IoT gadget may comprise a few interfaces for communications with different devices, both wired and remote. These incorporate the following (Elijah et al., 2018):

- Input and output interfaces for perceivable and propelled units
- Interfaces for transmitting real-time information
- Interfaces for storage
- Interfaces for images/videos transmission

IoT gadgets can have a diversified area of applications; for example, as the instance of brilliant cultivating, numerous sensors, various sorts of actuators, and microcontrollers are actualized truly to screen distinctive agricultural applications (Subhashini and Khang, 2024).

Numerous other system types of gear likewise executed like switches, gateways, and routers are incorporated. At this stage, all natural conditions are detected, and afterward, actions are incited per the predefined guidelines. The microcontroller assumes the manager's job and performs organizing related tasks and some other utilities, which are finished by sensors and actuators.

The improvement of a basic IoT-based framework should be possible by utilizing a basic microcontroller-based framework alongside a few subjective sensor units, and a transmission system could be created with the focal server framework (Rani et al., 2022).

7.2.1.2 Communication

The communication segment plays out the communication between the base station and the remote servers. The IoT communication hierarchy mostly works in the application layer, the analytical layer, the network, and the physical layer, which needs to carry all information from the bottom layer to the uppermost layer of the subsystem by any means of transmission.

The subtleties of different hierarchies are described in the next area in the block diagram of a layered system in Figure 7.2.

7.2.1.3 Services

An IoT framework serves different sorts of functions and provides the backbone for the foundation of a smart system. It does bind every system block by establishing the link between each of them. Its services are not limited to one boundary but perform activity for every unit and behave like blood in the human body (Khang et al., 2023c).

The services offered by such systems are system modeling, regulating system controls, distribution of information, information examination, and the disposal of information to every unit.

7.2.2 MANAGEMENT

Just like the services offered by an IoT framework, the management of the whole system is designed to assist the framework. Management segment is given various functions to administer in an IoT framework when looking for the basic administration of an IoT framework. The system is managed by following the coder protocols meant for every unit (Babasaheb et al., 2023).

The system is organized with a predefined protocol, prior to the development of the system, which makes the system smart by directing each unit to execute the task assigned to it, keeping forward and backward communication with neighboring units to monitor the right execution of the task assigned and hence making, on the whole, an intelligent system.

7.2.3 SECURITY

This section plays a crucial role in the proper functioning of the system as per the protocol developed for it by restricting the unidentified request that may make the system vulnerable to failure.

The security segment safeguards the IoT framework by giving capacities, for example, validation, approval, protection, message respectability, content uprightness, and information security (Khang et al., 2022b).

7.2.4 APPLICATION

This is the most significant as far as clients as it goes about as an interface that gives fundamental modules the ability to control and screen different parts of the IoT framework. Applications permit clients to envision and break down the framework status in the present phase of activity and at times forecast modern possibilities.

7.3 AN IOT STRUCTURE FOR AGRIBUSINESS APPLICATIONS

The IoT system requires the total arrangement of gathering crude information from the field, which passes through different stages to dissect and change over it into significant data, which will help the homestead administrator make important choices and move as needed (Fountas et al., 2015; Wu et al., 2015; Khanna et al., 2019). The system shown in Figure 7.3 is separated into four phases, which are additionally examined in detail.

7.3.1 DATA PERCEPTION UNIT

The data perception unit is the essential unit of a keen agriculture IoT network design framework that incorporates sensors, actuators, microcontroller modules, and system hardware (e.g., switches, routers, gateways, etc.) for information transmission of a receipted information to the primary base station. This unit detects and incites diverse agricultural parameters.

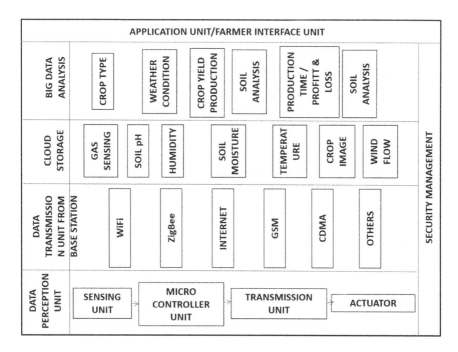

FIGURE 7.3 Block diagram of an Internet of Things structure for agribusiness.

The most significant part of this unit is to deploy those devices that use minimal power, since this unit is set on the field and must work on a battery or solar power source for a longer period. This unit is responsible for detecting and observing, which is done by applying distinctive detecting and checking gadgets. Sensors detect physical information and move the yield illness data. Information that is handled through different assets is consequently accomplished by the sensing framework.

Statistical investigation has been done on information received from sensors so as to impel the crop sickness. Ranchers get fundamental data, for example, soil pH, temperature, and soil dampness and moistness, through internet or mobile applications and message administrations (Muangprathub et al., 2018).

Real-time picture and video checking of the information encourages the rancher to get opportune and precise data as Figure 7.4.

The gadgets that can be utilized for this unit appear in Figure 7.4 and their capacity utilization in Table 7.1

7.3.2 DATA TRANSMISSION UNIT

The data transmission unit is a vital innovation for smart farming and dependent on transmitting horticulture data from a base station to the cloud or the investigation unit to perform information mining (Rani et al., 2023).

This layer includes web and other applicable communication innovations. Global System for Mobile Communication (GSM), Wi-Fi, Long-Term Evolution (LTE) (4G),

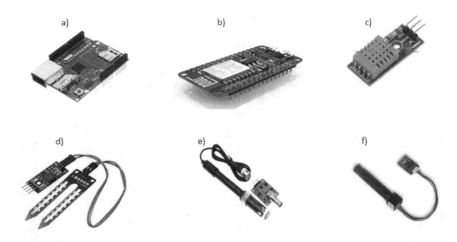

FIGURE 7.4 The gadgets in the data perception unit.

TABLE 7.1
Parameters of the Field Devices

Deployed unit and parameters	Type	Required power
Node MCU	8266	2.5–3.6V/170 mA
Arduino Uno Mega	2560	7–12 V
Temperature and Humidity	DHT22	2.5 mA
Gas Sensor	MH-Z16	3.7–5V
Soil PH Sensor	RC-A-41520	PH4= 1.5V, PH7 = 1.2V, PH9 = 2.5V.
Actuating coil	Plastic ¼	1.3 mA
Architecture	Crop field	—

and Code-Division Multiple Access (CDMA) are common to act in horticulture fields in a suitable way. ZigBee is known as the fundamental empowering agent for communication over significant distances when third-party servers, for example, LTE, CDMA, or GSM, are not accessible.

7.3.3 ANALYTICAL UNIT

This unit is used to perform examinations on the ground information that is stored in the cloud. Big data analysis is applied to the sensor information and the factual information about the different parameters identified with crops, and statistical investigation encourages the detecting, activating, and illness-distinguishing exercises.

In addition, field crop manor management and pesticide control administrations are intended to be an incentive from the horticulture information. The yield illness control and harvest development models expand based on ranch information. The big data analysis on the cloud-figured information is the foundation of shrewd horticulture. These are additionally clarified through cloud-based investigations (Pathak et al., 2019).

7.3.3.1 Cloud Infrastructure

A cloud infrastructure is developed to store the data on several virtual servers, which enhances the storage capacity to a greater extent. The data can be accessed online to perform all the fundamental activities.

Cloud storage is a centralized server for all agrarian-related information, which is collected from the field through the determination stage and passed on to the cloud server via the IoT network. Analytics assets and web administrations are likewise introduced into cloud or web, which is available by cloud administrations.

The overall system developed on the field has subsystems that are not accessible for gathering information via the network directly. This problem is resolved by developing a bridge network of local gateways that gather the information and pass it on to the cloud network, which makes the system fully controllable and secure. This whole setup helps the farmer improve the capability of the system, making the system reliable, with the expectation of maximum output.

A gateway system used in greenhouse farming (Hashem et al., 2015) resulted in better performance when compared to conventional means due to the real-time observation and regulation of parameters.

7.3.3.2 Data Analysis

Data analysis can straightforwardly be performed on the cloud server to discover the required and significant data from the huge measure of information from various information formats. When the data analysis is performed on bulk data normally a size of zettabyte, then it is called big data analysis.

The yield ailment control and harvest development models expand based on ranch information. Big data analysis likewise gives choice help administrations to ranchers for crop profitability and ideal cost investigation. This examination makes the entire environment more astute by investigating, exploring, and handling the horticultural data for advanced mindfulness. This investigation is made to predict the expected yield probability of harvest efficiency for the upcoming season.

Although various recognizing gadgets could be deployed for supervising parameters, such as climatic conditions, crop quality, or any pest attack or disease, at regular intervals, they can also gauge the benefit/misfortune-based yield efficiency.

Big data analysis applied in horticulture encourages the homestead to comprehend the ideal time for planting and reaping and different ranch management methods. Big data analysis limits the danger of yield obliteration per logical ways (Kamilaris et al., 2017).

7.3.4 Application Unit or Farmer Interface Unit

This is the last interface unit which is obvious to the rancher and is completely intended for the rancher's own understanding. This unit encourages the rancher to screen the harvest efficiency in numerous manners, for example, for viable development of yields ranchers are awarded by distinguishing the propriety of prolific determination.

Atmosphere conditions, crop development conditions, soil quality, or checking the well-being of cows encourage ranchers to follow the conditions of their business and alleviates lower creation dangers (Cadavid et al., 2018).

Ranchers are enabled with recognizing the fitness of the prolific choice for the successful development of the harvests. Ranchers can get data through web administration, message administration, and master administration.

7.4 THE GENERIC SYSTEM MODEL FOR SMART AGRICULTURE

The block diagram shown in Figure 7.5 works on the real-time data perceived through sensors from the farm field, which is further transmitted through the Node MCU 8266 Wi-Fi module to the base station at a set time interval depending on the programmer's instruction to the controller unit on the farm. These data are further sent to the cloud server using IoT techniques and are used for analytics by the data interpretation and data-mining methods.

The former idealistic big data available from the cloud and used to make decisions in comparison to the received data at regular intervals can be further refined and used to consider the various aspects of farm parameters and crop-related issues.

The analysis done by considering all the parameters will guide the farmer using any of the information tools designed to be used by the farmer for further measures. The information may be passed through a mobile app designed for this specific

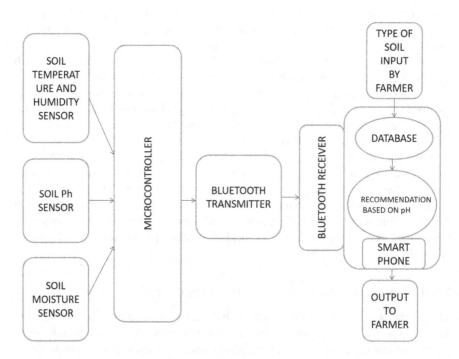

FIGURE 7.5   Block diagram of a real-time data perceivable system.

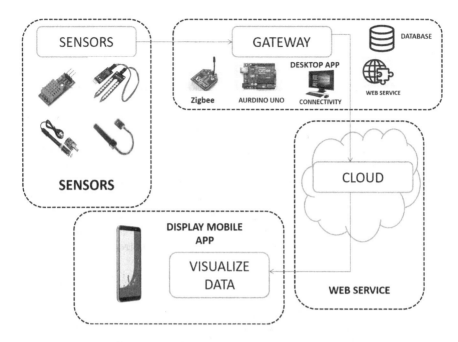

FIGURE 7.6   Ubiquitous sensor-based Internet of Things system for smart agriculture.

application; it may be through an SMS system (WhatsApp, email, etc.) per the convenience of the farm manager (Muangprathub et al., 2018).

The composite design of a ubiquitous sensor-based IoT system is shown in Figure 7.6, which is designed using the entire sensor and controller unit discussed so far in the former sections.

7.5   APPLICATIONS OF THE IOT IN AGRICULTURE

IoT agribusiness framework can be applied to a variety of wide assortment of fields, for example, precision cultivating, animal rearing, and nursery observing. Agribusiness applications have been classified into different segments depending on choices like smartphone-based applications designed for any segment of agribusiness.

IoT-based agribusiness applications are utilized to make progressively proficient assets for farming efficiency. The primary spaces of IoT agribusiness applications are smart farming, domesticated animal observation, nursery checking, and agrarian automatons (Hajimahmud et al., 2022). The accompanying subsection comprises different sorts of agrarian applications.

7.5.1   SA

SA causes the ranchers to improve, robotize, and upgrade every practical course so as to improve farming efficiency and make the crop framework brilliant (Khang et al., 2024).

Distinctive remote perceivable devices are set to estimate the various parameters relevant to the suitable environment for crop growth which enhances the harvesting methods and efficiency.

To ameliorate the yield production, the comparative results between horticultural surrounding data and a harvest factual investigation have to be interpreted for better crop information and production.

IoT-based stages have been created for SA for natural observation and plant disease detection by Basori et al. (2020). IoT-based climate figures assist with improving profitability and making expectant examinations to keep the harvest from harm.

Different checking gadgets/sensors are utilized to anticipate problem conduct and plant or yield development, which invokes a warning generated by the system on the basis of a machine learning algorithm before causing any harm to the field (Parameswaran et al., 2016). Farmers are utilizing water systems based on the IoT that oversees and examines crop water system necessities. A remote farming checking stage has been introduced based on observed information from Muhammad et al. (2016).

A theoretical engineering dependent on digital frameworks and the programming of the characterized networks has been introduced for accuracy cultivation.

IoT-based accuracy cultivation comprises various monitoring and handling applications, for example, finding a location for cultivation, soil analysis, climate SA, deciding the ideal time for seeding and weeding, water systems, disease detection, reaping, and more (Bastiaanssen et al., 2000).

7.5.1.1   Location Finding Based on Land Suitability

IoT gives educational information to agrarian organizations for better choices making, for example, arranging, overseeing, and associating with colleagues insightfully by setting aside cash and time. In developing conditions, the parameters that favor or hamper the crop quality are observed by RFID and GPS.

GPS is utilized to locate the specific area of the agribusiness field and screen different agrarian parameters by utilizing remote communication systems (Jayathilaka et al., 2012, and Mandal et al., 2020).

Devi et al. (2016) developed a significant model that screens the basic field parameters like soil pattern and its type remotely according to necessity of harvesting trends.

This model is designed with a wireless device (ZigBee) which is associated with different gadgets like CMS, GSM, and GPRS by utilizing ubiquitous sensor systems that screens and acknowledges the information of the field at a regular interval (Hahanov et al., 2022).

The GPS system can be easily interfaced with the Advanced RISC (reduced instruction set computer) Machine (ARM) cortex (a wise observing framework to accomplish capacities like SMS/MMS) which warns the cultivator remotely in the occurrence of any undesirable event and encourages the ranchers to make remedial moves. Despite the fact that the functional cost of such a system is quite high enough, still these are broadly utilized in farming because the true calculations for area checking and supervising the farm are precise.

7.5.1.2 Soil Analysis

Soil observing acquires the requesting practices for both enterprises and ranchers in the field of agribusiness. In the analysis of soil, there are several issues that influence or weaken crop growth or production. On the off chance that these sorts of issues are distinguished information precisely, at that point, cultivating examples and procedures can be seen without any problem.

The soil whose sample is being analyzed contains most of the parameters needed by a crop, such as moisture, humidity, temperature, nutrients, and so on. There are sensors available for monitoring the moisture and humidity in the soil that can transmit to a remote location.

The test conducted on soil for proper measurement of soil nutrients and other necessary parameters can provide advice to the farmer, who can further maintain the soil by adding the deficient parameter and simultaneously increasing the probability of a maximum crop yield (Dayoub, 2018). Also, recognizable proofs of sullied soil by utilizing IoT advances shield the field from overfertilization and yield misfortune.

7.5.1.3 Climate Smart Agriculture

While farming, the most critical situation is to screen climate conditions constantly with the goal that future exercises can be planned in a like manner. Climate stations would be the first well-known devices for the agriculture field, which are utilized to screen distinctive weather.

Climate concerns regarding any specific crop require observing important measures like moistness, wind bearing, gaseous tension, temperature, and so forth. Situated over the field, climate stations gather the ecological information that continuously needs to be stored in the cloud for further data analysis (Khang et al., 2023c).

The gathered information is utilized for climate investigations to outline conditions and gives new bits of knowledge to take expected activities to improve agrarian efficiency. There is an organization in the Food and Agriculture Organization of the United Nations that has designed a special program that relates the different climate strategies and methodologies, hence named climate SA, and it plays a major role in guiding the farm field managers by distinguishing atmosphere conditions (Lova et al., 2020).

The measurements of essential information, transmitted through the sensor network deployed at the farm field, can be analyzed at a remote server station, can guide the farmer on climatic change and weather forecasts with the use of an IoT-enabled sensor network at the farm (Salam, 2020; Kodali et al., 2013).

7.5.1.4 Ideal Time for Seeding and Weeding

IoT-enabled systems enhance operational proficiency and upgrade yield efficiency with the help of a sensor network on the field, which feeds the program at the cloud server with real-time data, which are analyzed to inform the farm manager about the suitable time for cultivation.

The system may use a variety of sensors and wireless transmission device on the field which works along with the IoT network to guide the farmer about the correct time of seeding and weeding (Hajimahmud et al., 2023).

7.5.1.5   Irrigation Observing Framework

The IoT improves the present water system framework in an increasingly creative manner. A rancher can advance a water system framework in numerous manners by checking climatic patterns, soil nutrients, and other constituents.

A system equipped with IoT services offers several ways of monitoring a water system framework, such as climate determination information, and handling and monitoring the farm, from any remote location.

The advancement in the sprinkling mechanism framework encourages ranchers by introducing different sensors, lowering ranchers' month-to-month water system costs, and cutting off water assets. Venkatra et al. (2020) introduced a keen water system management framework by utilizing the latest technique called machine learning, which analyzes distinctive climate and soil parameters.

Similarly, the quality of water and its dissolved mineral composition, pH values, and all its chemical properties that have been sent from the field by a variety of sensors deployed at the farm field, which transmit information through an IoT-enabled network, are also analyzed at a remote server.

The assembled information regarding the irrigation framework can be seen using a mobile application or on a website using a registered ID. These services are possible just because of the integration of the IoT services at the field through a sensor network and cloud computing that can manipulate the number of parameters at the field using different strategies, which are more suitable for a particular crop being sowed.

7.5.1.6   Disease Detection

An underlying driver of income and creation misfortunes is crop sickness. Because the blast of the IoT rural framework has been changed into a digital framework, this causes the rancher to settle on educated choices.

Forecasts about yield ailments at the beginning of a period cause the ranchers to create more income by defending the crop from insects and other pests, which hamper the crop quality severely. The technological merger of a sensor network, the IoT, and machine learning guides the farmer to nurture and protect their crop from disease in numerous ways by recognizing various ailments and keep crops from assaults from creatures.

Rittika et al. (2017) introduced a pest, weed, and wheat disease detection technique based on an IoT-enabled framework.

A reduction in the predicted amount of crop yield is a major issue of concern because of the limitations of developed land frequently place on the various wildlife. Johannes et al. (2017) introduced an observing and repulsing framework for the assurance of harvest against assault from wild creatures.

Recognizing a yield malady at the beginning is an exceptionally important reason for testing in the field of horticulture. Since expert advice and supervision are needed when distinguishing harvest or petal maladies, this may be costly and time-consuming, which is very difficult for a farmer to manage.

However, programming the location of illnesses is gainful, exact, and less expensive for ranchers when contrasted with manual observation by specialists. An image-processing strategy additionally assumes an essential job for identifying plant sickness (Dhingra et al., 2018; Foughali et al., 2018).

7.5.1.7   Agriculture Management Technique

Selecting a savvy cultivation method is a way of enhancing the degree of profitability while diminishing the ecological effect yet this keen cultivating method can be conceivable by means of agriculture management techniques (AMT).

AMT is the backbone of the system for preparing, arranging, and making choices, with the end goal being shrewd cultivation (Fountas et al., 2015). Incorporating an AMT system permits ranchers to screen all field parameters gathered by means of a system developed for this specific management system; this system comprises a ubiquitous sensor network working over an IoT-enabled microcontroller-based system that transmits information to the cloud for further analysis.

A machine learning algorithm working over the big data in the cloud provides the best possible information on fertilization and a complete climate information forecast at every fixed interval of time as required and instructed in the program by a rancher or farm manager. All these data are put away in the PC in a standard configuration and can be available by means of mobile phone or web for additional processing.

To enhance the utilization of water assets, a robotized water system and observing framework are utilized (Parameswaran et al., 2016). Aside from a water system framework, the ranch is additionally shielded from vermin and animal interruption (Foughali et al., 2018).

7.5.1.8   Unmanned Aerial Vehicles or Drones for Agricultural Work

This is a new technique of reducing the manpower required at the field as it works in any of the aerial direction while carrying the necessary pesticides or nutrients required to a specific portion of the field and can be easily sprayed at any time remotely. These unmanned vehicles are designed to carry small loads, which is managed by an RF controller from any place of work.

These drones are very useful in screening and recovery of deficiencies in agricultural procedures for farming. These can be helpful for showering the field, spraying pesticides, tracing crop well-being, and exploring the field report and the soil condition by receiving real-time video and images at a remote center.

This latest flying machine encourages the ranchers by reconciling GIS mapping and yield well-being imaging. These flyers are, for the most part, used on enormous ranches where identified microscopic organism growth issues are hard to deal with and require ordinary checking.

In the territory of agribusiness pesticides and manures are significant for crop yield (Boursianis et al., 2020). Agrarian automatons are completing this activity productively in view of their fast speed and viability in the showering activity.

Furthermore, they additionally conveyed to screen backwoods, domesticated animals, and aquaculture. An association Precision Hawk is utilizing rambles for significant information assembling through sensors for studying mapping, and imaging of agrarian land.

This continuous monitoring by flyers assists ranchers with gaining insight concerning the selected ground and the portion of the farm field being investigated. There is much research with IoT-based platforms where an executive data framework approach has been created to meet business needs.

Rural rambles are incorporated with GPS gadgets' cameras and sensors to screen crop well-being, such as seeding, splashing of crops, and analyzing and investigating the texture of the soil. This analysis has numerous different favorable circumstances for using automatons, such as harvest well-being imaging; plant numbering; measuring gas level rises in wheat fields, especially nitrogen fixation when necessary in such crops; weed growth; seepage; plant growth and size; and so forth.

7.5.2 GREENHOUSE MONITORING

This is another method of farming, where the small saplings are grown in a closed environment. Sometimes, they are formed to nurture the plants such as in a nursery, which is further used in horticulture.

However, this method is employed to develop the shrub level of farming where the plants that do not develop deep roots under the ground, such as cucumber, tomatoes, capsicum, bell peppers, and floriculture are very wisely nurtured under this environment. In nurseries, the plantation is done in a closed, shaded chamber, maintaining a suitable environment with a feedback control system.

The enclosed shaded chamber innovation gives advantages for the plants, which can develop anywhere just by providing the proper natural conditions suitable for their development. The development of a greenhouse is progressively extreme, when concerned with the required degree of accuracy in terms of controlling the parameters inside the chamber.

For screening natural and climatic changes, only a certain number of ubiquitous sensor-based systems are available to examine the greenhouse setup.

Introducing an IoT-enabled setup at a greenhouse farm can definitely ease service by reducing manpower and increasing the accuracy of conventional farming, a mass vitality that gives a direct connection of greenhouse from farmers to clients.

The greater part of the investigations have concentrated distinctly on remote observing and limitations (Dan et al., 2015; Reka et al., 2019). The whole setup of a greenhouse farming working over a fully automated IoT-enabled system, which can be controlled and operated from a remote location, can benefit the farm manager to a greater extent by providing accurate information regarding the plantation, seeding, weeding, water supply, nutrient deficiency, and qualified crop retention and can be made easily with 100% accuracy, which encourages the farmer to plant the crop of his choice in any season and can earn a greater profit from the market.

Under such systems, the entire unit is partitioned into multiple parts that form the information and continuously monitor the output. Information can be acquired by comparing sensors and indicators and afterward moved to the fundamental central server system for handling.

Actually, the execution of the significant parts of the system is basically the perceivable unit and the transmission system for sending information. Cultivators are arranged, diverse checking gadgets and sensors, as indicated by the particular necessities, and track or record the necessary data.

Agriculturists settle on better choices by breaking down the data and accomplishing explicit objectives by acquiring ideal information. There are numerous IoT-based nursery applications, for example, irrigation systems, plant observation, and atmosphere checking, among others.

7.5.2.1 Managing Irrigation

The main issue in the greenhouse is to maintain the water level as per the requirement. To cope up with this problem, specialized sensors can be deployed on the farm to properly monitor the level of water, which can be actuated by an IoT procedure and controlled by any remote location.

This complete setup can be programmed, and the threshold value can be set, which will actuate the watering system on crossing the threshold level as per the readings received through a soil moisture sensor (Parameswaran et al., 2016).

7.5.2.2 Observing Plant Growth

The camera for capturing videos or images can be used which is equipped with an IoT service so that the captured images or videos can be uploaded to the cloud for further analysis or proper observation, so that the farmer can be advised on the basis of studies made and any conspicuous activity can be resolved immediately.

The machine learning algorithm applied on the cloud for image or video analysis will properly guide the supervisor for any type of disease, dryness, fungus, or anything and assist the farm manager at all times, which is an assurance that the system works properly and without any problems (Cadavid et al., 2018).

7.5.2.3 In-House Atmospheric Observation

There are numerous parameters that are consolidated to keep up and make a perfect domain for plants inside severe cutoff points. The goal is to maintain the level of in-house atmosphere suitable for the plants to grow. This is only made possible by supervising everything through an automated sensory system, which measures the accurate level of temperature, CO_2, and O_2.

The IoT-enabled system can continuously store the series of real-time data in the cloud. The machine learning algorithm analyzes the data and sends the signal to operate the actuators to maintain the level if any of the parameter crosses the threshold level (Dan et al., 2015; Reka et al., 2019). The degree threshold that can be set depends on the ideal condition in nurturing the crop with 100% efficiency.

7.6 CONSTRAINTS OF THE IOT

The current arrangements have used more astute IoT-based applications for settling various difficulties in the agrarian and cultivating space. Here, the different possibilities of these applications are examined to improve the current arrangements as mentioned, although the accompanying segments will demonstrate the way to improve the present circumstance point-wise (Birla et al., 2015).

7.6.1 Cost Viability

Developers and analysts around the world are essentially interested in decreasing the use of machinery and application software charges in IoT deployments while lifting the structure output. People are looking for financially savvy devices, but auxiliary price is essential for the adoption because imported components for constructing the arrangement may be limited.

Yet universal ranches are at the forefront of the growing advancements in aforementioned way, but the problem of how to reduce the price further remains. Introduced works do require cost viability. Consequently, previously mentioned tip is purposely a requirement of time.

7.6.2 NORMALIZATION

The latest works do not fit in with the normalized arrangements of portraying information just as a procedure. Normalization is a main component that might unequivocally be worked out in the development of the IoT.

Normalization in the IoT means reducing the underlying hindrances for specialist co-ops or dynamic clients, extemporizing the interactivity affairs among various exertions or frameworks, and seeing a superior rivalry between the created items or administrations at the implementation level.

Safety principles, communication guidelines, and recognizable proof gauges must be expanded with the advancement of IoT when planning growing innovations at a parallel identicalness.

Likewise, individual analysts will record industry-explicit rules and determine the needed normalizations for the productive execution of the IoT. Farming-related normalization when utilizing the IoT ought to carefully be followed.

7.6.3 DIVERSIFIED SYSTEM

The IoT is at an entangled, diversified system stage. Yet, referenced works in agribusiness cannot cooperate with diversified modules or communication advances. This, indeed, upgrades the multifaceted nature between miscellaneous sorts of applications by miscellaneous communication innovations exhibiting inconsiderate performance of appliance to be false and postponed.

It has unmistakably referenced that the administration of associated protests by encouraging community efforts among various things (machinery parts or programming administrations) and controlling that in the wake of tending to, identifying, and streamlining at compositional and convention levels is a genuine investigation issue. In any case, to prevail in the farming area, the IoT should be rethought to sift through the exhaustion of the basic stage.

7.6.4 CONCERNED DATA

While tons of sensor-empowered objects are associated with the web, it might not be practical for a client to deal with all the information gathered by sensors. Concerned and mindful calculating strategies should be utilized in an improved manner to assist in choosing what information should be prepared.

Talks about agro-undertakings are bereft of concerned mindfulness. This appears to determine the refutation of data approval in a type of ceaseless disturbed strategy. Including natural parameters and self-estimation may move the limited setting to others while making an careful inside-out-related self-cum-periphery IoT organic framework.

7.6.5 VITALITY CONTROL

Vitality control is indeed a significant problem in IoT-based frameworks. Framework parts, for example, IoT gadgets, network antennas, and other ward inactive modules alongside the algorithms ought to appropriately be readdressed while reveling into the collecting of vitality. Or else, nontraditional wellspring of energy-gathering arrangements.

For example, sun-powered force, wind, biofuel, and electron clouds have also been tried when planning brilliant IoT-based agribusiness frameworks. Starting at now, sun-powered IoT frameworks are being used. Consequently, analysts may consider different sources in the future (Khang et al., 2023b).

7.6.6 ADAPTATION TO NONCRITICAL FAILURE

Fault resistance is generally missing in the previously discussed arrangements. To make an impeccable framework, the adaptation to internal failure level of the framework ought to be kept exceptionally high so that, despite a specialized blunder, the framework continues working. Equipment modules may flop because of exhausted batteries or some other explanation.

Essentially, a mistaken use of the sensor, flawed adjustment, and disappointment in communication may build to be issues of circumstance. While looking for an arrangement, sun-based energy may be an option in contrast to the battery-powered modules.

Utilizing a large number of communication conventions may maximize energy utilization however frequent the connection. Force usage in this case may be dropping the nearby sanctioning of one protocol to achieve the initiated change at any example. Legitimate standardization should be completed before conclusive insertion.

7.6.7 THE NEED FOR A CONTINUOUS ARRANGEMENT

Most of the introduced arrangements do not include genuine practicality in the record. Be that as it may, empowering exactness in farming, climatic data, and soil parameters is keenly coordinated with the present turn of events (Khang, 2023).

7.7 FUTURE SCOPE

Here, numerous potential utilizations of IoT in the agribusiness and cultivating region. The ebb-and-flow suggestions incorporate different gadgets, cloud arrangements, and frameworks that chip away at water system management, floriculture, vermiculture, crop illness expectations, climate smart forecasts, and aquaculture, among others.

The variables required for development related to the IoT that need further consideration later on follow:

- **Autonomy:** The upcoming applications would be completely self-ruling to get utilized along with particular requirements.

- **Price:** Minimum price arrangements are alluring for developing and utilizing IoT-based arrangements.
- **Client control board:** Generally, nonspecialized individuals utilize the IoT plus agriculture-based arrangements on ranches. Thus, it is smarter to plan an easy-to-understand interaction for the type of control board for productive uses.
- **Power:** Green registering procedures should be dispersed along the latest IoT-based horticulture when IoT gadgets will use less energy that may, in turn, boost their future and make them less defective and, henceforth, profoundly gainful.
- **Interactivity:** Interactivity problems are widely recognized in IoT gadgets. Gadgets must be sufficiently skilled to communicate along many more classifications such that a general framework would be filled in as a live, biological arrangement.
- **AI:** Machine learning and AI strategies must be actualized jointly to adapt to the future and conduct investigation performances by using cutting-edge decisions to greatly nurture networks and continue evaluations.
- **Maintenance:** IoT frameworks will be structured in a way that supports time duration and price; these two are decreased to a low threshold of acknowledgment for approaching innocent clients.
- **System motility:** The likelihood of present framework designs should be upgraded to make them requested enough with specific viewpoints toward agrarian necessities.
- **Robustness:** IoT design should be powerful and issue-lenient so its uses can be assured to be supported at their respective work.
- **Weather, loam, and water:** While structuring an IoT-dependent framework for farming, the main trouble is, by all accounts, the distinctive warmth, loam, and aqua properties across the world. Ranchers will be furnished along nearby climate sensors that may communicate with federal climate communities to make them aware of natural situations preceding undesirable circumstances. This could build yield creation and oppose the demolition of agro-items in an exact way.
- **Segmented land structure:** Partitioned land cultivating is an issue in numerous nations around the world. Reasonable IoT engineering will be created to provide food regarding this particular issue. A suitable approach and appropriate arranging ought to be considered for handling the problem in advance.
- **Minimal support:** Participation in upkeep at the same time is a tremendous issue. Consequently, it is important to plan a low-upkeep framework that could play out the tasks naturally without human mediation in an advanced manner.
- **Flexibility:** Flexibility is a significant element that might create trouble when taking care of the framework. Along these lines, compact-measure types of equipment, for example, built-in Session Initiation Protocol (SIP) and so on, can be utilized to convey the last item.

7.8 CONCLUSION

The consideration of IoT is alleged to be treasured for growing agrarian and fertilizing enterprises through imparting new measures. In the present chapter, an exhaustive analysis of IoT organizations for cutting-edge farming use was discussed (Jaiswal et al., 2023).

An IoT-dependent horticultural system was presented to apply a simple blend of horticulture and the IoT. To start with, a clear IoT scheme and its qualities were proposed. After that, the attributes of miscellaneous important utilizations of IoT in horticulture area were discussed.

Then the hardware parts on hand in the market and areas where they are presently being deployed in harvesting areas were presented. Miscellaneous unwired communication advances that can be equitable for agribusiness programs were, moreover, presented.

IoT–cloud specialist organizations are not widely known in farming areas. A list of cloud professional agencies in this area of utilization depending on key qualities was presented. The list was determined by using an assessment of IoT-dependent systems being with the aid of and via sent in miscellaneous places all around the globe.

A scanty of provisional investigations had been delineated along formation particulars in different uses, likewise, apiculture, orchard observing, exactness harvesting, stream aqua attribute checking, and so forth.

Ultimately, the problems of the present applications were discussed. A short list of headings for subsequent innovation and applications was conceived. In particular, minimal cost, independent, power effective, interoperable, normalized, diversified, and vigorous elucidation with properties such as AI, decision-making sustenance structures, and low upkeep are popular.

All in all, the miscellaneous parts of the IoT must provide food so that horticulture can be keen and universal.

REFERENCES

Babasaheb J, Sphurti B, Khang A. "Industry Revolution 4.0: Workforce Competency Models and Designs," *Designing Workforce Management Systems for Industry 4.0: Data-Centric and AI-Enabled Approaches* (1st Ed., pp. 14–31). CRC Press, 2023. https://doi.org/10.1201/9781003357070-2

Basori AH et al. "SMARF: Smart Farming Framework Based on Big Data, IoT and Deep Learning Model for Plant Disease Detection and Prevention," *ACRIT 2019: Communications in Computer and Information Science* (vol. 1174). Springer, 2020. https://doi.org/10.1007/978-3-030-38752-5_4

Bastiaanssen W. et al. "Remote Sensing for Irrigated Agriculture: Examples from Research and Possible Applications," *Agricultural Water Management*, 46(2), 137–155, 2000. www.sciencedirect.com/science/article/pii/S0378377400000809

Bauer M. et al. "IoT Reference Architecture," *Enabling Things to Talk*. Springer, 2013. https://doi.org/10.1007/978-3-642-40403-0_8

Bhambri P, Rani S, Gupta G, Khang A. *Cloud and Fog Computing Platforms for Internet of Things*. CRC Press, 2022. https://doi.org/10.1201/9781003213888

Birla, A. et al. "The Challenges of M2M Massive Access in Wireless Cellular Networks," *Digital Communication Network*, 1(1), 1–19, 2015. www.sciencedirect.com/science/article/pii/S2352864815000005X

Boursianis D. et al. "Internet of Things (IoT) and Agricultural Unmanned Aerial Vehicles (UAVs) in Smart Farming: A Comprehensive Review," *Sciencedirect Vol. 9 Internet of Things* (pp. 100–187). 2020, ISSN: 2542–6605. https://doi.org/10.1016/j.iot.2020.100187

Cadavid H. et al. "Towards a Smart Farming Platform: From IoT-Based Crop Sensing to Data Analytics," *Advances in Computing. CCC 2018. Communications in Computer and Information Science* (vol. 885). Springer, 2018. https://doi.org/10.1007/978-3-319-98998-3_19

Foughali K. et al. "Using Cloud IoT for Disease Prevention in Precision Agriculture," *Procedia Computer Science*, 130, 575–582, 2018. https://doi.org/10.1016/j.procs.2018.04.106

Dan L. et al. "Intelligent Agriculture Greenhouse Environment Monitoring System Based on IoT Technology," *In: 2015 International Conference on Intelligent Transportation, Big Data and Smart City (ICITBS)*. IEEE, 2015. https://ieeexplore.ieee.org/abstract/document/7384072/

Dayoub M. "Factors Affecting the Soil Analysis Technique Adopted by the Farmers," *Advances in Time Series Data Methods in Applied Economic Research. ICOAE 2018. Springer Proceedings in Business and Economics*. Springer, 2018. https://link.springer.com/chapter/10.1007/978-3-030-02194-8_19

Devi MPK. et al. "Enhanced Crop Yield Prediction and Soil Data Analysis Using Data Mining," *International Journal of Modern Computer Science*, 4(6), 2016. https://doi.org/10.1201/9781003145011-2

Dhingra G. et al. "Study of Digital Image Processing Techniques for Leaf Disease Detection and Classification," *Multimedia Tools and Applications*, 77, 19951–20000, 2018. https://doi.org/10.1007/s11042-017-5445-8

Elijah O. et al. "An Overview of Internet of Things (IoT) and Data Analytics in Agriculture: Benefits and Challenges," *IEEE Internet Things*, 5, 3758–3773, 2018. https://doi.org/10.1109/JIOT.2018.2844296

Fountas S. et al. "Farm Management Information Systems: Current Situation and Future Perspectives," *Computers and Electronics in Agriculture*, 115, 40–50, 2015. https://doi.org/10.1016/J.COMPAG.2015.05.011

Gebbers R, Adamchuk V. "Precision Agriculture and Food Security," *Science*, 828–883, 2010. www.science.org/doi/abs/10.1126/science.1183899

Hahanov V, Khang A, Litvinova E, Chumachenko S, Hajimahmud VA, Alyar AV. "The Key Assistant of Smart City—Sensors and Tools," *AI-Centric Smart City Ecosystems: Technologies, Design and Implementation* (1st Ed.). CRC Press, 2022. https://doi.org/10.1201/9781003252542-17

Hajimahmud VA, Khang A, Hahanov V, Litvinova E, Chumachenko S, Alyar AV. "Autonomous Robots for Smart City: Closer to Augmented Humanity," *AI-Centric Smart City Ecosystems: Technologies, Design and Implementation* (1st Ed.). CRC Press, 2022. https://doi.org/10.1201/9781003252542-7

Hajimahmud VA. et al. (Eds.). "The Role of Data in Business and Production," *AI-Aided IoT Technologies and Applications in the Smart Business and Production*. CRC Press, 2023. https://doi.org/10.1201/9781003392224-2

Hashem IAT. et al. "The Rise of 'Big Data' on Cloud Computing: Review and Open Research Issues," *Information Systems*, 47, 98–115, 2015. ISSN 0306-4379. https://doi.org/10.1016/j.is.2014.07.006

Jaiswal N, Misra A, Misra PK, Khang A (Eds.). "Role of the Internet of Things (IoT) Technologies in Business and Production," *AI-Aided IoT Technologies and Applications in the Smart Business and Production*. CRC Press, 2023. https://doi.org/10.1201/9781003392224-1

Jayathilaka PMS. et al. "Spatial Assessment of Climate Change Effects on Crop Suitability for Major Plantation Crops in Sri Lanka," *Regional Environmental Change*, 12, 55–68, 2012. https://doi.org/10.1007/s10113-011-0235-8

Johannes A. et al. "Automatic Plant Disease Diagnosis Using Mobile Capture Devices, Applied on a Wheat Use Case," *Computer Electronic Agriculture*, 138, 200–209, 2017. ISSN: 0168-1699. https://doi.org/10.1016/j.compag.2017.04.013

Kamilaris A. et al. "A Review on the Practice of Big Data Analysis in Agriculture," *Computers and Electronics in Agriculture*, 143, 23–37, 2017. ISSN 0168-1699. https://doi.org/10.1016/j.compag.2017.09.037

Kang H. et al. "A Design of IoT Based Agricultural Zone Management System," *Information Technology Convergence, Secure and Trust Computing, and Data Management. Lecture Notes in Electrical Engineering* (vol. 180). Springer, 2012. https://doi.org/10.1007/978-94-007-5083-8_2

Khang A (Ed.). *AI-Oriented Competency Framework for Talent Management in the Digital Economy: Models, Technologies, Applications, and Implementation*. CRC Press, 2023. https://doi.org/10.1201/9781003440901

Khang A. *Advanced Technologies and AI-Equipped IoT Applications in High-Tech Agriculture* (1st Ed.). IGI Global Press, 2024. https://doi.org/10.4018/978-1-6684-9231-4

Khang A, Gupta SK, Shah V, Misra A (Eds.). *AI-aided IoT Technologies and Applications in the Smart Business and Production*. CRC Press, 2023b. https://doi.org/10.1201/9781003392224

Khang A, Hahanov V, Abbas GL, Hajimahmud VA. "Cyber-Physical-Social System and İncident Management," *AI-Centric Smart City Ecosystems: Technologies, Design and Implementation* (1st Ed.). CRC Press, 2022b. https://doi.org/10.1201/9781003252542-2

Khang A, Rani S, Gujrati R, Uygun H, Gupta SK (Eds.). *Designing Workforce Management Systems for Industry 4.0: Data-Centric and AI-Enabled Approaches*. CRC Press, 2023c. https://doi.org/10.1201/9781003357070

Khang A, Rani S, Sivaraman AK. *AI-Centric Smart City Ecosystems: Technologies, Design and Implementation* (1st Ed.). CRC Press, 2022a. https://doi.org/10.1201/9781003252542

Khang A, Shah V, Rani S. *AI-Based Technologies and Applications in the Era of the Metaverse* (1st Ed.). IGI Global Press, 2023c. https://doi.org/10.4018/9781668488515

Khanh HH, Khang A. "The Role of Artificial Intelligence in Blockchain Applications," *Reinventing Manufacturing and Business Processes Through Artificial Intelligence* (pp. 20–40). CRC Press, 2021. https://doi.org/10.1201/9781003145011-2

Khanna A. et al. "Evolution of Internet of Things (IoT) and Its Significant Impact in the Field of Precision Agriculture, Elsevier," *Computers and Electronics in Agriculture*, 157, 218–231, 2019. ISSN: 01681699. https://doi.org/10.1016/j.compag.2018.12.039

Kodali RK. et al. "An IoT Based Weather Information Prototype Using WeMos," *2nd International Conference on Contemporary Computing and Informatics*, 612–616, 2013. https://ieeexplore.ieee.org/abstract/document/7918036/

Kranenburg RV. *The Internet of Things: A Critique of Ambient Technology and the All Seeing Network of RFID*. Institute of Network Cultures, 2008. https://mediarep.org/handle/doc/20469

Lova Raju K. et al. "IoT Technologies in Agricultural Environment: A Survey," *Wireless Personal Communication*, 2020. https://doi.org/10.1007/s11277-020-07334-x

Mandal VP. et al. "Land Suitability Assessment for Optimal Cropping Sequences in Katihar District of Bihar, India Using GIS and AHP," *Spatial Information Research*, 2020. https://doi.org/10.1007/s41324-020-00315-z

Muangprathub J. et al. "IoT and Agriculture Data Analysis for Smart Farm, Elsevier," *Computers and Electronics in Agriculture*, 156, 467–474, 2018, January 2019. https://doi.org/10.1016/j.compag.2018.12.011

Muhammad A. et al. "IoT Enabled Analysis of Irrigation Rosters in the Indus Basin Irrigation System," *Proceeding Engineering*, 154, 229–235, 2016. www.sciencedirect.com/science/article/pii/S187770581631846X

Parameswaran G. et al. "Arduino Based Smart Drip Irrigation System Using Internet of Things," *International Journal Engineering Science Computer*, 6, 5518–5521, 2016. https://doi.org/10.4010/2016.1348

Pathak A. et al. "IoT Based Smart System to Support Agricultural Parameters: A Case Study," *Procedia Computer Science*, 155, 648–653, 2019. https://doi.org/10.1016/j.procs.2019.08.092

Pritiprada P, Satpathy I, Patnaik BCM, Patnaik A, Khang A (Eds.). "Role of the Internet of Things (IoT) in Enhancing the Effectiveness of the Self-Help Groups (SHG) in Smart City," *Smart Cities: IoT Technologies, Big Data Solutions, Cloud Platforms, and Cybersecurity Techniques*. CRC Press, 2024. https://doi.org/10.1201/9781003376064-14

Rana G, Khang A, Sharma R, Goel AK, Dubey AK (Eds.). *Reinventing Manufacturing and Business Processes Through Artificial Intelligence*. CRC Press, 2021. https://doi.org/10.1201/9781003145011

Rani S, Bhambri P, Kataria A, Khang A. "Smart City Ecosystem: Concept, Sustainability, Design Principles and Technologies," *AI-Centric Smart City Ecosystems: Technologies, Design and Implementation* (1st Ed.). CRC Press, 2022. https://doi.org/10.1201/9781003252542-1

Rani S, Bhambri P, Kataria A, Khang A, Sivaraman AK. *Big Data, Cloud Computing and IoT: Tools and Applications* (1st Ed.). Chapman and Hall/CRC Press, 2023. https://doi.org/10.1201/9781003298335

Rani S, Chauhan M, Kataria A, Khang A (Eds.). "IoT Equipped Intelligent Distributed Framework for Smart Healthcare Systems," *Networking and Internet Architecture*. CRC Press, 2021. https://doi.org/10.48550/arXiv.2110.04997

Reka SS. et al. "A Novel Approach of IoT-Based Smart Greenhouse Farming System," *Green Buildings and Sustainable Engineering: Springer Transactions in Civil and Environmental Engineering*. Springer, 2019. https://doi.org/10.1007/978-981-13-12021_20.

Rittika R. et al. "On Analysis of Wheat Leaf Infection by Using ImageProcessing," *Proceedings of the International Conference on Data Engineering and Communication Technology. Advances in Intelligent Systems and Computing* (vol. 468). Springer, 2017. https://link.springer.com/chapter/10.1007/978-981-10-1675-2_56

Salam A. "Internet of Things for Environmental Sustainability and Climate Change," *Internet of Things for Sustainable Community Development. Internet of Things (Technology, Communications and Computing)*. Springer, 2020. https://link.springer.com/chapter/10.1007/978-3-030-35291-2_2

Sethi P. et al. "Internet of Things: Architectures, Protocols, and Applications," *Hindawi Journal*, 2017. https://doi.org/10.1155/2017/9324035

Subhashini R, Khang A (Eds.). "The Role of Internet of Things (IoT) in Smart City Framework," *Smart Cities: IoT Technologies, Big Data Solutions, Cloud Platforms, and Cybersecurity Techniques*. CRC Press, 2024. https://doi.org/10.1201/9781003376064-3

Venkatra K. et al. "Optimized Water Scheduling Using IoT Sensor Data in Smart-Farming," *Smart Technologies in Data Science and Communication: Lecture Notes in Networks and Systems* (vol. 105). Springer, 2020. https://doi.org/10.1007/978-981-15-24073_26

Wu Z. et al. "The Actuality of Agriculture Internet of Things for Applying and Popularizing in China," *Proceedings of the International Conference on Advances in Mechanical Engineering and Industrial Informatics (EII'15)*, 2015. www.atlantis-press.com/article/22019.pdf

8 Implementation of a Smart Vehicle Parking System Using the Internet of Things

Anuj Kumar Goel and Alex Khang

8.1 INTRODUCTION

Today, in our day-to-day lives, a major problem we face on the roads is free spaces available for parking. A system of entitizing (to convert abstract idea into an entity; to perceive as tangible or alive) and modeling must be executed with the help of smart tech-based devices that can accommodate all the needs that modern-day drivers deserve.

Since the turn of the 20th century, we have invented and stumbled on many things, one among them being vehicles (Karbab et al., 2015a).

As the years pass by, we, as people of India, are getting influenced by Western culture; cars and other motor vehicles that were seen as a luxury in the past can now be found in abundance in specific families (Chou et al., 2008).

With the growing population, the number of vehicles on the road is increasing. In the transportation sector, vehicles brought a drastic change in the time to mobilize goods and boost the economy for good (Karbab et al., 2015b).

As we all know, there are two aspects of the same coin. The plethora of vehicles has resulted in increased traffic, emissions like carbon monoxide and carbon dioxide and noise pollution, as well as environmental degradation and waste energy (Goel, 2017).

As the transition in motor speed came in vehicles in the current era, resulted in more and more man-made occurrences like road mishaps. Increased traffic issues gave birth to economic dilemmas that are somewhat common in most cities (Goel et al., 2019a).

The real problem today is that the cities are becoming sizable, yet people cannot find parking spaces (Khang et al., 2023a). Cars are gradually challenging the public engineers, architects and government. A smart and efficient parking system will favor both city planners and drivers by reducing parking time and increasing the use of parking spaces, thereby increasing the revenue from drivers parking in a city (Goel et al., 2019b).

The need for available parking spaces varies, that is, the price changes over time, so it is logically necessary for real-time data about vacant parking locations and conditions of parking provided to car drivers in an instant (Rahayu et al., 2013).

The parking challenge is vital, as people across the nation account for a personal vehicle identity, something that has become part of people's lives. Still, vehicles require a parking space. Because the number of vehicles is increasing on the road, the need for parking areas has also (Sakthivel et al., 2022).

Consistent with data released by the Department of Road Transport and Highways, India released in 2018, the number of motorists in India has risen by approximately 400 %/r from 5.5 crores in 2001 to 21 crores in the fiscal year 2015, as depicted in Figure 8.1 (Khang et al., 2022a).

The speed of motor growth is comparatively higher in metropolitan cities like Mumbai, Bangalore and National Capital Region (NCR), which is around 1%, the expansion number is predicted to hoop from 1 % to 7% based on 2018 data from the Directorate of Economics and Statistics (Goel et al., 2022) as shown in Figure 8.1.

Hence, there is a need to include parking spaces in the design. Parking is a crucial problem within the field, and level of designing strategies lies one of the foremost stressful belongings; if left unmanaged, they will stray and lose track, also causing traffic problems, accidents, and a waste of money and importantly time. Some previous research on parking indicates that parking problems are sometimes caused by people who want to park directly ahead of their destination (Goel et al., 2016).

People park vehicles close to their destination when feasible, which leads to a rise in drift when looking for parking spaces, which leads to choosing curbside parking. On-street/curbside parking is dangerous for through traffic, and due to curbside parking, delays in traffic clearance are created. Many curbside parking locations with high traffic in metropolitan regions create problems in clearing local traffic problems (Orrie et al., 2015).

In Masmoudi et al. (2014) proposed a "division, construction, adapt, and share" strategy for parking in China's Wu Jiang area. Box (2004) found that casualties and

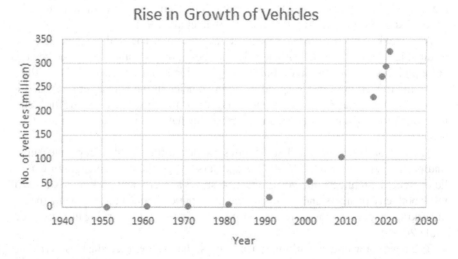

FIGURE 8.1   Rise in growth of vehicles.

Source: Masmoudi et al. (2014).

accidents arose from curbside parking, particularly vehicles parked at angles. The news writer suggested that curbside parking should be restricted on key city roads and demonstrated that the number of injuries and accidents from vehicles parked at an angle is more than that of vehicles parked parallel (Yang et al., 2012).

IoT-based smart parking system is to be designed to solve these problems. There are many ways available to resolving this issue. Magnet-based sensors, sound-based sensors, radio-frequency identification (RFID), ultrasound, "inductive loop", vision cameras and more are different technologies that are available and can be used for monitoring parking (Rani et al., 2022).

There are many solutions that have been developed in recent times for the smart street, parking detection, and monitoring. One such example includes crowdsourcing. It is based on a device, that is, an ultrasonic range finder, that is installed over the vehicles to detect and identify space near an already-parked vehicle (Jayashree et al., 2024).

Information that is collected will be aired to assist drivers of other vehicles. They get these free slots (Geng et al., 2012). This chapter focuses more on the point of smart parking. As seen in past decades, the rise in demand models and studies regarding the behavior of parking and its characteristics has led to much research in this context. The things that were common in most of this research are the importance of parking, distance to be traveled by a vehicle and its time of travailing (Sushil et al., 2011).

The preceding points not only influence the nature of parking but also bring us an idea of how modernization can play a big role in the "parking system" as shown in Figure 8.2. Many such factors have been considered in the research that evaluates and compares the earlier parking models to the existing demand models to search for obstacles that a person can face (Khang et al., 2023b).

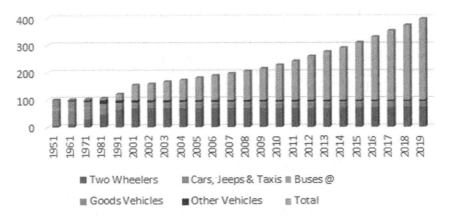

FIGURE 8.2   Vehicle composition data.

Source: Bagula et al. (2015).

8.2 PROBLEM STATEMENT

The main task is to create a smart parking system using automated transactions:

- To make software that will be applied to the parking lifts and accordingly park the cars
- To calculate fares automatically according to the entry time and exit time noted with the registration number of the car by using algorithms
- To provide the best possible places for parking according to the time requested by the user by using algorithms

8.3 REQUIREMENTS FOR AN AUTOMATED SMART PARKING SYSTEM

This project implements an automated smart parking system that will resolve existing problems in reserved vehicle parking stands in various malls, buildings, airports, and other public and private vehicle parking stands (Rana et al., 2021).

It will be more time-efficient, hassle-free, and cost-effective and will provide a better software infrastructure than the existing service system, which involves a lot of human work, thus making it much less effective; involves a scope for errors; and is exhausting. It will be developed to reduce human effort and implement it in a way where we can use digitization for the advancement of day-to-day activities, such as reserved vehicle parking.

8.4 PROPOSED WORK

A smart car parking system is really in demand, and being a handler of a real-life day-to-day application that everyone uses, the demand is increasing day by day as most builders want to equip their projects with more advanced parking systems.

Thus, considering quality and consistency as mandatory factors, we carried out some testing experiments simulating the prototype system to evaluate its dependability as Figure 8.3. The test area consisted of three parking slots equipped with infrared (IR) sensors, microcontrollers, and a display board (Hahanov et al., 2022).

After having the information for all free parking slots, it is the driver's choice as to which slot they want to park their car. Then our algorithm calculates the parking charges according to the duration of car parking and the place they choose to park their car.

8.5 RESULTS AND DISCUSSION

8.5.1 WHEN THE PARKING AREA IS CLOSED

There are several reasons why a parking area might be closed temporarily like maintenance, among others, and it can be considered as closed when a single car is not there.

Keeping these things in mind and to reduce electricity consumption, we designed a system that will turn off all the lights, sensors and the display board as well,

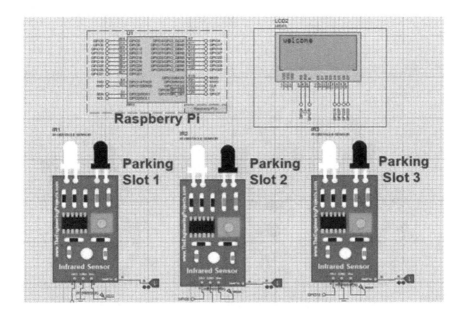

FIGURE 8.3   Initial display of smart parking system.

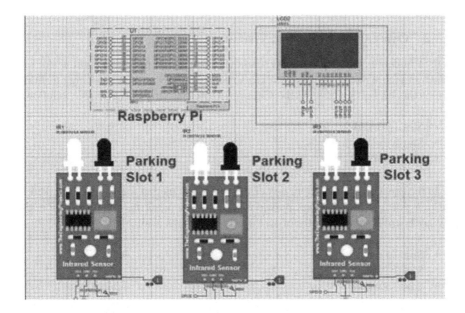

FIGURE 8.4   Display of closed smart parking system.

although there is an option to turn on lights and sensors manually but not the display board as shown in Figure 8.4.

The lights are also checked during this routine maintenance checkup, and faulty lights are replaced so that movement can be easy.

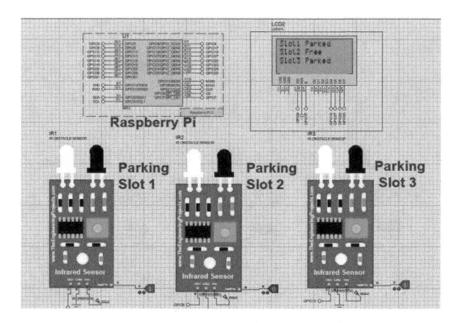

FIGURE 8.5 Smart parking with free slot 2.

8.5.2 OTHER CASES OF PARKING SYSTEMS

In the simulation, it is clearly visible to us that there is a car parked in parking slot 1 as shown in Figure 8.6, as well as a car is also parked in parking slot 3, taking this information the display board shows the status of all three parking slots available as follows:

- Slot 1 Parked
- Slot 2 Free
- Slot 3 Parked

In the preceding simulation, it is clearly visible to us that there is a car parked in parking slot 2, as well as a car parked in parking slot 3; taking this information, the display board shows the status of all three parking slots available as follows:

- Slot 1 Free
- Slot 2 Parked
- Slot 3 Parked

In the preceding simulation, it is clearly visible to us that there is a car parked in all three parking slots; taking this information, the display board shows the status of all three parking slots available as follows:

- Slot 1 Parked
- Slot 2 Parked
- Slot 3 Parked

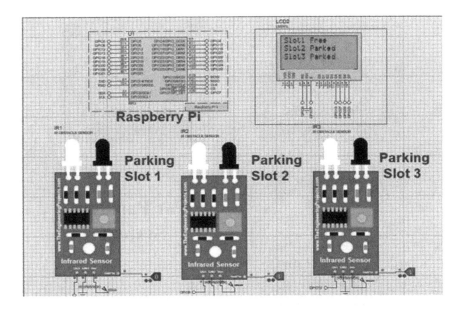

FIGURE 8.6   Smart parking with free slot 1.

LCD D7	LCD D6	LCD D5	LCD D4		LCD D7	LCD D6	LCD D5	LCD D4
Byte 7	Byte 6	Byte 5	Byte 4		Byte 3	Byte 2	Byte 1	Byte 0
0	0	1	1		0	0	1	1
0	0	0	1		0	0	0	0
0	0	0	1		0	0	0	0

FIGURE 8.7   Smart parking liquid crystal display (LCD) arrangement.

This is the truth table of IR sensors on which sensors work, sensing the car and, after that, giving the information to the display board controller, which then analyzes the data and displays the status of the parking slots on the display board (Bhambri et al., 2022).

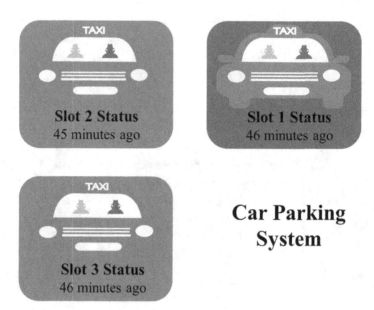

FIGURE 8.8   Smart parking mobile application.

Source: Khang (2021).

8.5.3 SMARTPHONE APPLICATION

Figure 8.8 represents the user interface of our mobile application where users can select the location where they want to park their car and then the application shows them the status of all the slots available also being a safety measure added by us.

This app notifies the user whenever the car is moved from the parking slot in which they parked their car, and if it is not happening according to their will, they can contact us, and as an action, we will call the police and try to stop the car from being stolen.

8.6   CONCLUSION

This smart parking system will provide several free slots according to the hours of stay of the customer. Furthermore, this application will automatically transact the respective amount to be paid by the customer as per the hours for which the car was parked. It will help in decreasing the rush at the cash collection counters (Khang et al., 2022b).

This project will help with people not blocking traffic while looking for parking places when on roads. This project implements an automated smart parking system, which will resolve existing problems in reserved vehicle parking stands in various malls, buildings, airports, and other public and private vehicle parking stands (Rani et al., 2021).

It will be more time-efficient, hassle-free, and cost-effective and will provide a better software infrastructure than the existing service system, which involves a lot of human work, thus making it less effective; involves scope for errors; and is exhaustive. It will be developed to reduce human effort and implement it in a way where we can use digitization for the advancement of day-to-day activities such as reserved vehicle parking (Jayashree et al., 2024).

8.7 FUTURE SCOPE

The futuristic approach toward the project includes implementation and updating of the database and making the smart parking system implementation more effective with a wider range of systems in metropolitan cities starting with establishing systems in places like malls, government buildings and airports.

The project would help solve a lot of problems like traffic congestion in parking lots, saving a lot of time and money.

The smart parking system will also be able to solve conflicts between the authorities and the common people for parking in crowded parking lots. This will, in turn, improve the functioning of parking lots and work toward building smarter cities (Shah et al., 2024).

REFERENCES

Bagula L, Castelli MZ. "On the Design of Smart Parking Networks in the Smart Cities: An Optimal Sensor Placement Model," *Sensors (Switzerland)*, 15(7), 15443–15467, June 2015. https://doi.org/10.3390/s150715443

Bhambri P, Rani S, Gupta G, Khang A. *Cloud and Fog Computing Platforms for Internet of Things*. CRC Press, 2022. https://doi.org/10.1201/9781003213888

Chou SY, Lin SW, Li CC. "Dynamic Parking Negotiation and Guidance Using an Agent-Based Platform," *Expert Systems with Applications*, 35(3), 805–817, 2008. https://doi.org/10.1016/j.eswa.2007.07.042

Geng Y, Cassandras CG. "A New 'Smart Parking' System Infrastructure and Implementation," *Procedia — Social and Behavioral Sciences*, 54, 1278–1287, October 2012. https://doi.org/10.1016/j.sbspro.2012.09.842

Goel AK. "Analytical Modeling and Simulation of Microcantilever Based MEMS Devices," *Wulfenia Journal*, 24(1), 79–91, 2017. www.academia.edu/download/34914378/anuj_012.pdf

Goel AK. "Design of Passive Fuel Cell for Even Current Density and Heat Profile," *JAC: A Journal of Composition Theory*, 15(12), December 2022a. https://doi.org/10.1201/9781003145011

Goel AK. "Integration of MEMS Sensors for Advanced IoT Applications," *Electronic Devices and Circuit Design: Challenges and Applications in the Internet of Things* (pp. 33–50). CRC Press, 2022b. www.taylorfrancis.com/chapters/edit/10.1201/9781003145776-3/integration-mems-sensors-advanced-iot-applications-anuj-kumar-goel

Goel AK. et al. "Design and Simulation of Microcantilevers for Sensing Applications," *International Journal of Applied Engineering and Research*, 11(1), 501–503, 2016. www.researchgate.net/profile/Anuj-Goel-3/publication/298711066_Design_and_simulation_of_microcantilevers_for_sensing_applications/links/5f0c45a34585155a55250100/Design-and-simulation-of-microcantilevers-for-sensing-applications.pdf

Goel AK. et al. "Performance Analysis of SAW Gas Sensors with Different Number of Electrodes," *International Journal of Recent Technology and Engineering*, 8(3), 4397–4401, 2019a. https://doi.org/10.35940/ijrte.C5530.098319

Goel AK. et al. "Sensitivity Enhancement and Comparison of MEMS/NEMS Cantilevers," *Journal of Mechanics of Continua and Mathematical Sciences*, 14(1), 414–421, 2019b. https://doi.org/10.26782/jmcms.2019.02.00029

Hahanov V, Khang A, Litvinova E, Chumachenko S, Hajimahmud VA, Alyar AV. "The Key Assistant of Smart City—Sensors and Tools," *AI-Centric Smart City Ecosystems: Technologies, Design and Implementation* (1st Ed.). CRC Press, 2022. https://doi.org/10.1201/9781003252542-17

Jayashree M. et al. (Eds.). "Vehicle and Passenger Identification in Public Transportation to Fortify Smart City Indices," *Smart Cities: IoT Technologies, Big Data Solutions, Cloud Platforms, and Cybersecurity Techniques*. CRC Press, 2024. https://doi.org/10.1201/9781003376064-13

Karbab El M, Djenouri D, Boulkaboul S, Bagula A. "Car Park Management with Networked Wireless Sensors and Active RFID," *IEEE International Conference on Electro/Information Technology (EIT)*, 2015a. https://doi.org/10.1109/EIT.2015.7293372

Karbab El M, Djenouri D, Boulkaboul S, Bagula A. "Car Park Management with networked Wireless Sensors and Active RFID," *IEEE International Conference on Electro Information Technology*, 373–378, 2015b. https://doi.org/10.1109/eit.2015.7293372

Khang A. "Material4Studies," *Material of Computer Science, Artificial Intelligence, Data Science, IoT, Blockchain, Cloud, Metaverse, Cybersecurity for Studies*, 2021. www.researchgate.net/publication/370156102_Material4Studies

Khang A, Gupta SK, Rani S, Karras DA (Eds.). *Smart Cities: IoT Technologies, Big Data Solutions, Cloud Platforms, and Cybersecurity Techniques*. CRC Press, 2023b. https://doi.org/10.1201/9781003376064

Khang A, Gupta SK, Shah V, Misra A (Eds.). *AI-Aided IoT Technologies and Applications in the Smart Business and Production*. CRC Press, 2023a. https://doi.org/10.1201/9781003392224

Khang A, Hahanov V, Abbas GL, Hajimahmud VA. "Cyber-Physical-Social System and İncident Management," *AI-Centric Smart City Ecosystems: Technologies, Design and Implementation* (1st Ed.). CRC Press, 2022a. https://doi.org/10.1201/9781003252542-2

Khang A, Rani S, Sivaraman AK. *AI-Centric Smart City Ecosystems: Technologies, Design and Implementation* (1st Ed.). CRC Press, 2022b. https://doi.org/10.1201/9781003252542

Masmoudi A, Wali A, Jamoussi AMA. "Vision Based System for Vacant Parking Lot Detection: VPLD," *VISAPP 2014—Proceedings of the 9th International Conference on Computer Vision Theory and Applications*, 2, 526–533, 2014. https://doi.org/10.5220/0004730605260533

Orrie O, Silva B, Hancke GP. "A Wireless Smart Parking System," *IECON 2015–41st Annual Conference of the IEEE Industrial Electronics Society*, 4110–4114, 2015. https://doi.org/10.1109/iecon.2015.7392741

Rahayu Y, Mustapa FN. "A Secure Parking Reservation System Using GSM Technology," *International Journal of Computer and Communication Engineering*, 518–520, 2013. https://doi.org/10.7763/IJCCE.2013.V2.239

Rana G, Khang A, Sharma R, Goel AK, Dubey AK (Eds.). *Reinventing Manufacturing and Business Processes Through Artificial Intelligence*. CRC Press, 2021. https://doi.org/10.1201/9781003145011

Rani S, Bhambri P, Kataria A, Khang A. "Smart City Ecosystem: Concept, Sustainability, Design Principles and Technologies," *AI-Centric Smart City Ecosystems: Technologies, Design and Implementation* (1st Ed.). CRC Press, 2022. https://doi.org/10.1201/9781003252542-1

Rani S, Chauhan M, Kataria A, Khang A (Eds.). "IoT Equipped Intelligent Distributed Framework for Smart Healthcare Systems," *Networking and Internet Architecture*. CRC Press, 2021. https://doi.org/10.48550/arXiv.2110.04997

Sakthivel M, Gupta SK, Karras DA, Khang A, Dixit CK, Haralayya B. "Solving Vehicle Routing Problem for Intelligent Systems using Delaunay Triangulation," *2022 International Conference on Knowledge Engineering and Communication Systems (ICKES)*, 2022. https://ieeexplore.ieee.org/abstract/document/10060807/

Shah V, Jani S, Khang A (Eds.). "Automotive IoT: Accelerating the Automobile Industry's Long-Term Sustainability in Smart City Development Strategy," *Smart Cities: IoT Technologies, Big Data Solutions, Cloud Platforms, and Cybersecurity Techniques*. CRC Press, 2024. https://doi.org/10.1201/9781003376064-9

Sushil KG, Goel A, Gupta M. "Microstrip Filter Designing by Using Split Ring Resonator Metamaterial," *International Journal of Applied Engineering and Research*, 6(5), 709–712, 2011. https://ieeexplore.ieee.org/abstract/document/10009575/

Yang J, Portilla J, Riesgo T. "Smart Parking Service Based on Wireless Sensor Networks," *IECON 2012–38th Annual Conference on IEEE Industrial Electronics Society*, 6029–6034, 2012. https://doi.org/10.1109/iecon.2012.6389096

9 Artificial Intelligence Algorithms for Unmanned Aerial Vehicles and Wireless-Based Sensor Technologies

Mohammed Qayyum, Mohammad Adam Baba, and Nidhya M. S.

9.1 INTRODUCTION

9.1.1 FLYING AD HOC NETWORKS

Organizations can be framed by little automated aerial vehicles (UAVs), some of which are alluded to as flying specially appointed organizations (flying ad hoc networks or FANETs) is a part of the versatile impromptu organizations (mobile ad hoc networks or MANETs) and vehicle impromptu organizations (vehicle ad hoc networks or VANETs) gatherings.

Portable hubs, remote media, decentralized control, and multi-jump correspondence are some of their common qualities. Be that as it may, FANETs are recognized for their quick hubs, frequent topology changes, and low network density.

UAVs have become more popular because of the wide range of industries they are used in, including the military, surveillance, agriculture, healthcare, traffic management, inspection, and public safety (Shah et al., 2023).

Ad hoc refers to a network's capacity to create connections between nodes. Although this network offers dependability, simplicity in deployment, and relatively low operating costs, there are also a number of difficulties. Based on configuration, drones may be split into two categories: rotary-wing UAVs (RW-UAVs) and fixed-wing UAVs (FW-UAVs).

9.1.2 RW-UAVs

These drones are able to take off and land vertically and may be fixed in the air. They are thus more stable and well suited for interior spaces.

 DOI: 10.1201/9781003392224-9

In contrast to FW-UAVs, these drones have smaller capacities, slower speeds, and stricter energy limits. Because their flight time depends on a number of variables, including path planning, speed, weight, and energy source, these features have an impact.

9.1.2.1 FW-UAVs

These drones outperform RW-UAVs in every metric, including flight duration, top speed, and aerodynamic design. The usage of these drones for aerial surveillance is possible. They come in two sizes (small and giant) and have one body and two wings.

Additionally, they typically weigh more than RW-UAVs. They are like normal planes and cannot be fixed in midair. As a result, these drones should not be used in fixed locations. The picture shows FW-UAVs.

* UAVs are further classified into two categories, completely autonomous and remote control, depending on their level of independence from the pilot.
* Remotely piloted UAVs, sometimes known as "drones," are operated either by a human pilot inside the drone's line of sight (LoS) or in reaction to data gathered by the drone's sensors.

9.1.2.2 An Entirely Automated UAV

When confronted with unforeseeable operational and climatic situations, since they run the flight operations themselves without any help from humans, there is still hope that they can complete their mission.

9.2 RELATED WORK

9.2.1 Applications

Mobile ad hoc networks have several applications. We describe these applications in the sections that follow.

Rescue and search efforts: Due to their quick and simple deployment capabilities, in times of natural disaster, the utilization of portable impromptu organizations in the air might be the first line of defense.

Drones: Drones additionally help human troops in hazardous places and seek after unambiguous objectives, for example, pinpointing the whereabouts of survivors and setbacks.

Monitoring of wildfires: Ad hoc networks in the air might be used to check forest temperatures, identify issues, and put out forest fires.

Traffic observation: One FANET application is highway traffic monitoring. This traffic management data reporting and gridlock detection duty can be readily handled by drones. This is a sensible and affordable choice. Furthermore, these networks allow for the implementation of various real-time security solutions, making roads and trains safer places to travel.

Patrolling: In aerial surveillance applications, drones fly in a fixed pattern to locate a certain area without the need for human intervention. When conducting a surveillance operation over a large area, drones can take photographs of the targets and locations of interest. These readings are processed rapidly and sent to an intelligent command center. In order to check and keep track of their security objectives, drones that are in charge of a certain target or region patrol it frequently. For instance, using a FANET, the border patrol may spot illicit border crossing.

Agricultural surveillance: Precision agriculture is the term for this application. It encompasses all information technology-based approaches and techniques used to keep track of the well-being of agricultural output. FANETs may be used to update this program to address any current issues. Information on plant development, agricultural product quality, and chemical fertilizers may be collected and analyzed rapidly by employing drones in this context due to the use of exact scales and criteria.

Remote sensing (RS): Recently, wireless sensor networks (WSNs) have been integrated with FANETs. Drones with sensors and other equipment are used in its implementation. These drones automatically scan the region to gather data about the target environment.

Networks for relaying: Drones are used as airborne relays safely in VANETs and WSNs and efficiently transmit data from ground nodes to base stations. UAVs are also utilized to extend the ground relay nodes' communication range.

9.2.2 CHALLENGES IN FANETs

Mobility: In a FANET, UAVs are frequently mobile and have the ability to move quickly and erratically. Establishing and maintaining dependable communication channels between nodes may be challenging as a result.

Topology of a dynamic network: The topology of a FANET is dynamic because of the mobility of UAVs and the addition and removal of nodes. This may result in routing problems such as packet loss, congestion, and connection failures.

Constrained Bandwidth: FANETs often use bandwidth-constrained wireless communication routes. This may cause traffic jams and delays in data transmission between UAVs.

Security: Eavesdropping, tampering, and denial-of-service assaults are just a few of the security risks that FANETs are susceptible to. A significant difficulty is protecting the network and guaranteeing the accuracy of the data delivered.

Energy Limitations: Due to their limited power, UAVs in a FANET must conserve energy in order to increase their flight time. The energy consumption of routing algorithms must be kept to a minimum while ensuring the integrity of the communication channels.

Scalability: A FANET's complexity rises with the number of UAVs it contains, making it more challenging to properly manage and route traffic. To guarantee that the network can support several nodes, scalable routing methods are required.

Insufficient Infrastructure: FANETs are often deployed in underdeveloped or undeveloped areas. For this reason, it is necessary for the network to use decentralized routing protocols and provide self-organization and configuration.

Scanty Computer Resources: The computing resources and storage space available to UAVs in a FANET are limited. This may put a limit on the sophistication of the routing protocols that may be employed.

Interference: FANET node-to-node connections may suffer from operating in a crowded wireless band due to interference from other devices.

Level of Service: There are certain criteria for the quality of service (QoS) that must be supplied in various FANET applications, such as minimum data throughput, latency, and dependability. The design of routing protocols must take into account these QoS specifications while also maximizing other factors like energy use and network performance.

9.2.3 Architecture for Communications

Two networking modes need to be activated in accordance with the functions that UAVs carry out in a FANET architecture:

1. UAV to UAV (U2U)
2. UAV to Infrastructure (U2I)

To this end, one of the UAVs is known as the "spine UAV"; it probably goes at the middle point of the FANET structure. The UAV filling in as the organization's spine uses U2U correspondence to gather data from the individual UAVs that make up the organization and then uses U2I communication to send that information to the ground station (GS).

Both centralized and decentralized approaches to integrating several UAVs were proposed by Li et al. (2013). Centralized designs, like the one seen in Figure 9.1, have all UAVs link directly to the GS. In order to send data between any two UAVs, a GS is required.

However, UAVs in the decentralized architecture may interact directly or indirectly without a GS. Three other forms of distributed architecture were also defined

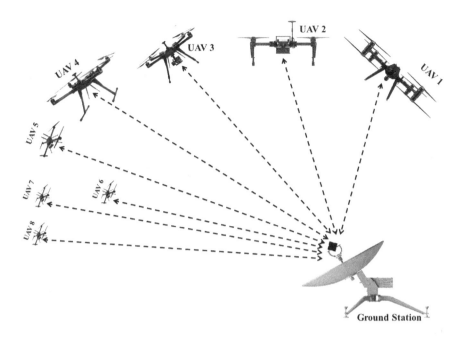

FIGURE 9.1   Centralized UAV network architecture.

Source: Khang (2021).

by the authors: the multilayer UAV ad hoc network, the multigroup UAV ad hoc network, and the UAV ad hoc network with multiple layers.

Similar to Khan et al. (2017), the authors looked at these designs for FANET deployment. The three architectures are described in the following subsections.

9.2.3.1   UAV Ad Hoc Network

In a UAV impromptu organization, as portrayed in Figure 9.2, each UAV adds to the information sent to the other UAVs. The "spine," or the UAV in the center, associates the GS to each UAV.

Two radios, one with low power for short reach and one with high power for long reach, commonly make up the standard UAV setup. A powerful long-range radio is expected to speak with the GS, although a low-controlled short-range radio is used for correspondence between UAVs.

The UAV specially appointed systems administration idea significantly extends the organization's inclusion region by interfacing only one spine UAV to the GS. In addition, because of the UAVs' close proximity, the transceiver put on the UAV may be cheap and light, which is especially desired for tiny UAVs.

However, for the network to be more stable, all the linked UAVs' motion patterns, speeds, altitudes, and directional velocities must be consistent. Surveillance, monitoring operations, and similar missions are ideal for this kind of networking architecture since only a small number of similarly configured UAVs need to be deployed.

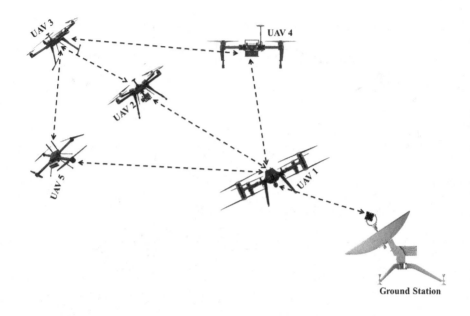

FIGURE 9.2   Unmanned aerial vehicle ad hoc network architecture.

Source: Khang (2021).

9.2.3.2   Multigroup UAV Ad Hoc Network

Figure 9.3 outlines how a multi-bunch UAV impromptu organization consolidates components of both decentralized and concentrated network designs. The UAVs in each group operate independently of one another, forming a network on the fly, while each gathering's spine UAV is formally associated with the GS.

While the spine UAV is utilized for between-bunch correspondences, a separate ad hoc network of UAVs handles intragroup communication independently of the GS. The missions that call for the deployment of several heterogeneous UAVs are best suited for this architecture. However, the decentralized nature of this network design makes it vulnerable.

9.2.3.3   Multilayer UAV Ad Hoc Network

The multifaceted UAV specially appointed network is a design, as displayed in Figure 9.4, that associates a few groups of UAVs with fluctuating levels of similitude. In this architecture, ad hoc UAV networks (made up of UAVs that are not part of any larger network) form the network's bottom layer.

The top layer incorporates the spine UAVs, everything being equal, which are interconnected. The gathering's "spine" is one UAV, the sole UAV that communicates with the GS. The data sent to the GS are solely stored at the GS to reduce the amount of data transfer and processing required by the GS. Therefore, missions requiring a large number of uniquely configured UAVs will benefit from this design.

To sum up, decentralized communication architecture is appropriate for establishing a connection between several UAVs in a FANET. This communication architecture may provide greater coverage for data transfer thanks to its multi-hop structure.

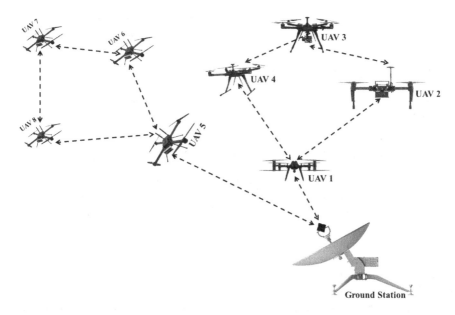

FIGURE 9.3 Multigroup unmanned aerial vehicle network architecture.

Source: Khang (2021).

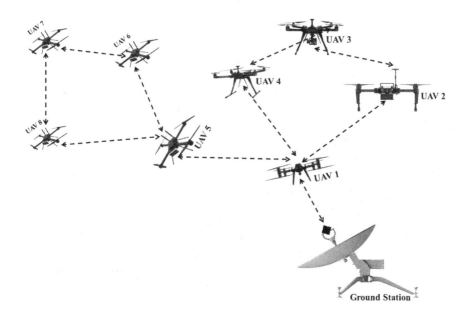

FIGURE 9.4 Multilayer unmanned aerial vehicle ad hoc network architecture.

Source: Khang (2021).

Homogeneous, sparsely dispersed UAVs also perform well with the ad hoc network architecture for UAVs. Nonetheless, when countless heterogeneous UAVs are required, multi-bunch and multifaceted models can be used. Furthermore, because of its decentralized nature, a multifaceted UAV impromptu organization might be linked to its multi-bunch partner. A multifaceted impromptu UAV organizing configuration is likewise more adaptable as far as giving an "on-the-fly" correspondence network since it is more impervious to a weak link.

9.3 TECHNOLOGIES FOR WIRELESS COMMUNICATIONS

We concentrated on several networking topologies between UAVs and GS in the preceding section. According to the topologies shown in Figure 9.4, a wide variety of wireless communication technologies may be used to quickly establish FANET by offering durable and flexible communication connections.

The kind of mission and the nature of the application determine the best technology to use. All wireless technologies can support the low demand for FANET control traffic, and there are two main categories of wireless technology: short-range and long-range communications.

However, when there is a significant amount of short- range and long-range communication, multigroup and multilayer architectures can be utilized. Short-range correspondence advancements incorporate any semblance of Wi-Fi, ZigBee, and Bluetooth. A few instances of long-range correspondence innovations that may be utilized to cover bigger regions incorporate cell, WiMAX, and satellite.

9.3.1 Wi-Fi (IEEE 802.11)

A bunch of principles for laying out wireless local area networks (WLANs; remote neighborhood) on the radio frequencies of 5 GHz, 60 GHz, 3.6 GHz, and 2.4 GHz is known as Wi-Fi, which means "remote loyalty." Because of the great data transmission rates they give, IEEE 802.11a/b/g/n/ac varieties may be the best choice for by far most FANET applications.

A Wi-Fi framework regularly has a transmission scope of 100 meters. However, the communication range between UAVs could be extended by several kilometers through ad hoc networking. Clients discover and join WLANs in a standard 802.11 network based on announcements sent by access points (Joh et al., 2015).

The device's role is clearly defined as either client or access point (AP). Furthermore, these functions can be performed concurrently by the same device and are dynamically assigned. Throughput, received signal strength indication, and range were all taken into account when evaluating a wireless connection between a UAV and the GS by Cheng et al. (2006). In a networking setup based on UAVs, it is recommended that an 802.11a remote connection between the GS and the UAV may be helpful.

9.3.2 Bluetooth (IEEE 802.15.1)

Bluetooth uses the unlicensed 2.4 GHz radio band and has an extent of 10 to 200 meters for data transmission. Bluetooth comes in a wide variety of flavors, with data transfer speeds anywhere from 1–3 Mbps depending on the model. Nonetheless, a maximum throughput of 24 Mbps is feasible.

Bluetooth Low Energy (BLE) was proposed for consideration in the Bluetooth 4.0 particulars by the Bluetooth Special Interest Group (SIG). At this time, the most recent specifications are the core ones for Bluetooth 5 (Bluetooth Core, 2016). Speed, transmission range, energy productivity, and interoperability with other short-range innovations are the essential objectives of Bluetooth 5.

In addition to the standard location data, Bluetooth 5 can transmit more complex data like URL files and multimedia. Given the significant advancements, Bluetooth 5 gives off an impression of being a serious competitor for its minimal expense and low-power sending of a future FANET.

The 802.15.1 hybrid method proposed by Khan et al. (2018) was specifically designed for FANET. The proposed method relied on 802.15.1 for the data transmission between the UAVs. The authors presented their work on a round-robin scheduling-based Bluetooth-enabled wireless network platform in (Afonso et al., 2006).

The ongoing UAV network works utilizing a Bluetooth convention and comprises one expert station and anything from one to seven slave stations. The attestation that the proposed stage offers trustworthy correspondence while requiring fewer computational resources is supported by the findings.

In the Hoffmann et al. (2004) testbed, multiple UAVs were used. The testbed opens the door to actual Bluetooth-enabled UAV network applications in the real world.

9.3.3   ZIGBEE (IEEE 802.15.4)

Due to its long battery life, secure networking, and low power consumption, the ZigBee innovation is oftentimes used in low-information rate applications. It could go similar to 100 yards. It is more affordable and less complex than Bluetooth and Wi-Fi. It communicates information at a pace of 250 kbps and works on the 2.4 GHz band. Its 16 channels are totally dispensed 5 MHz of data transfer capacity each.

ZigBee-based indoor quadcopter localization was described by Yut et al. (2013). The findings confirm the ZigBee localization system's viability, effectiveness, and ease of use. Jiang et al.'s (2006) analysis of UAV landings included tests of ZigBee's communication and location estimation capabilities.

The research shows a considerable reduction in inaccuracy and an accurate calculation of the location as a result. In their hybrid design, Zafar et al. (2017) recommended ZigBee for inter-cluster data transmission. ZigBee might be a decent choice for low-information rate FANET applications, according to simulation results.

9.4   TECHNOLOGIES FOR LONG-RANGE COMMUNICATION

As a backhaul, vast-distance communication technology can provide data communication services over vast distances between two locations. Aerial vehicles may potentially benefit from these technologies by being able to interface directly with U2U- and U2I-enabled stationary infrastructure. In this section, the strengths of today's localized communication technologies are examined.

9.4.1 WiMAX IEEE 802.16

With the use of point-to-point connections and complete mobile cellular connectivity, WiMAX is a technical standard with the goal of extending high-speed Internet connectivity across vast distances and in a variety of ways.

The applications of both fixed and mobile broadband are supported by this technology. It allows for download speeds of up to 30 Mbps (3–5 Mbps per subscriber) for mobile apps and up to 75 Mbps (20–30 Mbps per subscriber) for stationary ones.

WiMAX was developed so that high-quality speech and video could be streamed without interruptions. WiMAX is thought to be the most suited technology for UAV-based rescue systems in dangerous situations (Rahman et al., 2014).

Dalmasso et al. (2012) proposed a technique for network design that took into account the quantity and location of UAVs. The outcomes of the simulation show that it is possible to determine each UAV's dimensions, including location and altitude, and ensure a certain QoS using WiMAX.

9.4.2 Long-Term Evolution

Beyond use cases requiring visual LoS, the secure wireless control and the security of Long Term Evolution (LTE) may benefit greatly from increased connection, portability, and data throughput. LTE is optimized for IP and offers bandwidths as low as 5 MHz and as high as 20 MHz (Khang et al., 2022a).

Recurrence division duplexing (both matched and unpaired) and time division duplexing (TDD) are additionally upheld. In spite of the fact that it performs splendidly inside 30 km and at an OK level up to 100 km, the ideal cell size is 5 km. Unsurprisingly, there has been a significant increase in activity around using an existing LTE network for UAVs during the last 3 years.

9.4.3 Iteration Five

After the first 2G (GSM), the 3G (UMTS), and the 4G (LTE/WiMAX) ages of cell portable correspondence, the most up to date is iteration 5 (5G). High information rates, decreased dormancy, energy investment funds, expanded framework limit, and broad network are among its most eminent benefits.

The debut of 5G mobile networks is anticipated for 2020, according to the International Telecommunication Union (ITU). Such systems will have a capacity that can span up to 1000 times and a speed per user of 100 GB/s (Jiang et al., 2017). These qualities make 5G technology well suited for UAV communication systems and herald the arrival of novel uses. In the case of UAVs (5G), for instance, the core network may be connected to less active parts of the FANET architecture. This will make it easier to provide services like streaming multimedia for surveillance.

UAVs in a 5G network have more transparency than ever before, but continuous connection requires backhauling (Bor-Yaliniz et al., 2016; Dong et al., 2018).

Including a large-scale cell base station (mobile base station, MBS) and interfacing UAVs to a base band unit (BBU) are two examples of what can be done to meet the backhauling requirement (Sharma et al., 2017).

9.4.4 SATELLITE COMMUNICATION

Including a large-scale cell base station (MBS) and interfacing UAVs to a BBU. Different satellites use various frequency bands in satellite communication (SATCOM). The C-bands use a 4GHz downlink and a 6-GHz uplink. They continue to have ties to many networks. In contrast, X-Bands require 7 GHz for downlink and 8 GHz for uplink and are hence primarily employed by government and military systems.

The downlink and uplink frequencies used by the so-called Ku-Bands are 11–12 GHz and 14 GHz, respectively. Furthermore, these bands are becoming saturated, leading to a shift to Ka-Bands. Ka-Gatherings use an uplink of 30 GHz and a downlink of 20 GHz.

The makers of (Ma et al., 2004) focused on critical components such the satellite downlink, the UAV uplink transmission strength, and the picture transmission rate. It has been expressed that satellite hand-off can broadcast high-quality images over a large area and has a wider overlay range.

In addition, Ma et al. (2004) detailed the main difficulties of using SATCOM for real-time picture and video communications with micro and tiny unmanned aerial system (UAS). The significant expense of information transmission is one of the main pressing concerns, and another is the absence of sufficient transfer speed.

Due to the range and throughput requirements, it is apparent from the previous depiction that medium-range FANET applications might need to consider short-range correspondence innovations like Bluetooth, ZigBee, or Wi-Fi.

Notwithstanding, long-range correspondence advances like WiMAX, LTE, 5G, and SATCOM might be considerably more reasonable in the event that the inclusion region is enormous, and these short-range advances cannot uphold the important throughput prerequisites.

For our proposed system, we look into transmission properties-based short-range communication options. These systems use unlicensed spectrum, do not need a strict LoS, and meet the requirements for acceptable data rate and coverage. Likewise, little UAVs may effectively incorporate them into their frameworks (Khang et al., 2023a).

Besides, among the previously mentioned short-range remote advancements, Bluetooth 5 is the most ideal choice because of huge upgrades in speed, power utilization, limit, and inclusion. However, our proposed approach combines Wi-Fi with Bluetooth 5 to maximize the benefits of both technologies (Khang et al., 2023b).

9.5 ANALYSIS OF AI ALGORITHMS APPLIED IN UAVS

9.5.1 LINEAR REGRESSION

Linear regression means that the number of independent points is marked in a single line, or the number of events happening in an environment that can be grouped and combined as a single event. During search-and-rescue efforts, UAVs can pinpoint survivors and setbacks (Rana et al., 2021).

Direct relapse is utilized to decide the idea of the association between the autonomous and subordinate factors. This line, which is also referred to as the regression line, is represented by the linear Equation 9.1.

$$Y = a * X + b. \tag{9.1}$$

This formula includes

- Y as the dependent variable,
- a as the slope,
- X as the independent variable, and
- b as the intercept.

The a and b coefficients are found by minimizing the sum of the squared deviations from the regression line to the data points.

9.5.2 LOGICAL DISCRIMINATION

Strategic relapse can be utilized to gauge discrete qualities, regularly twofold qualities like 0/1. Utilizing it to fit information to a logit capability makes it easier to predict an event's likelihood.

This technique is also known as logit regression. Logistic regression models are commonly refined using the following techniques:

- Including interaction words
- Getting rid of extras
- Imposing order on processes
- Incorporating a nonlinear model

9.5.3 DECISION TREE

The decision tree method is a common supervised learning technique for issue classification in the field of machine learning. It is applicable to the categorization of both discrete and continuous outcomes. Using the most important characteristics as the dividing line, the population is split into two or more groups using this method.

9.5.4 SUPPORT VECTOR MACHINE ALGORITHM

An example of a grouping strategy that works by extending crude information is focusing it in an n-layered space, the support vector machine (SVM) algorithm (where n is the number of attributes). After assigning a numerical value to each attribute, the data may be easily sorted into categories. Data may be partitioned and plotted on a graph using lines called classifiers (Khanh and Khang, 2021).

9.5.5 ALGORITHM OF NAIVE BAYES

In the event that we utilize a Naive Bayes classifier, we will be working under the bogus presumption that the presence of one part in a class does not make any difference to the presence of some other element.

A Credulous Bayes classifier would work out the likelihood of an occasion regardless of any connections between the elements being thought of a Naive Bayesian model can be built quickly and works well with large datasets. It has been demonstrated to perform better than complex categorization methods and is simple to implement (Hajimahmud et al., 2022).

9.5.6 CALCULATION FOR TRACKING DOWN THE k-NEAREST NEIGHBORS

Both characterization and relapse issues might be settled utilizing this methodology. It appears to be used more frequently in the data science industry to address categorization issues. It is a straightforward approach that assigns classes to any new examples based on the consensus of its k-nearest neighbors (KNN) and keeps track of all the cases that are available (Hajimahmud et al., 2023).

After that, the case is placed in the category that most closely matches it. Using a distance function, it is measured. By drawing parallels to everyday situations, KNN becomes intuitively clear.

9.5.6.1 KNN Algorithm Selection Criteria

* KNN has high computational costs.
* Higher-range variables might skew the algorithm; thus, it is important to normalize them.
* Pre-handling of information is as yet required.

9.5.6.2 k-Means

Unsupervised learning underpins this method for resolving clustering issues. A cluster's data points are similar to each other but distinct from those in other clusters; this is the process by which informational indexes are gathered into a specific number of groups, or k. The clustering process in k-means happens as follows:

* k-means selects a set number of cluster centers, or "centroids," for each group.
* Each data point groups with its nearest neighbors to generate k clusters.
* It generates new cluster centers using the preexisting nodes in the cluster as input.
* The new centers are used to calculate the nearest distance between each pair of data points. This procedure is continued until there is no longer any movement in the centroids.

9.5.7 THE ALGORITHM OF RANDOM FORESTS

A group of decision trees is called a random forest. Each tree is doled out a class and afterward "votes" for that class when another thing is being sorted in light of its properties. The most famous order is chosen by the woodland (from among every one of the trees).

Here's how each tree is planted and developed:

- An irregular example of *N* occurrences is chosen in the event that the all-out number of cases in the preparation set is *N*. The tree will be developed utilizing this informational collection as the preparation set.
- In the event that there are *M* info factors, the best parted on these *M*s is utilized to partition the hub at each node and is chosen at random from the *M*. We keep the value of m constant throughout this procedure.

The maximum feasible size of each tree is achieved. It is not pruned in any way.

9.5.7.1 Algorithms for Reducing Dimensions

Today, organizations, state-run administrations, and scholastic establishments gather, store, and investigate monstrous volumes of information. As an information researcher, we know that this crude information contains an abundance of data; try to extract the most useful patterns and variables.

Decision trees, factor analysis, missing value ratios, and random forests are all examples of dimensionality reduction methods that may aid in the discovery of important information.

9.5.7.2 Gradient Boosting Algorithm and Ada Boosting Algorithm

At the point when a lot of information should be handled to create exact forecasts, supporting calculations like the Gradient Boosting algorithm and the Adaptive Boosting algorithm come into play. The ensemble learning process known as "boosting" utilizes the strengths of many "base estimators" to produce more accurate predictions.

In a nutshell, it takes many mediocre or weak predictors and merges them into one robust one. Data science contests like Kaggle portal, AV Hackathon program, and CrowdAnalytix platform have consistently found success with these boosting methods. In the field of machine learning, these are the top choices. For reliable results, combine them with Python and R codes.

9.6 CONCLUSION

In this chapter, we discussed how UAVs are essentially flying robots that are controlled remotely or can fly self-directed with software controlled flight plans embedded in its system that work in conjunction with sensors and global positioning systems (Hahanov et al., 2022).

The four main types of UAVs are there multi-rotor, fixed wing, single rotor, and fixed-wing hybrid, as well as some other types.

We analyzed a number of UAV types and the wireless technologies used in UAV and AI-driven applications like monitoring, imaging, and communication. The AI algorithms applied to UAVs were also analyzed (Khang et al., 2022b).

REFERENCES

Afonso JA, Coelho ET, Carvalhal P, Ferreira MJ, Santos C, Silva LF, Almeida H. "Distributed Sensing and Actuation Over Bluetooth for Unmanned Air Vehicles," *In Proceedings of the International Conference on Robotics and Automation*, May 15–19, 2006. https:// ieeexplore.ieee.org/abstract/document/8528677/

Bluetooth Core Specification, Bluetooth Special Interest Group (SIG), 2016. www.bluetooth. com/specifications/bluetooth-core-specification. Last visit October 6, 2018.

Bor-Yaliniz I, Yanikomeroglu H. "The New Frontier in Ran Heterogeneity: Multi-Tier Drone-Cells," *IEEE Communications Magazine*, 54, 48–55, 2016. https://ieeexplore.ieee.org/ abstract/document/7744808/

Cheng C-M, Hsiao PH, Kung HT, Vlah D. "Performance Measurement of 802.11a Wireless Links from UAV to Ground Nodes with Various Antenna Orientations," *Proceedings of the 15th International Conference on Computer Communications and Networks*, 303–308, October 9–11, 2006. https://ieeexplore.ieee.org/abstract/document/4067672/

Dalmasso I, Galletti I, Giuliano R, Mazzenga F. "WiMAX Networks for Emergency Management Based on UAVs," *Proceedings of the 2012 IEEE First AESS European Conference on Satellite*. https://ieeexplore.ieee.org/abstract/document/6400206/

Dong Y, Hassan M, Cheng J, Hossain M, Leung V. *An Edge Computing Empowered Radio Access Network with UAV-Mounted FSO Fronthaul and Backhaul: Key Challenges and Approaches*, 2018. arXiv:1803.06381. https://ieeexplore.ieee.org/abstract/document/8403965/

Hahanov V, Khang A, Litvinova E, Chumachenko S, Hajimahmud VA, Alyar VA. "The Key Assistant of Smart City—Sensors and Tools," *AI-Centric Smart City Ecosystems: Technologies, Design and Implementation* (1st Ed., vol 17, p. 10). CRC Press, 2022. https:// doi.org/10.1201/9781003252542-17

Hajimahmud VA, Khang A, Hahanov V, Litvinova E, Chumachenko S, Alyar V. A., "Autonomous Robots for Smart City: Closer to Augmented Humanity," *AI-Centric Smart City Ecosystems: Technologies, Design and Implementation* (1st Ed., vol. 7, p. 12). CRC Press, 2022. https://doi.org/10.1201/9781003252542-7

Hajimahmud VA. et al., "The Role of Data in Business and Production," *AI-Aided IoT Technologies and Applications in the Smart Business and Production*. CRC Press, December 2023. https://doi.org/10.1201/9781003392224-2

Hoffmann G, Rajnarayan DG, Waslander SL, Dostal D, Jang JS, Tomlin CJ. "The Stanford Testbed of Autonomous Rotorcraft for Multi Agent Control (STARMAC)," *In Proceedings of the 23rd Digital Avionics Systems Conference* (IEEE Cat. No.04CH37576), October 28, 2004. https://ieeexplore.ieee.org/abstract/document/1390847/

Jiang D, Liu G. *An Overview of 5G Requirements—5G Mobile Communications* (pp. 3–26), Springer, 2017. https://link.springer.com/content/pdf/10.1007/978-3-030-58197-8.pdf

Jiang Y, Cao J, Du Y. "Unmanned Air Vehicle Landing Based on Zigbee and Vision Guidance," *Proceedings of the 2006 6th World Congress on Intelligent Control and Automation, Dalian, China*, 10310–10314, June 21–23, 2006. https://ieeexplore.ieee.org/abstract/document/1714021/

Joh H, Yang I, Ryoo I. "The Internet of Everything Based on Energy Efficient P2P Transmission Technology with Bluetooth Low Energy," *Peer-to-Peer Networking and Applications*, 9, 520–528, 2015. https://link.springer.com/article/10.1007/s12083-015-0377-4

Khan MA, Khan IU, Qureshi IM, Alam MK, Shah SB, Shafiq M. "Deployment of Reliable, Simple, and Cost-Effective Medium Access Control Protocols for Multi-Layer Flying Ad-Hoc Networks," *Proceedings of the 2nd International Conference on Future Networks and Distributed Systems (ICFNDS '18), Amman, Jordan*, 49, June 26–27, 2018. https://dl.acm.org/doi/abs/10.1145/3231053.3231115

Khan MA, Qureshi IM, Safi A, Khan IU. "Flying Ad-Hoc Networks (FANETs): A Review of Communication Architectures, and Routing Protocols," *Proceedings of the 2017 First International Conference on Latest Trends in Electrical Engineering and Computing Technologies (INTELLECT), Karachi, Pakistan*, 692–699, November 15–16, 2017. https://ieeexplore.ieee.org/abstract/document/8277614/

Khang A. "Material4Studies," *Material of Computer Science, Artificial Intelligence, Data Science, IoT, Blockchain, Cloud, Metaverse, Cybersecurity for Studies*, 2021. www.researchgate.net/publication/370156102_Material4Studies

Khang A, Gupta SK, Rani S, Karras DA (Eds.). *Smart Cities: IoT Technologies, Big Data Solutions, Cloud Platforms, and Cybersecurity Techniques*. CRC Press, 2023a. https://doi.org/10.1201/9781003376064

Khang A, Hahanov V, Abbas GL, Hajimahmud VA. "Cyber-Physical-Social System and İncident Management," *AI-Centric Smart City Ecosystems: Technologies, Design and Implementation* (1st Ed., vol. 2, p. 15), CRC Press, 2022a. https://doi.org/10.1201/9781003252542-2

Khang A, Rani S, Gujrati R, Uygun H, Gupta SK (Eds.). *Designing Workforce Management Systems for Industry 4.0: Data-Centric and AI-Enabled Approaches*. CRC Press, 2023b. https://doi.org/10.1201/9781003357070

Khang A, Rani S, Sivaraman AK. *AI-Centric Smart City Ecosystems: Technologies, Design and Implementation* (1st Ed.). CRC Press, 2022b. https://doi.org/10.1201/9781003252542

Khanh HH, Khang A. "The Role of Artificial Intelligence in Blockchain Applications," *Reinventing Manufacturing and Business Processes through Artificial Intelligence* (vol. 2, pp. 20–40. CRC Press, 2021. https://doi.org/10.1201/9781003145011-2

Li J, Zhou Y, Lamont, L. "Communication Architectures and Protocols for Networking Unmanned Aerial Vehicles," *Proceedings of the 2013 IEEE Globecom Workshops (GC Workshops), Atlanta, GA*, 1415–1420, December 9–13, 2013. https://ieeexplore.ieee.org/abstract/document/6825193/

Ma D, Yang S. "UAV Image Transmission System Based on Satellite Relay," *Proceedings of the ICMMT 4th International Conference on, Proceedings Microwave and Millimeter Wave Technology, Nanjing, China*, 874–878, August 18–21, 2004. https://ieeexplore.ieee.org/abstract/document/1411670/

Rahman MA. "Enabling Drone Communications with WiMAX Technology," *Proceedings of the IISA 2014, the 5th International Conference on Information, Intelligence, Systems and Applications, Chania, Greece*, 323–328, July 7–9, 2014. https://ieeexplore.ieee.org/abstract/document/6878796/

Rana G, Khang A, Sharma R, Goel AK, Dubey AK (Eds.). *Reinventing Manufacturing and Business Processes Through Artificial Intelligence*. CRC Press, 2021. https://doi.org/10.1201/9781003145011

Shah V, Khang A. "Internet of Medical Things (IoMT) Driving the Digital Transformation of the Healthcare Sector," *Data-Centric AI Solutions and Emerging Technologies in the Healthcare Ecosystem* (1st Ed., p. 1). CRC Press, August 24, 2023. https://doi.org/10.1201/9781003356189-2

Sharma V, Song F, You I, Chao H-C. "Efficient Management and Fast Handovers in Software Defined Wireless Networks Using UAVS," *IEEE Network*, 31, 78–85, 2017. https://ieeexplore.ieee.org/abstract/document/8120266/

Yut L, Fei Q, Geng Q. "Combining Zigbee and Inertial Sensors for Quadrotor UAV Indoor Localization," *Proceedings of the 2013 10th IEEE International Conference on Control and Automation (ICCA), Hangzhou, China*, 1912–1916, June 12–14, 2013. https://ieeexplore.ieee.org/abstract/document/6565087/

Zafar W, Khan BM. "A Reliable, Delay Bounded and Less Complex Communication Protocol for Multicluster FANETs," *Digital Communications and Networks*, 3, 30–38, 2017. www.sciencedirect.com/science/article/pii/S2352864816300256

10 Management and Analysis of House Prices Using the Python Platform

Radha Agrawa, Anuradha Misra, and
Praveen Kumar Misra

10.1 INTRODUCTION

The need for data is increasing day by day. It is not easy to analyze big data. Data analysis helps us understand large data sets in an efficient way. Python language, which is rich with libraries, helps us gain information from big data. It is more useful than traditional methods and technologies.

Data analysis helps companies analyze their work and hence helps in making decisions that will result in the betterment of the company. For example, YouTube, after reviewing the interests of customers in a field, displays the ads related to that particular field only. Data analysis is easy to apply in most of domains if we can understand the tools and needs.

An e-commerce company analyzes the feedback of customers using proper visualization techniques. Exploratory data analysis (EDA) is an approach that concludes data based on key characteristics and visualizes them with appropriate descriptions.

- Row/column record counts, missing data, data types, and previews are quickly explained by EDA. Delete any error data. Invalid data types, unlawful values, and missing data should be addressed.
- Data distribution is visualized by EDA, such as by using histograms, boxplots, and bar charts.
- Calculate and display the associations (correlations) between different variables.

10.2 SAS OVERVIEW

SAS is a software program used to examine statistical data. SAS's primary functions are data retrieval, reporting, and analysis. In the SAS environment, every statement must conclude with a semicolon; otherwise, an error message will be displayed. It is an effective tool for running SQL queries and using macros to automate user tasks.

DOI: 10.1201/9781003392224-10

Aside from this, SAS offers descriptive graph visualization and offers reporting for machine learning (ML), data mining, time series, and other things in different SAS versions. There are two different types of statements that can be utilized with SAS. Data stages and procedures are two categories into which the statements in SAS programs are frequently subdivided (Rana et al., 2021).

10.3 METHODS FOR EDA

Exploratory data analysis is primarily a method to learn more about the data's potential to inform our thinking outside of formal modeling or hypothesis-testing tasks. These essential EDA components are described in the following subsections. Every stage of the ML process, as depicted in Figure 10.1, makes considerable use of data analysis and visualization tools. Some techniques are covered in the following subsections.

10.3.1 EXPLORATION OF DATA

This is the initial phase of data analysis. Here, we can learn more about the data set's characteristics and content. It provides information on the data's size. The data's missing values can be located (Khanh and Khang, 2021).

We can discover any potential relationships between the data. By using tabular data and comprehending the qualities as Figure 10.1, data are visualized according to Khang (2021).

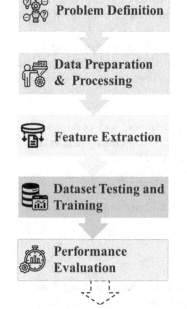

Problem Definition	• *Automatic detection of phishing emails*
Data Preparation & Processing	• *Data (ham and phishing mail) collection* • *Data formatting and representation* • *Data shuffling / randomizing*
Feature Extraction	• *Outline features to be used* • *Automatic feature extraction*
Dataset Testing and Training	• *Cross validation to test and train dataset* • *Prediction of result*
Performance Evaluation	• *Evaluating performance metrics of algorithm* • *Accuracy, recall, precision, f1-score, ROC, AUC*

FIGURE 10.1 Steps of data exploration.

Source: Khang (2021).

10.3.2 DATA CLEANING

It entails identifying faulty data, removing extraneous information from the data, and then replacing the incorrect data. The actual procedure for cleaning data entails removing errors and confirming information. Data cross-checking can be used to discover and correct inaccuracies (Bhambri et al., 2022).

10.3.3 MODAL DESIGN

It is feasible to use supervised and unsupervised models. We can use a classification or regression model to get the results. A model allows us to visualize the results. The model then has to be assessed (Hajimahmud et al., 2023).

10.3.4 CURRENT RESULT

By using charts, graphs, and tables, we may view vast amounts of complex data. It can pinpoint the areas that require improvement. It effectively explains the factor.

10.3.5 GRAPHICAL EDA WITH ONE VARIABLE

The raw data set's fields are statistically summarized by univariate GEDA, or the summary is limited to one variable. We can see the distribution of numerical data by using a histogram. Two variables are not necessary for the histogram.

In this case, the entire value range can be divided into several intervals. Histograms are frequently used to represent continuous data. A rectangle can be used to represent the frequency distribution of a histogram, with the width standing in for the class interval and the area for the frequencies.

10.3.6 PLOT BOXES

A box plot can be used to show the data concentration in a meaningful visual way. The primary trends, symmetry, skew, and outliers are all clearly shown. It can be produced using five values: the minimum, first quartile, median, third quartile, and maximum values. These numbers are compared to show how similar other data values are to them.

10.3.7 STEM PLOTS

A bit of extra information is represented in a stem plot compared to a histogram. It also functions as a visualization tool. Data comparison is much easier in this case. The numbers are arranged according to place value. They mainly help to draw attention to the mode. They are used with little data sets.

10.3.8 GRAPHICAL MULTIVARIATE EDA

To comprehend the relationships between several fields in the data set or identify the relationships between more than two variables, multivariate GEDA is used. Pair plots and three-dimensional (3D) scatter plots are examples of these kinds of GEDA.

The bar graph plot is the most used graphical method. Today, box plots are widely used to show how two sets of data relate to one another. To show the relationship between all variables and their views, pair graphs are occasionally utilized (Rani et al., 2023).

10.3.9 COMPARATIVE BOX PLOTS

We utilize a box plot that compares all potential values' levels side by side. Two data sets are compared using it. The data for each instance of a category variable are essentially summarized.

10.3.10 SURFACE PLOTS IN 3D AND HEAT MAPS

With the entire feature variable, we can produce a heat map. Row and column headers, as well as the variable, pitted against itself on the diagonal are used to identify feature variables. Visualizing the link between variables in a high-dimensional space is highly helpful.

10.4 PANDAS

The most effective data analysis program is this one. The information can be modified, cleansed, and examined. Data can be kept in CSV format on a computer. The data can be organized, visualized, and saved. It is constructed on top of the NumPy module. EDA applications contain the following features:

- Using EDA, errors and anomalies can be found.
- We can learn new things about different kinds of data.
- Data outliers can be identified.
- Using EDA, we may test an assumption.
- It can be used to identify significant factors.
- The connections between various data are clear to us.
- Utilizing visualization techniques, data can speak for itself.

10.5 WORKING WITH THE DATA SETS

It is time to research the facts and gain additional knowledge. Data from the housing data set is what we are using. We will use a variety of methods to analyze the data. In the first step, we imported the pandas and NumPy libraries as Figure 10.2.

A substantial housing CSV file was then imported as a data frame (df). Rows and columns are used to present the data sets. Our CSV file has 20 columns and 5 rows as shown in Figure 10.3. We used the head () method to get the top five rows of the data frame or series.

Both buyers and sellers are very interested in property values, and house prices are a vital indicator of the state of the economy. Real estate is one of the least open industries in our economy. This study's main goal is to forecast property values using current variables.

```
[1] import pandas as pd
    import numpy as np

    path='/housing.csv'
    df=pd.read_csv(path)
    df.head()
```

index	longitude	latitude	housing_median_age	total_rooms	total_bedrooms	population	households	median_income	median_house_value	ocean_proximity
0	-122.23	37.88	41.0	880.0	129.0	322.0	126.0	8.3252	452600.0	NEAR BAY
1	-122.22	37.86	21.0	7099.0	1106.0	2401.0	1138.0	8.3014	358500.0	NEAR BAY
2	-122.24	37.85	52.0	1467.0	190.0	496.0	177.0	7.2574	352100.0	NEAR BAY
3	-122.25	37.85	52.0	1274.0	235.0	558.0	219.0	5.6431	341300.0	NEAR BAY
4	-122.25	37.85	52.0	1627.0	280.0	565.0	259.0	3.8462	342200.0	NEAR BAY

Show 25 per page
Like what you see? Visit the data table notebook to learn more about interactive tables.

FIGURE 10.2   Adding the pandas library and displaying the first five rows of the data frame.

Source: Output of executable code.

```
[9] import matplotlib.pyplot as plt
    import seaborn as sns
    %matplotlib inline
```

```
df.dtypes

longitude              float64
latitude               float64
housing_median_age     float64
total_rooms            float64
total_bedrooms         float64
population             float64
households             float64
median_income          float64
median_house_value     float64
ocean_proximity         object
dtype: object
```

FIGURE 10.3   Displaying the types of data for each column in the data frame.

Source: Output of executable code.

We must choose the best visualization method. Before we can view a variable, it is essential to identify the kind of variable we are dealing with as with 10.4.

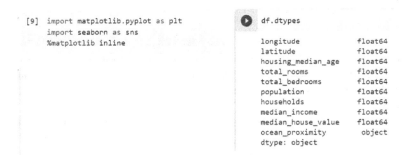

```
df.corr()
```

	longitude	latitude	housing_median_age	total_rooms	total_bedrooms	population	households	median_income	median_house_value
longitude	1.000000	-0.924664	-0.108197	0.044568	0.069608	0.099773	0.055310	-0.015176	-0.045967
latitude	-0.924664	1.000000	0.011173	-0.036100	-0.066983	-0.108785	-0.071035	-0.079809	-0.144160
housing_median_age	-0.108197	0.011173	1.000000	-0.361262	-0.320451	-0.296244	-0.302916	-0.119034	0.105623
total_rooms	0.044568	-0.036100	-0.361262	1.000000	0.930380	0.857126	0.918484	0.198050	0.134153
total_bedrooms	0.069608	-0.066983	-0.320451	0.930380	1.000000	0.877747	0.979728	-0.007723	0.049686
population	0.099773	-0.108785	-0.296244	0.857126	0.877747	1.000000	0.907222	0.004834	-0.024650
households	0.055310	-0.071035	-0.302916	0.918484	0.979728	0.907222	1.000000	0.013033	0.065843
median_income	-0.015176	-0.079809	-0.119034	0.198050	-0.007723	0.004834	0.013033	1.000000	0.688075
median_house_value	-0.045967	-0.144160	0.105623	0.134153	0.049686	-0.024650	0.065843	0.688075	1.000000

FIGURE 10.4   Showing the data frame's pairwise correlations.

Source: Output of executable code.

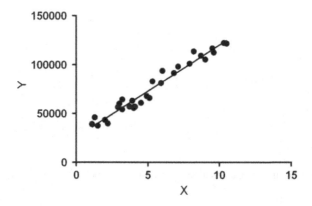

FIGURE 10.5 The linear relationship observed between *x* and *y*.

Source: Output of executable code.

This report's evaluations are intended to be based on certain fundamental factors that go into calculating a home's cost. Real-time housing data for Boston has been manually gathered to conduct the real-time research. Due to the fact that we are working with a continuous outcome variable, the regression technique for ML was applied to the project.

The objective of this study is to develop an efficient ML model that can precisely predict the cost of a property based on provided features and implement the ML model as a website to interact with people (Misra et al., 2023).

In recent years, ML has shown that it can address problems in the real world using various techniques. Business analysis, security alerts, spam and fraud detection, the automobile industry, and medical imaging all rely on it heavily (Jaiswal et al., 2023).

In order to determine the price of a property based on several characteristics, we will be evaluating a house price prediction data set. We will split the training and testing sets of data, perform exploratory data analysis, and make predictions using the model as Figure 10.5.

10.5.1 HOUSE PREDICTION DATA SET

Problem Statement: Real estate agents need assistance forecasting home prices across the US. You chose to employ the linear regression model when handed the data set to work with (Khang et al., 2022a).

Make a model that will assist him in estimating the price at which the house would sell. The data set has 5000 rows and 7 columns with a CSV extension. The following columns are present in the data:

- "**Avg. Area Income**" refers to the average annual household income in the city where the house is located.
- "**Avg. Area House Age**" refers to the average age of homes inside a city.
- "**Avg. Area Number of Rooms**" refers to the typical number of rooms in a given city's houses.

- "**Average Area Bedroom Count**" refers to the average number of bedrooms in a city's homes.
- "**Area Population**" refers to the populace of the city.
- "**Price**" refers to the price at which the house was sold.
- "**Address**" is the home's address.

An example is utilizing NumPy, pandas, seaborn, and scikit-learn to predict housing prices using linear regression.

10.5.2 IMPORT LIBRARIES

scikit-learn, pandas, seaborn, matplotlib, and NumPy will all be imported. Plots are added using the "%matplotlib inline" command in your Jupyter Notebook as shown in Figures 10.6, 10.7, 10.8, and 10.9.

10.5.3 DATA IMPORT AND CHECKOUT

```
import pandas as pd
import numpy as np
import seaborn as sns
import matplotlib.pyplot as plt

%matplotlib inline
```

FIGURE 10.6 Imported important libraries.

Source: Output of executable code.

```
df = pd.read_csv('/USA_Housing.csv')
df.head()
```

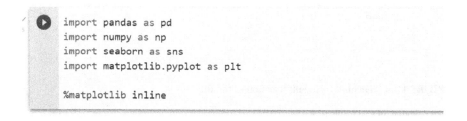

	Avg. Area Income	Avg. Area House Age	Avg. Area Number of Rooms	Avg. Area Number of Bedrooms	Area Population	Price	Address
0	79545.458574	5.682861	7.009188	4.09	23086.800503	1.059034e+06	208 Michael Ferry Apt. 674\nLaurabury, NE 3701...
1	79248.642455	6.002900	6.730821	3.09	40173.072174	1.505891e+06	188 Johnson Views Suite 079\nLake Kathleen, CA...
2	61287.067179	5.865890	8.512727	5.13	36882.159400	1.058988e+06	9127 Elizabeth Stravenue\nDanieltown, WI 06482...
3	63345.240046	7.188236	5.586729	3.26	34310.242831	1.260617e+06	USS Barnett\nFPO AP 44820
4	59982.197226	5.040555	7.839388	4.23	26354.109472	6.309435e+05	USNS Raymond\nFPO AE 09386

FIGURE 10.7 Importing the Excel file and displaying the records of the file.

Source: Output of executable code.

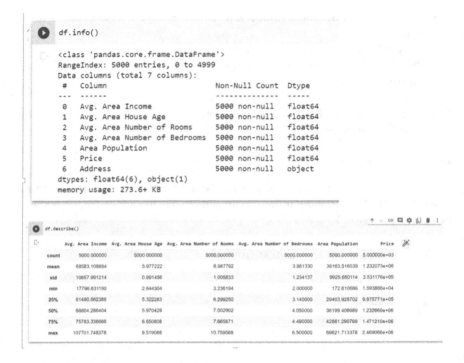

FIGURE 10.8　Displaying the information of the file.

Source: Output of executable code.

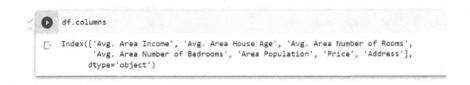

FIGURE 10.9　Reading the columns of the file.

Source: Output of executable code.

10.5.4　EXPLORATORY DATA ANALYSIS FOR HOUSE PRICE PREDICTION

We will make a straightforward plot to display the data by using sns.pairplot (df) function and output results as Figures 10.10 and 10.11.

10.5.5　TRAINING A LINEAR REGRESSION MODEL

Let's start building the regression model immediately. The Address column will be disregarded because it contains only text, which is useless for linear regression modeling. Here is the list object of *X* and *y* as Figure 10.12.

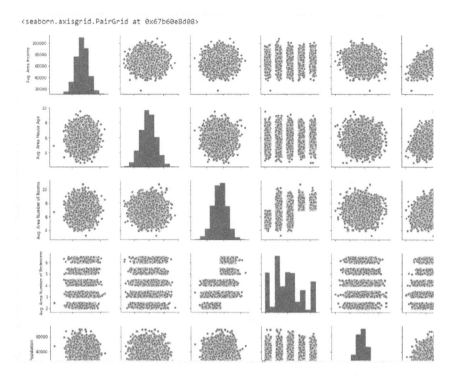

FIGURE 10.10 Displaying the different plots.

Source: Output of executable code.

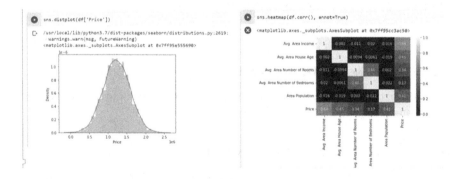

FIGURE 10.11 Patterns observed between the data.

Source: Output of executable code.

The training model's data are in the files X_train and y_train. The testing model's data are in the files X_test and y test. Names for the characteristics and target variables are X and y as Figure 10.13.

```
X =df[['Avg. Area Income', 'Avg. Area House Age', 'Avg. Area Number of Rooms',
                'Avg. Area Number of Bedrooms', 'Area Population']]

y = df['Price']
```

FIGURE 10.12   The attributes of *x* and *y*.

Source: Output of executable code.

```
from sklearn.model_selection import train_test_split

X_train, X_test, y_train, y_test = train_test_split(X, y, test_size=0.4, random_state=101)
```

FIGURE 10.13   Imported the split to split the data.

Source: Output of executable code.

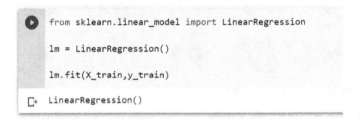

```
from sklearn.linear_model import LinearRegression

lm = LinearRegression()

lm.fit(X_train,y_train)

LinearRegression()
```

FIGURE 10.14   Imported linear regression to train the data.

Source: Output of executable code.

10.5.6   CREATING AND TRAINING THE LINEAR REGRESSION MODEL

We will import the training data set, build a sklearn linear model with a linear regression object, and fit it to it as shown in Figure 10.14.

10.5.7   LINEAR REGRESSION MODEL EVALUATION

Let's now assess the model by looking at its coefficients and discussing how to interpret them as shown in Figure 10.15.

- When all other factors are held constant, a rise of $21.52 is associated with a unit increase in average area income.
- When all other factors are held constant, an increase of $164,883.28 is associated with a 1-unit increase in the average area house age.
- All other characteristics remain constant as the average increases by 1 unit. The number of rooms has increased by $122,368.67 in relation to the area.

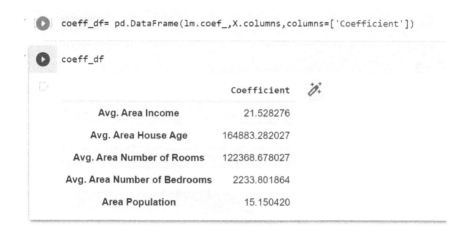

```
coeff_df= pd.DataFrame(lm.coef_,X.columns,columns=['Coefficient'])
```

```
coeff_df
```

	Coefficient
Avg. Area Income	21.528276
Avg. Area House Age	164883.282027
Avg. Area Number of Rooms	122368.678027
Avg. Area Number of Bedrooms	2233.801864
Area Population	15.150420

FIGURE 10.15   The coefficients of data.

Source: Output of executable code.

- When all other factors are held equal, an increase of 1 unit in the average area number of bedrooms leads to an increase of $2233.80.
- Keeping all other factors unchanged, an increase in the area population of 1 unit translates to an increase of $15.15.

10.5.8   PREDICTIONS FROM OUR LINEAR REGRESSION MODEL

Discover the predictions made by our test set and assess its performance as Figure 10.16.

The fact that the data in the scatter plot is in a line form indicates that our model has made accurate predictions as shown in Figure 10.17.

The bell-shaped (normally distributed) data in the preceding histogram plot indicates that our model has made accurate predictions as Equations 10.1, 10.2, and 10.3.

$$\frac{1}{n}\sum_{i=1}^{n}|y_i - \hat{y}_i| \tag{10.1}$$

$$\frac{1}{n}\sum_{i=1}^{n}|y_i - \hat{y}_i|^2 \tag{10.2}$$

$$\sqrt{\frac{1}{n}\sum_{i=1}^{n}|y_i - \hat{y}_i|^2} \tag{10.3}$$

10.5.9   COMPARING THESE MEASUREMENTS

Because it is the average mistake, mean absolute error (MAE) is the most straightforward to comprehend. Due to its ability to be interpreted in "y" units, the root mean

```
[29] predictions = lm.predict(X_test)
```

```
plt.scatter(y_test,predictions)
```

<matplotlib.collections.PathCollection at 0x7ff95be314d0>

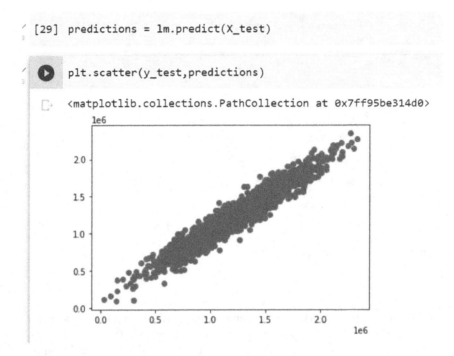

FIGURE 10.16   Observed linear relationship between variable data.

Source: Output of executable code.

```
sns.distplot((y_test-predictions),bins=50);
```

/usr/local/lib/python3.7/dist-packages/seaborn/distributions.py:
2619: FutureWarn: warnings.warn(msg, FutureWarning)

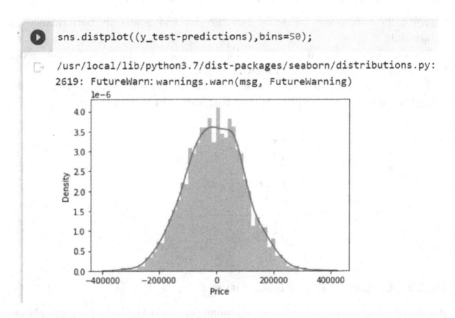

FIGURE 10.17   The normal distribution of the data.

Source: Output of executable code.

```
from sklearn import metrics
print('MAE:', metrics.mean_absolute_error(y_test, predictions))
print('MSE:', metrics.mean_squared_error(y_test, predictions))
print('RMSE:', np.sqrt(metrics.mean_squared_error(y_test, predictions)))
```

```
MAE:  82288.22251914942
MSE:  10460958907.208977
RMSE: 102278.82922290897
```

FIGURE 10.18 Calculated different metrics to compare the result.

Source: Output of executable code.

squared error (RMSE) is even more widely used than the mean squared error (MSE). These are all loss functions since we aim to reduce them as shown in Figure 10.18.

10.6 IMPROVING THE ACCURACY OF MODEL

10.6.1 MORE INFORMATION

Usually, having more information is a good thing. It relies less on shaky correlations and presumptions and more on letting the "evidence speak for itself." More data translate into better models with higher accuracy.

We are aware that there is no option to include further data. We are not given the option, for instance, to increase the quantity of the training data.

10.6.2 INCORPORATE MISSING VALUES AND OUTLIERS

The unintentional inclusion of missing and outlier values reduces model accuracy or produces biased models rather frequently. It leads to inaccurate forecasts. Inappropriate behavior and relationship analysis regarding other variables are the causes of this. It is imperative to handle missing and outlier values carefully.

10.7 COMPARISON OF ML ALGORITHMS FOR HOUSE PRICE PREDICTION USING REAL-TIME DATA

Real estate price prediction using regression and classification: In this study, explanatory variables that cover a wide range of residential house characteristics are used to forecast house prices. Numerous regression techniques, such as Lasso, Ridge, support vector machine (SVM) regression, and random forest, are used to estimate home prices.

The best-performing model for a regression problem, according to this chapter, is support vector regression (SVR) with a Gaussian kernel, which has an RMSE of 0.5271. However, due to SVR's large dimensionality, visualization is challenging. Its investigation shows that the neighborhood, roof material, and living area square feet have the most statistical relevance in determining a house's sale price.

A method for predicting real estate prices based on SVR: Classification, clustering, and forecasting have all been effectively accomplished using the SVM. SVR is suggested in this study as a method for predicting Chinese real estate prices. This study looked at the viability of SVR in predicting real estate prices.

The development of house price prediction models in this work involved the application of machine learning methods. In this study, 5359 townhouses in Fairfax County, Virginia, have their housing data analyzed. C4.5, RIPPER, Bayesian, and AdaBoost were all subjected to a 10-fold cross-validation.

Neural networks and ML for predicting home prices: This chapter seeks to evaluate each fundamental factor taken into account when calculating price. The findings of this model were not decided by just one regression technique but, rather, by the weighted means of several procedures, which produced the most accurate results.

The results showed that, as compared to using individual methods, this strategy produces the least error and maximum accuracy.

Predicting house prices using ML algorithms: This study employs machine learning algorithms to forecast house prices. The lasso regression method, decision tree, and logistic regression algorithms are used in this study to construct a prediction model. It took 3000 properties' housing data into account.

R-squared values of 0.98, 0.96, 0.81, and 0.99 are displayed for logistic regression, SVM, lasso regression, and decision tree, respectively. Additional comparisons of these algorithms are based on metrics like accuracy, MAE, MSE, and RMSE.

10.7.1 DESCRIPTION OF THE DATA SET

Real-time details on homes for sale in Pune are included in the data set. There are more than 1635 entries in this manually collected data, sometimes referred to as manual web scraping. It has been collected from real estate websites like 99acres, magic bricks, and housing.co.

10.7.2 REGRESSION METHODS

The two learning techniques available with ML are supervised learning and unsupervised learning. The algorithm being trained must already have some examples of successful outcomes in the data being utilized for supervised learning.

The most common method is this one since it increases the possibility of getting findings that are far more accurate. We adopted a supervised learning strategy for this project. Following supervised learning are the two categories of regression and classification.

10.7.3 REGULAR REGRESSION

There are two types of linear regression: basic and multiple, which vary depending on the number of independent variables. The linear relationship between the single independent variable and the dependent variable must be established in the simple linear regression model.

Multiple linear regression, by contrast, makes use of a number of independent variables to determine the relationship between the dependent and independent variables.

10.7.4 FLASK FRAMEWORK

We require flask, a framework, for creating a working webpage for our developed model, in order to launch our machine learning model.

Although there are other possibilities. Additionally, flask integration with the model is simpler. We can build an interface for our model using flask. We may create a web application using the tools, frameworks, and technologies offered by flask.

10.8 RESULTS AND DISCUSSION

Finding the best-fitting algorithm for the model has been demonstrated to be possible through cross-validation of various algorithms. The estimation of home prices provided by the linear regression algorithm is extremely accurate. It is providing significantly more precise estimates for various locations (Khang et al., 2023a).

Additionally, linear regression provides almost correct predictions, according to the confusion matrix. Our data set is fit by linear regression, which yields the maximum accuracy of 85.64%. Decision trees have the lowest accuracy (56.02%). SVR has an accuracy rate of 62.81%.

10.9 CONCLUSION

By employing an Android phone's PIC microcontroller, global positioning system (GPS), and Global System for Mobile Communication (GSM), the suggested system will help us understand how to identify coastal states for sailors with the suggested system present.

Fishermen lose their livelihoods when they accidentally violate the maritime boundary while navigating since they cannot see it in the ocean. The fisherman are given a security system based on an Android application through this system, which allows them to understand when they are in danger (Khang et al., 2022b).

Because they can clearly detect the sea's boundaries, fishermen are prevented from allowing others into their territory and can safeguard their lives and foster friendly relations with the bordering countries (Khang et al., 2023b). This verdict will prevent many sailors from dying while crossing borders. Our project is primarily concerned with the good relations between the two nations. Fishermen's average life span will improve while the death rate will drop.

REFERENCES

Bhambri P, Rani S, Gupta G, Khang A. *Cloud and Fog Computing Platforms for Internet of Things*. CRC Press, 2022. https://doi.org/10.1201/9781003213888

Hajimahmud VA, Khang A, Gupta SK, Babasaheb J, Morris G. *AI-Centric Modelling and Analytics: Concepts, Designs, Technologies, and Applications* (1st Ed.). CRC Press, 2023. https://doi.org/10.1201/9781003400110

Jaiswal N, Misra A, Misra PK, Khang A (Eds.). "Role of the Internet of Things (IoT) Technologies in Business and Production," *AI-Aided IoT Technologies and Applications in the Smart Business and Production*. CRC Press, 2023. https://doi.org/10.1201/9781003392224-1

Khang A. "Material4Studies," *Material of Computer Science, Artificial Intelligence, Data Science, IoT, Blockchain, Cloud, Metaverse, Cybersecurity for Studies*, 2021. www.researchgate.net/publication/370156102_Material4Studies

Khang A, Gupta SK, Shah V, Misra A (Eds.). *AI-Aided IoT Technologies and Applications in the Smart Business and Production*. CRC Press, 2023a. https://doi.org/10.1201/9781003392224

Khang A, Hahanov V, Abbas GL, Hajimahmud VA. "Cyber-Physical-Social System and Incident Management," *AI-Centric Smart City Ecosystems: Technologies, Design and Implementation* (1st Ed.). CRC Press, 2022a. https://doi.org/10.1201/9781003252542-2

Khang A, Rani S, Sivaraman AK. *AI-Centric Smart City Ecosystems: Technologies, Design and Implementation* (1st Ed.). CRC Press, 2022b. https://doi.org/10.1201/9781003252542

Khang A, Shah V, Rani S. *AI-Based Technologies and Applications in the Era of the Metaverse* (1st Ed.). IGI Global Press, 2023b. https://doi.org/10.4018/9781668488515

Khanh HH, Khang A. "The Role of Artificial Intelligence in Blockchain Applications," *Reinventing Manufacturing and Business Processes Through Artificial Intelligence* (pp. 20–40). CRC Press, 2021. https://doi.org/10.1201/9781003145011-2

Misra PK, Kumar N, Misra A, Khang A. "Heart Disease Prediction using Logistic Regression and Random Forest Classifier," *Data-Centric AI Solutions and Emerging Technologies in the Healthcare Ecosystem* (1st Ed., p. 6). CRC Press, 2023. https://doi.org/10.1201/9781003356189-6

Rana G, Khang A, Sharma R, Goel AK, Dubey AK. *Reinventing Manufacturing and Business Processes through Artificial Intelligence*. CRC Press, 2021. https://doi.org/10.1201/9781003145011

11 Heart Disease Classification and Its Early Prediction Using Machine Learning

Soumya Chandra and Anuradha Misra

11.1 INTRODUCTION

The heart, a fist-sized organ whose role is to pump sufficient blood throughout the body, is undoubtedly among the most important human organs. The broader structure of this organ consists of four chambers that are further separated by valves that divide it into two sections: the atrium, which accumulates blood, and the ventricle, which pushes blood out.

Heart disease is a general term that refers to all the malfunctions of the heart like heart attacks and heart failures. One of the common causes of the disease is the development of plaques in the arteries and blood vessels that reach heart. Plaque is a substance made up of cholesterol, fatty molecules and minerals (Dangare et al., 2012).

Heart disease is noted to be the leading cause of mortality. According to the World Health Organization, 18 million deaths were reported from cardiovascular disease in 2016, that is, approximately 30% of all deaths each year. As per the statistics provided by Centers for Disease Control and Prevention, about 6.7% (almost equal to 18.2 million) of adults aged 20 or older have a chance of coronary artery disease (CAD), and out of 10 adults, two die from CAD.

The American Heart Association laid out some symptoms that a person may undergo experience. These symptoms are an increase or decrease in heart rate and swelling in the legs; a person may have difficulty sleeping, or in some cases, there is a rapid weight gain of, say, 1 or 2 kg per day. But these indicators are quite similar to those of many other diseases that might come with age. So, in order to control the mortality rate from heart disease in the near future, a correct diagnosis needs to be made (Sai Kumar et al., 2023).

Since humans have a complex biology with variations among each patient, a healthcare system equipped with field specialist and integrated with the technology will be helpful in diagnosing the disease early, providing better informed treatment and hence reducing the space of errors. Medical data of patients that are generated on a large scale can be used for analysis supported by computer technology. This may provide us with critical inferences that can be consolidated with the presence of a domain expert (Rana et al., 2021).

DOI: 10.1201/9781003392224-11

Machine learning (ML) algorithms that consider physiological factors (including age and sex), lifestyle factors (smoking, physical activity, alcohol, stress), metabolic syndrome factors (insulin resistance), dyslipidemia (e.g., obesity, high blood pressure), and dietary, among others, can be used for analysis. To assess the condition of the patient, their exposure to these constituents would determine the severity of cardiovascular disease.

11.2   OBJECTIVE

There happens to be numerous amounts of medical tests whose outcomes are regularly made under the supervision of specialists. This study focuses on classifying heart health issues by analyzing the various ML algorithms for accurately discovering and identifying issues that will help medical analysts.

With ML, the principle point should be confined to not only the prediction and detection of disease but also the aspect of how to provide better medical treatment, patient care, allocation of medical resources as per need, management of hospital volume, policymaking, and much more (Khang et al., 2022a).

The objective of this chapter is to work on these ML-supervised algorithms for increased precision in diagnosis. Affecting features are analyzed and worked to identify which algorithm would be suitable for the prediction process. This would act as an additional support in cross-verifying the diagnosis suggested by doctors (Khang et al., 2023a).

11.3   RELATED WORK

The work by Dangare et al. (2012) is based on a study using a higher number of attributes for analyzing prediction systems. Along with 13 attributes, such as age, blood pressure, cholesterol, and others, two more factors, namely, smoking and obesity, are considered.

The algorithms included naïve Bayes (NB), with 90.74% accuracy; decision tree, with 99.62% accuracy; and neural networks, with 100% accuracy. Puyalnithi et al. (2016) used supervised learning algorithms: NB, random forest, support vector machine, and random decision forest classification tree for analysis using a data-mining tool called oranges. The result of this study is that the precision of the classification tree is higher than the other aforementioned algorithms (Shashi et al., 2023).

Muhammad et al. (2020) used an intelligent computational system for identification and treatment. Distinctive types of ML algorithms have been reviewed, and for removing outliers, four different feature selection algorithms are implemented, along with classifiers that have been analyzed. Parameters like accuracy, specificity, area under the ROC curve (AUC), F1 score, and receiver operating characteristic (ROC) curves have been plotted to better understand the reinforcing factors.

With each feature selection method, p-values and chi-square values have also been computed. Fatima et al. (2017) dealt with frequently used computer-aided diagnosis in the medical field that can further be enhanced using ML to avoid inaccuracies and miscalculations.

The authors did a comparative study on various kinds of ML techniques for diagnosing liver diseases, dengue diseases, and others.

Nashif et al. (2018) aimed to predict heart disease in a structured and systematic manner. They authenticated the two most widely accepted databases. A support vector machine (SVM) attained accuracy of 97.53% along with factors like sensitivity 97.50% and specificity 94.54% using 10-fold cross-validation. For real-time monitoring, Arduino was also deployed. The report by Shanta et al. (2011) aims to model a framework for predicting heart disease by considering the main factors that affect the heart.

Classifier algorithms like NB, Bayesian optimized SVM (BO-SVM), k-nearest neighbors (KNN), and Salp swarm optimized neural network (SSA-NN) were applied, and it was reported that the BO-SVM had the highest accuracy at 86.7%, precision at 100%, and a sensitivity of about 80%. The next highest was SSA-NN, with an accuracy of about 86.7%, a precision of 100%, and a sensitivity of 60%.

The focus of Patro et al. (2021) was to gauge a persons' risk of heart disease by working on rules like original rules, pruned rules, rules without duplicate values, Polish rules, classified rules, and sorted rules that showcased the extraordinary level of their precision, accuracy, and prediction.

The paper by Saxena et al. (2016) used an intelligent computational prediction technique modeled for diagnoses of cardiovascular disease. Along with machine learning algorithms and outlier removal, four feature selection methods were applied to the available data set. The output of the feature selection algorithm for AUC, F1 score, Matthews correlation coefficient (MCC), and ROC, along with classifier were examined. p-values and chi-square values were also used, with a classifier for better feature selection.

Palaniappan et al. (2008) attempted to identify not-so-prominent information and design a high-accuracy framework to exploit it to get the model with improved decision-making capabilities.

Using intelligent heart disease prediction system (IHDPS) along with the neural network, NB, and decision trees, a prototype model was prepared. IHDPS helps answer complex queries that are usually unanswerable by a conventional decision support system.

Repaka et al. (2019) used a standardized data format for data mining. Some attributes like age, blood pressure, gender, fasting blood sugar, cholesterol, and others were used to check the possibility of a patient having a heart disease. These mentioned attributes have been passed as input to the NB classification for predicting disease. The data set was divided into 80-to-20 ratios for training and testing, respectively.

The methodology followed was the collection of data followed by the registration of users and their login, which was application-based. Then NB was applied for classification and reliable data transfer with the help of attention-based feature selection (AFS), and eventually, the result was predicted. It has also demonstrated varied type of abstraction techniques in order to make the data-mining techniques more adoptable in a real-world scenario (Misra et al., 2023).

11.4 PROPOSED WORK

The importance of correct diagnosis is unparalleled. Being able to separate cardiovascular disease from any other disease having almost similar symptoms is still an intricate part that needs to be performed with utmost care and responsibility. An intelligent learning-based computational framework would help in refining research as well as ameliorating medical considerations.

In this chapter, our focus is on analyzing the constituting factors that are present in the data set and, among those, identifying which factors are most prominently affecting. Supervised ML algorithms are applied to the data set, and thereafter, the algorithm with the highest accuracy would further be improved using hyper-parameter tuning (Misra et al., 2022a).

The algorithms that have been taken into consideration for the prediction purpose are logistic regression, KNN, random forest, and NB.

11.5 PROPOSED METHODOLOGY

11.5.1 DATA COLLECTION

The foremost step of data analysis or machine learning pipeline is collecting data that should be relevant and reliable (Khanh and Khang, 2021). The quality of predictions directly depends on the quality of the data set. Data collection is a tedious job, and the few factors that might cause trouble include the following:

1. **Inaccuracy in data:** This means that the data set collected does not align with the provided problem statement.
2. **Missing values:** There can be columns with missing data, or it can be images for prediction purposes.
3. **Disproportioned data:** Some attributes or categories of the data set might have a higher or lower number of imbalances to the corresponding samples. This can lead to under-representation in our model.
4. **Biased data:** There can happen a data set where the attributes and labels chosen could pipe bias toward, perhaps, a gender or an age or a region, for example. It is usually difficult to identify bias at first and then remove it.
5. **Techniques:** There are some techniques that may come in handy while dealing with these problems.
6. **Data sets:** Using pre-cleaned freely available data sets if the requirement of the problem can be satisfied with a clean, well-formulated data set.
7. **Tools:** Web scraping using automated tools and headless browsers can scrape the data from websites. They are customizable depending on what factor can require scraping based on those labels.
8. **ML:** Building your own data set that is for practice by private ML engineers in order to create their data, but this is more suitable if the problem statement is very specific and it can be generalized for a freely available data set. This method can also be efficiently used in case the amount of data for training our model is small.
9. **Corporations:** Agencies or big corporate firms usually crowdsource data for a fee.

The source of the data set taken for analysis was downloaded from the University of California, Irvine (UCI) Machine Learning Repository, which is a collection of data

TABLE 11.1

Data Set Attributes Description.

Data set Characteristics	Multivariate	Number of Instances	303	Area	Life
Attribute Characteristics	Categorical, Integer, Real	Number of Attributes	75	Date Donated	1988-07-01
Associated Tasks	Classification	Missing Values?	Yes	Number of Web Hits	1,704,520

sets, algorithms, and tools for machine learning research of the UCI, as shown in Table 11.1. It is the Public Health Dataset, which dates from 1988 and is a combination of four different databases extracted from Cleveland, Hungary, Switzerland, and VA Long Beach (Misra et al., 2022b).

UCI is an ML repository that has collections for many domain-based theories, extensive collections of databases, and data generators that are extensively used by people in the ML field, educators, and researchers all over the world (Rani et al., 2021).

This was a project in collaboration with Rexa.info at the University of Massachusetts. It was also provided funding support from the National Science Foundation.

This heart disease data set contains the records of 303 patients and 76 attributes along with the target column as shown in Table 11.1. But for experimental purposes, among those 76 attributes, a subset of 14 features is considered. In the target field, 0 indicates 'no disease', and 1 indicates 'disease'. List of the attributes before preprocessing used for the analysis are the following:

1. **Age:** Patient's age in terms of year, sex: here, 0 is for female while 1 indicates male. Age is very strongly associated with an elevated risk of heart-related complications. According to empirical estimates, over 80% of those who die from heart disease happen to be 65 years or old in age.
2. **Classification of Chest Pain**: Abbreviated as CP, where 1 is typical, 2 is atypical, 3 is pain, and 4 is asymptotic. It has been shown to be closely related to heart-related risk factors.
3. **Trestbps:** This is the blood pressure when resting (in mm Hg). The normal range is 120/180. A relationship between elevated resting heart rate and heart-related disease in both sexes even after controlling for potential confounders, such as general and abdominal obesity, has been found. The control here is important as resting blood pressure is highly correlated to other factors that also contribute to the risk of heart disease (Hajimahmud et al., 2022).
4. **Chol:** Serum cholesterol provides the number of triglycerides.
5. **Fbs:** This is the abbreviation for fasting blood sugar, greater than 120 mg/dL (1 true), less than 100mg/dL (5.6 mmol/L) is the normal range, and

100–125 mg/dL is the prediabetic stage. Relationships between fasting glucose levels and heart-related disease risks generally follow a J-shape curve. Lowest risk in the glucose range is 85–99 mg/dL. Patients with fasting glucose levels above 110 mg/dL are prone to heart diseases.

6. **Restecg:** Electrocardiographic (ECG) results from when a patient is resting. Symptomatic heart disease subjects have delineated ECG preternaturalness (both ECG and EKG stand for electrocardiogram), thereby making it a good indicator for detection. In men with symptomatic heart diseases, the resting EGC can be crucial in identifying a patient in imminent danger who might progress by intervention.

7. **Thalach:** This is the max heart rate. For any person, the max heart rate is 220 – (age). The max heart rate is also an indicator of fitness, which is negatively correlated with the presence of access adipose tissue. Hence, it is an effective tool, presumably with an inverse relation to heart disease risk.

8. **Exang:** Exercise-induced chest pain that happens due to reduced blood flow. Angina is a symptom of heart-related disease.

9. **Oldpeak:** ST depression is caused by exercise with respect to rest (The ST segment is the line between the "S" and the "T" on the readout of an ECG).

10. **Slope:** This is the slope of the peak exercise ST segment.

11. **Fluoroscopy Colored Vessels (Ca):** This ranges from 0 to 3 major vessels, which are colored by fluoroscopy.

12. **Thal:** Although there is no attached explanation, it probably is thalassemia (7: reversible defects, 6: fixed defects, 3: normal).

13. **Target (T):** This indicates the status of disease using angiography (0: no disease, 1: disease).

The attributes of the data set are shown in Tables 11.2a, 11.2b, and 11.2c.

TABLE 11.2A
Cleaned-Up UCI Data Set

	age	sex	chest_pain	blood_pressure	cholesterol
count	303.000000	303.000000	303.000000	303.000000	303.000000
mean	54.366337	0.683168	0.966997	131.623762	246.264025
std	9.082101	0.466011	1.032052	17.538143	51.830751
min	29.000000	0.000000	0.000000	94.000000	126.000000
25%	47.500000	0.000000	0.000000	120.000000	211.000000
50%	55.000000	1.000000	1.000000	130.000000	240.000000
75%	61.000000	1.000000	2.000000	140.000000	274.500000
max	77.000000	1.000000	3.000000	200.000000	564.000000

TABLE 11.2B
Cleaned-Up UCI Data Set

	blood_sugar	restecg	thalach	exang	oldpeal
count	303.000000	303.000000	303.000000	303.000000	303.000000
mean	0.148515	0.528053	149.646865	0.236733	1.039604
std	0.356198	0.525860	22.905161	0.469794	1.161075
min	0.000000	0.000000	71.000000	0.000000	0.000000
25%	0.000000	0.000000	135.500000	0.000000	0.000000
50%	0.000000	1.000000	153.000000	1.000000	0.800000
75%	0.000000	1.000000	166.000000	1.000000	1.600000
max	1.000000	2.000000	202.000000	1.000000	6.200000

TABLE 11.2C
Cleaned-Up UCI Data Set

	slope	vessel	Thal	target
count	303.000000	303.000000	303.000000	303.000000
mean	1.339934	0.729373	2.313531	0.544554
std	0.616226	1.022606	0.612277	0.498835
min	0.000000	0.000000	0.000000	0.000000
25%	1.000000	0.000000	2.000000	0.000000
50%	1.000000	0.000000	2.000000	1.000000
75%	2.000000	1.000000	3.000000	1.000000
max	2.000000	4.000000	3.000000	1.000000

11.5.2 DATA PREPROCESSING

On preprocessing, no null values were mentioned in the data set; however, the outliers present demanded careful attention and handling. The data set was not properly distributed. So, in such a situation, there are two alternatives (Tailor et al., 2022).

The first is to ignore outliers and work on crucial features of the data set, followed by feature selection, which then would have the algorithms applied. Another method would be overcoming the overfitting of the data and then applying the algorithm to tackle the outliers, this would yield more promising outcomes. Plotting the correlation heat map to understand the relation between the key features is shown in Figure 11.1.

11.5.3 EXPLORATORY DATA ANALYSIS

The data set has a varied range of features that might or might not be the factors that are causing the development of heart disease in a person. An exploratory data analysis would be a perfect fit in these situations (Geetha et al., 2024).

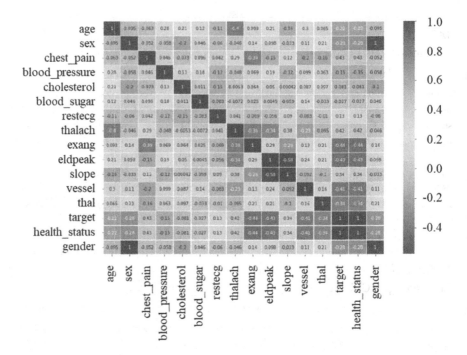

FIGURE 11.1   Correlation heat map between important features.

```
     gender          health_status
     0               0                          24
                     1                          72
     1               0                          114
                     1                          92
     Name: gender, dtype: int64
```

FIGURE 11.2   Health description based on gender.

Based on gender, the number of males was more than twice than that of females. Among 303, 206 were males while 96 were females as shown in Figure 11.2.

Out of 206 males, 114 do not have heart disease while 92 are suffering from some kind of heart disease as shown in Figures 11.2 and 11.3. In females, 24 do not have heart disease while 72 are suffering from it as shown in Figure 11.3.

Angina is a medical term for any kind of discomfort felt by a person when the heart muscle does not get oxygenated blood, thus leading to uneasiness in arms, shoulders, and so on, as shown in Figure 11.4.

Usually, women and older adult patients are seen to be atypical symptomatic with a history of disease. This parameter does not confirm with certainty that the chest pain felt is caused by heart disease.

Next factor is the slope of the peak exercise ST segment. This is divided into three categories wherein 0 represents up-sloping (uncommon). Next is flat-sloping, which is an indication of a healthy heart. Finally, down-sloping indicates an unhealthy heart (represented by 2) in Figure 11.5.

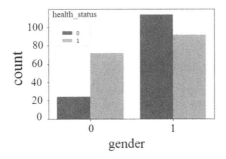

FIGURE 11.3 Histogram.

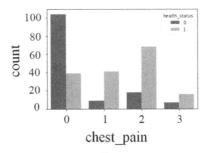

FIGURE 11.4 Bar plot for the number of people who faced chest pain.

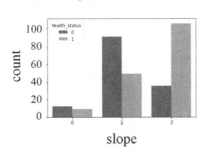

FIGURE 11.5 Health status based on the slope of the ST segment.

Thalach can help in understanding the situation of a person's heart health. It depicts the maximum heart rate achieved. The Figure 11.6 plot is right-skewed.

"Fixed defect" means that a person at some point in their life has suffered from this defect but has remained the same over time, and "reversible defect" shows that there is no movement of hemoglobin in the body but that any kind of physical activity such as exercising returns back to its normal state.

Category 3 denotes the classified part normal; 6, a fixed defect; and 7, a reversible defect. It was seen in Figure 11.7 that more than 120 people suffer from category 2, which is a fixed defect, which might be correlated with a heart disease.

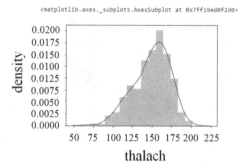

FIGURE 11.6   Density of people and their heart rates.

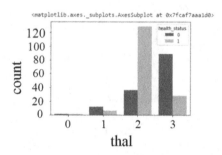

FIGURE 11.7   Bar plot for health of patients based on thalassemia.

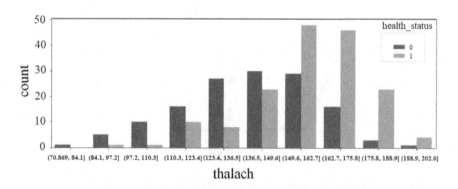

FIGURE 11.8   Health status based on maximum heart rate.

A maximum number of people have a maximum heartbeat of 150–160 in Figure 11.8. A heartbeat above 140 can be risky and indicate a person being more prone to heart disease. Next feature is thalach (thal).

When compared with nondiabetic class, it is seen that the true class is lower. But on a closer look, we can make out that a higher number of people who have or had heart disease are not diabetic as shown in Figure 11.9.

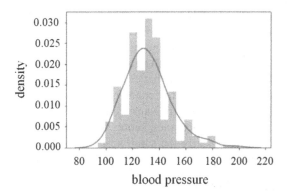

FIGURE 11.9   Density of population affected by blood pressure.

11.5.4   MODELING

Supervised learning algorithms are those that take up the already-known data set for learning with its pre-known results, with a target of building a model that can make projections for the newer input values. The goal of supervised learning algorithms is to find the relations between the input variable (denoted by X) and output attribute (denoted by Y) that would otherwise might not get noticed.

Working: In this kind of learning procedure, our model is fed with a labeled data set so that the model can learn about each kind of data. After, it is provided with a set of test data values for it to predict the outcome.

Algorithms used:

- Logistic regression
- K-NN
- Random forest
- NB

11.5.4.1   Logistic Regression

Logistic regression is a classification algorithm that is helpful for a problem statement with binary outcomes. This algorithm is essential for describing the data and helps with understanding the underlying relationship between one dependent binary variable and one or more nominal independent variables. The model uses logistic function to find the output of the equation that varies between 0 and 1. In vector form, logistic regression, as shown in Equation 11.1, is written as

$$p = h_\theta(x) = \sigma(X^T\theta), \tag{11.1}$$

where θ is the vector parameter for training and
$\sigma(x)$ denotes a sigmoid function. Its value can range from 0 to 1.

The value of the dependent variable should be binary. Outliers should not be available in the data, and there should be no high correlations. These were the few assumptions that were checked before applying this model.

For the accuracy score, the model will be asked to predict the target, and the value of the target will later be compared with the actual target values. The predictions for trained data are stored in X_train_prediction.

Later, following the same steps, the values for the X_test are predicted and then passed the predicted value to accuracy_score, which turned out to be 88%. Our model is supposed to have nearly the same accuracy score for the predicted and trained data. If the gap is huge, it presents that our model is overfitted.

11.5.4.2   Random Forest Classifier

A decision tree is the basic building block of a random forest classifier, so it is important to cover few features of a decision tree, which is a tree-like structure with three components: decision nodes, leaf nodes, and a root node.

Basically, this algorithm works by dividing the training data set into branches and then further separating them into more branches which continues further until a leaf node is attained as it cannot be further segregated. The attributes in the decision tree are represented in the form of nodes, which are used to forecast the result, while the nodes in the decision tree form a path or link to the next leaf.

11.5.4.3   Random Forest Classifier

Random forest classifier is an exemplar of a supervised learning algorithm, which is helpful in both classification- and regression-based problems. Random forest is an analogy to forest; the higher the number of trees in a forest, the more robust the forest.

This algorithm works by creating randomly choosing data points, then trying to achieve some type of predication from each of those trees, and, later on, selecting the best solution by voting. This algorithm lies under both the bagging and feature randomness (Bharti et al., 2021).

The random forest classifier handles the decision tree classifier's drawback of overfitting the data. However, another way of solving overfitting is by pruning, but that, in turn, reduces the quality of classification.

The sklearn library is a great tool that uses features to reduce impurity across the trees in the forest and scores them on its own against all the features present in the tree, and then, after the model is done with its training, it scales the result so that the sum of all the importance ends up equal to 1.

This helps identify what features do not contribute to the prediction process and therefore can be dropped. The accuracy score of the random forest model on passing the target test data set comes out to be 83.6%.

11.5.4.4   KNN

KNN is a nonparametric classifier used in supervised learning. This algorithm works on the idea of proximity, which means that it makes use of distance in order to group a single data point, classify it, and make predictions for it. It can be suitable for both classification and regression. It would work best with data sets that are well defined and nonlinear in nature.

It works by assuming that similar things are located in close proximity or a region. So, here, we calculate the distance between the mathematical values of each data point. After evaluating the distance between each point and the test data, it finds the possibility of similar valued data points to the test set data. It is based on the highest probability value of points. One of three methods— Euclidean, Minkowski, and Hamming—is adopted to find the distance.

11.5.4.4.1 Euclidean Distance

This is the distance calculated between two points no matter what the dimension is. It is the most widely used method among the three. Equation 11.2 of Euclidean distance calculation between two data points A(x,y) and B(a,b) will be calculated by

$$\text{Distance} \left(A \left(x, y \right), B \left(a, b \right) \right) = \sqrt{(x - a)^2 + (y - b)^2} \tag{11.2}$$

So, for, say, a value of K, the algorithm will find the k-nearest neighbor of the point, and then it will designate the appropriate class to which it has the highest number of points out of all the classes of the k-neighbors.

11.5.4.4.2 Minkowski Distance

It is a generalized version of Euclidean distance. Here, we find the distance between two points that lie in an n-dimensional real space or normed vector space. For example, there are two points P1(X1, X2, XN) and P2 (Y1, Y2, . . . , YN) Minkowski distance = (|X1 − Y1|p + |X2 − Y2|p + |X2 − Y2|p)1/p.

11.5.4.4.3 Hamming Distance

This is a metric for contrasting two binary data strings. In cases of equal-length strings, a comparison will be done based on the number of positions in which the two bits are different. For calculating distance between two strings, the distance between a and b is denoted as distance (a. b).

This algorithm is suitable when the data set worked on is labeled, free from outliers, and small in number. Applying the KNN model requires three sets of input that are from the sample data, the data point at which the model should be evaluated, and finally, the value of k, which is the number of neighbors we want to select.

Successively, we would now calculate the distance between the reference point and the point of the learning data set, and if it turns out to be less than the distances stored already in the neighbors list, it is added, and the next point is moved to (Bhambri et al., 2022).

Another thing to note while using this algorithm is that if the number of items present in the list is seemingly more important than the value of k itself, then it would be advisable to neglect and remove the last value from the neighbor's list.

After the training phase, the next step is to initialize the value of K for choosing the number of neighbors and then calculating the distance from any point to the data point using the distance calculating methods (usually Euclidean distance). The last step is picking the K entries from the sorted list and getting the most common class of the chosen values. To determine the value of K, a graph comparing the number of neighbors with accuracy was plotted and is shown in Figure 11.10.

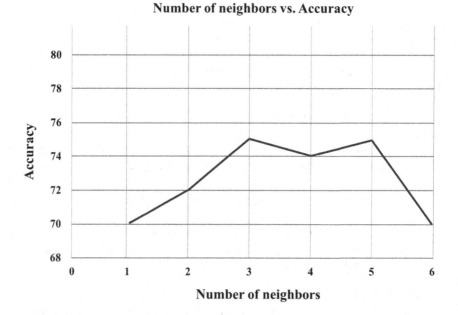

FIGURE 11.10   Comparison between number of neighbors and accuracy.

Source: Khang (2021).

Practically, if the value of *K* is less than 5, then 5 is chosen as *K*, which is also the default for sklearn. However, it was found that the accuracy of the KNN algorithm turned out to be lowest so far, which is about 68%.

11.5.4.4   NB

Sometimes the simplest of solutions make the most powerful ones, and NB perfectly holds this statement true. This algorithm is not only simple but also very fast, error-free, and reliable. Most particularly, it is used with natural language processing.

NB theorem is a supervised machine learning algorithm based on a conditional probability theorem known as Bayes theorem. Conditional probability, to put it into simple words, is the chance of an event happening because presumably that other event has already taken place. The Equation 11.3 of Bayes theorem is

$$P\left(A|B\right) = P\left(B|A\right) * P\left(A\right) / P\left(B\right). \tag{11.3}$$

This here tells how frequently does A happen with respect to B, which has happened. But this algorithm is functional, with an assumption that all the attributes of the data set are independent and equally responsible for the contribution of prediction.

NB algorithm is of three types: multinomial NB, Bernoulli NB, and the Gaussian NB. We used Gaussian NB as it is very appropriate when the predictor's values are continuous rather than discrete. So, in such scenarios, all the values are sampled from a Gaussian distribution.

Based on the type of value that is present in the working data set, the formula of conditional probability also changes. The model was trained pretty well, and the correctness of the model turned out to be 86%, which was the next highest, right after logistic regression.

11.6 IMPROVING THE ACCURACY

A model that is used in a practical way needs to check two conditions optimization and accuracy. The performance of a model can be tricky at times (Khang et al., 2023b).

Only applying the algorithms is never enough to make conclusions about the model as there are various other factors that may cause lower accuracy, and it should be a target to make the model predict accurately to the most of its capacity as shown in Figure 11.11.

Hence, the accuracy scores we got so far also are shown in Figure 11.11:

- Logistic regression: 88%
- KNN: 68%
- Random forest: 83%
- NB: 86%

We worked further on improving the two models with the highest accuracies, that is, logistic regression and NB. Theoretically, accuracy can be improved by following any of these eight steps, which have proven to be practically efficient in creating a robust model as shown in Figure 11.12.

It is important to understand the impact of the question statement to be handled and understand the intricacies of the attributes to get a rough idea of what would impact our results and then cross-verify it mathematically through the feature

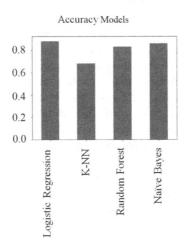

FIGURE 11.11 Accuracy of the models.

Accuracy Comparison

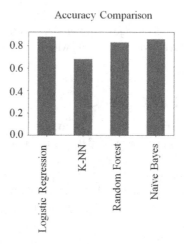

FIGURE 11.12   Accuracy comparison.

selection process. This would help develop crucial points before taking all possible permutations and combinations. Some ways that improve accuracy include the following:

1. **Adding more data so that we do not make assumptions on a weak correlation.** In our case, we took all the possible data that have been collected by the hospital over the years to form the data set.
2. **Treating missing values and the noisy unwanted values.** This can make the model biased toward a factor. For continuous-value data set, missing values can be computed with either the mean or the median of that same column or a fill-up value that will not bother the outcome like 0. For outliers, the observation can be removed followed by transformations, binning, and imputation. Another alternative can be to treat them separately. The outliers of our data set were treated at the preprocessing stage.
3. **Feature engineering:** Its extraction of information from the already-existing data and the extracted information are in terms of new attributes. Thus, they reflect the variance in the training data. It is further categorized into feature transformation and feature creation. The former one can be implemented through normalization or by normally distributing the data to remove any kind of skewness while the latter requires deriving a new attribute from the older attributes. This highlights the underlying relationship between the data points.
4. **Feature selection:** The theory behind this is that instead of working with all the attributes that have less or almost no impact on the outcome or prediction of the model should be removed. This is done by selecting the subset of best features that would help explain the interrelation between the independent and the target variable. In our modeling and working with the data set, out of 76 attributes, only 14 were considered for the analysis.

5. **Multiple algorithms:** In the pool of algorithms, identifying the most suitable can be a tedious job, and among them, understanding which would yield a higher accuracy can be difficult. For this, one needs to understand the mathematics behind the formulation of these algorithms and the kind of data that would be suitable for it. A better way to determine this is to apply multiple algorithms and compare their accuracy rates.

6. **Algorithm tuning:** The objective is to find the optimal value against each feature and then work to improve its accuracy further. We would follow this further to improve accuracy. We would apply the Grid Search CV to logistic regression and NB.

11.6.1 GRIDSEARCH COMPUTER VISION

This is a known process by which hyper-parameter tuning is performed so as to get the optimum values to a model for a given data set. The efficacy of the algorithm may vary significantly based on how the hyper-parameters contain the values (Khang et al., 2024).

There is no hard-and-fast rule to know what hyper-parameter tuning would be suitable so ideally all needs to be applied to see which works best. Grid Search Computer Vision (CV) works by passing an already-existing set of values to tune the function. Then it tries all the possible permutations for the value passed in the dictionary and then gauges the combination by making use of a method known as cross-validation (Gao et al., 2021).

The parameters of Grid Search CV differ based on the algorithm that has been applied. A brief list of them follows:

1. **Estimator:** It is an object that is supposed to be used to implement the scikit-learn estimator interface, which needs a function known as score.

2. **Param grid:** It takes a list of dictionaries or dictionaries as a parameter. The elements in the dictionary are in a string form as keys, and all the different of parameters in a list help find a pattern or sequence in the parameter situation.

3. **Scoring:** It can take strings, lists, dictionaries, and tuples to gauge the precision of the cross-validated framework.

4. **N jobs:** It takes an integer as an input parameter and helps, in parallel, execute those *n* jobs together. The usual symbolism is when 1 is passed, it means none; however, if a −1 is passed, it depicts all processors. The CV takes up the integer value whose aim is to strategize the cross-validation.

5. **Verbose:** It focuses on controlling the verbosity, which means more messages. If it is greater than 1, the time spent on computation for each fold and their parameter would be displayed. Another case is if it is greater than 2, then the score gets displayed, but if it is greater than 3, then the fold and indices of the candidate parameter get displayed along with the start time of the computation.

6. **Pre-dispatch:** It controls the count of jobs that would get sent during the time of parallel execution. Reducing it would be beneficial in avoiding an

explosion of memory consumption when the job that has been dispatched is far more than the capacity of a central processing unit (CPU) can process.

7. **Error score:** In the case of an error occurrence, values are assigned to the particular score. If it gets set to 'raise', then it raises the error. But a Fit-FailedWarning gets raised if a number is passed.

8. **Return train score:** By default, its value is false. It is used to gather valuable insights into how different parameter affects the under-fitting or over-fitting trade-off.

So, now the focus was to work on the models that produced the highest accuracies to get more precision.

1. **Logistic regression:** The GridSearch CV function is passed on the values of the logistic regression function; the param grid is also taken to consideration, and assigning the value of cross-validation to 5 and changing verbose to true to get the detailed print at the time, we fit our whole data set to the Grid Search CV. It performed five folds of fitting for each of the 30 candidates, which produced 150 fits. The accuracy remained at 88%.

2. **NB:** For this algorithm, the parameters have been chosen differently; the first parameter is an estimator that is equal to the Gaussian NB function. This would order the implemented scikit learn interface, and then we took param grid; verbose, whose value was assigned 1; and cross-validation, whose value was assigned to 5; and number of jobs represented as n jobs equal to −1. The accuracy achieved showed a slight improvement of 2%, which made it to 88%.

11.6.2 Evaluation Metrics of Model

As it can be seen statistically that the accuracy of logistic regression and NB was 88% and 86%, respectively, and furthermore, when both models underwent the Grid Search CV so as to enhance the feature selection process and make the models better, again, it turned out that NB improved from 86% to 88%. So, we need some measures to decipher between our two best models and conclude which one would be better compared to the other as shown in Figure 11.13.

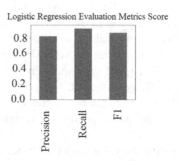

FIGURE 11.13   Logistic regression evaluation metrics score.

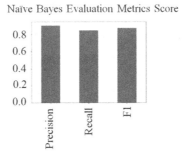

FIGURE 11.14 Naïve Bayes evaluation metrics score.

In order to find a good fit involves methods and process that provide the necessary balance between bias and variance. In our case, the only parameter left for gauging both models is understanding the performance metrics of each of them. In this chapter, we have focused on three evaluation metrics: precision, recall, and F1 score. These are sorting algorithms commonly used in search engines as shown in Figure 11.14.

11.7 RESULTS AND DISCUSSION

This chapter focused on extracting medical records of heart disease patients and the related symptoms and possible causes. Some of the features such as chest pain and their fasting blood sugar, age, gender, heart rate, and many more have been taken into account. Initially, it was 76 but was cut down to 14. The four algorithms that were used for prediction were logistic regression, random forest classifier, KNN, and NB.

The accuracy of each model came out as follows:

- Logistic regression: 88%
- Random forest classifier: 83%
- KNN: 68%
- NB: 86%

11.8 CONCLUSION

Based on accuracy, logistic regression and NB have been found to be the best-performing algorithms and the most suitable out of the four algorithms that have been worked on. To further improve the accuracy of the two, more features were extracted, and Grid Search CV was used to refine these models.

The accuracy improved for NB, rising to 88%, while it remains the same (88%) for logistic regression. Since the accuracy value of the two turned out to be same, different parameters were considered to evaluate the metrics of the algorithm (Khang et al., 2022b).

Precision, recall, and F1 scores were calculated for both logistic regression and NB algorithms shown in Table 11.3. It can be concluded that although the logistic

TABLE 11.3

Evaluation of Algorithms

Algorithm	Precision	Recall	F1 Score
Logistic Regression	82%	92%	86%
Naive Bayes	90%	84%	87%

regression and NB were the most suitable, the performance metric of the logistic regression was slightly better (Khang et al., 2023c).

From a long-term perspective, the work can be extended in an alliance with some medical organizations. The work focused just on discrete values, so formulating a framework that can work with a range of continuous data would make its application more realistic and relevant. Deep learning models may be used to capture the underlying relationship better and evaluate the performance (Rani et al., 2023).

REFERENCES

Bhambri P, Rani S, Gupta G, Khang A. *Cloud and Fog Computing Platforms for Internet of Things*. CRC Press, 2022. https://doi.org/10.1201/9781003213888

Bharti R, Khamparia A, Shabaz M, Dhiman G, Pande S, Singh P. "Prediction of Heart Disease Using a Combination of Machine Learning and Deep Learning," *Computational Intelligence and Neuroscience*, 2021. www.hindawi.com/journals/cin/2021/8387680/

Dangare CS, Apte SS. "Improved Study of Heart Disease Prediction System Using Data Mining Classification Techniques," *International Journal of Computer Applications*, 47(10), 44–48, 2012. https://citeseerx.ist.psu.edu/document?repid=rep1&type=pdf&doi=f4de0213b4a5777ff39d5a94cd574713799ca221

Fatima M, Pasha M. "Survey of Machine Learning Algorithms for Disease Diagnostic," *Journal of Intelligent Learning Systems and Applications*, 9(1), 1, 2017. www.scirp.org/html/1-9601348_73781.htm

Gao XY, Amin Ali A, Shaban Hassan H, Anwar EM. "Improving the Accuracy for Analyzing Heart Diseases Prediction Based on the Ensemble Method," *Complexity*, 2021. www.hindawi.com/journals/complexity/2021/6663455/

Geetha C, Neduncheliyan S, Khang A (Eds.). "Dual Access Control for Cloud Based Data Storage and Sharing," *Smart Cities: IoT Technologies, Big Data Solutions, Cloud Platforms, and Cybersecurity Techniques*. CRC Press, 2024. https://doi.org/10.1201/9781003376064-17

Hajimahmud VA, Khang A, Hahanov V, Litvinova E, Chumachenko S, Alyar AV. "Autonomous Robots for Smart City: Closer to Augmented Humanity," *AI-Centric Smart City Ecosystems: Technologies, Design and Implementation* (1st Ed.). CRC Press, 2022. https://doi.org/10.1201/9781003252542-7

Khang A. "Material4Studies," *Material of Computer Science, Artificial Intelligence, Data Science, IoT, Blockchain, Cloud, Metaverse, Cybersecurity for Studies*, 2021. www.researchgate.net/publication/370156102_Material4Studies

Khang A, Gupta SK, Shah V, Misra A (Eds.). *AI-aided IoT Technologies and Applications in the Smart Business and Production*. CRC Press, 2023c. https://doi.org/10.1201/9781003392224

Khang A, Hahanov V, Abbas GL, Hajimahmud VA. "Cyber-Physical-Social System and İncident Management," *AI-Centric Smart City Ecosystems: Technologies, Design and*

Implementation (1st Ed., vol. 2, p. 15). CRC Press, December 30, 2022b. https://doi. org/10.1201/9781003252542-2

Khang A, Hahanov V, Litvinova E, Chumachenko S, Triwiyanto T, Hajimahmud VA, Ali RN, Alyar AV, Anh PTN. "The Analytics of Hospitality of Hospitals in Healthcare Ecosystem," *Data-Centric AI Solutions and Emerging Technologies in the Healthcare Ecosystem* (p. 4). CRC Press, 2023b. https://doi.org/10.1201/9781003356189-4

Khang A, Hrybiuk O, Abdullayev V, Shukla AK. *Computer Vision and AI-integrated IoT Technologies in Medical Ecosystem* (1st Ed.). CRC Press, 2024. https://doi. org/10.1201/9781003429609

Khang A, Ragimova NA, Hajimahmud VA, Alyar AV. "Advanced Technologies and Data Management in the Smart Healthcare System," *AI-Centric Smart City Ecosystems: Technologies, Design and Implementation* (1st Ed.). CRC Press, 2022a. https://doi. org/10.1201/9781003252542-16

Khang A, Rana G, Tailor RK, Hajimahmud VA (Eds.). *Data-Centric AI Solutions and Emerging Technologies in the Healthcare Ecosystem.* CRC Press, 2023a. https://doi. org/10.1201/9781003356189

Khanh HH, Khang A. "The Role of Artificial Intelligence in Blockchain Applications," *Reinventing Manufacturing and Business Processes through Artificial Intelligence* (pp. 20–40). CRC Press, 2021. https://doi.org/10.1201/9781003145011-2

Misra A, Chandra S. "Covid-19 Analysis and Future Outbreak Prediction," *Design Engineering*, 2021(9), January 2022, ISSN: 0011-9342. https://doi.org/10.1201/9781003392224-1

Misra PK, Kumar N, Misra A, Khang A. "Heart Disease Prediction using Logistic Regression and Random Forest Classifier," *Data-Centric AI Solutions and Emerging Technologies in the Healthcare Ecosystem* (1st Ed., p. 6). CRC Press, 2023. https://doi. org/10.1201/9781003356189-6

Muhammad Y, Tahir M, Hayat M, Chong KT. "Early and Accurate Detection and Diagnosis of Heart Disease Using Intelligent Computational Model," *Scientific Reports*, 10(1), 1–17, 2020. www.nature.com/articles/s41598-020-76635-9

Nashif S, Raihan MR, Islam MR, Imam MH. "Heart Disease Detection by Using Machine Learning Algorithms and a Real-Time Cardiovascular Health Monitoring System," *World Journal of Engineering and Technology*, 6(4), 854–873, 2018. www.scirp.org/ journal/paperinformation.aspx?paperid=88650

Palaniappan S, Awang R. "Intelligent Heart Disease Prediction System Using Data Mining Techniques," *2008 IEEE/ACS International Conference on Computer Systems and Applications* (pp. 108–115). IEEE, March 2008. https://ieeexplore.ieee.org/abstract/ document/4493524/

Patro SP, Nayak GS, Padhy N. "Heart Disease Prediction by Using Novel Optimization Algorithm: A Supervised Learning Prospective," *Informatics in Medicine Unlocked*, 26, 100696, 2021. www.sciencedirect.com/science/article/pii/S2352914821001805

Puyalnithi T, Viswanatham VM. "Preliminary Cardiac Disease Risk Prediction Based on Medical and Behavioural Data Set Using Supervised Machine Learning Techniques," *Indian Journal of Science and Technology*, 9(31), 1–5, 2016. www. researchgate.net/profile/Thendral-Puyalnithi/publication/307531416_Preliminary_ Cardiac_Disease_Risk_Prediction_Based_on_Medical_and_Behavioural_Data_Set_ Using_Supervised_Machine_Learning_Techniques/links/57d61c8c08ae601b39aa734c/ Preliminary-Cardiac-Disease-Risk-Prediction-Based-on-Medical-and-Behavioural-Data-Set-Using-Supervised-Machine-Learning-Techniques.pdf

Rana G, Khang A, Sharma R, Goel AK, Dubey AK (Eds.). *Reinventing Manufacturing and Business Processes Through Artificial Intelligence.* CRC Press, 2021. https://doi. org/10.1201/9781003145011

Rani S, Bhambri P, Kataria A, Khang A, Sivaraman AK. *Big Data, Cloud Computing and IoT: Tools and Applications* (1st Ed.). Chapman and Hall/CRC Press, 2023. https://doi. org/10.1201/9781003298335

Rani S, Chauhan M, Kataria A, Khang A (Eds.). "IoT Equipped Intelligent Distributed Framework for Smart Healthcare Systems," *Networking and Internet Architecture*. CRC Press, 2021. https://doi.org/10.48550/arXiv.2110.04997

Repaka AN, Ravikanti SD, Franklin RG. "Design and Implementing Heart Disease Prediction Using Naives Bayesian," *2019 3rd International Conference on Trends in Electronics and Informatics (ICOEI)* (pp. 292–297). IEEE, April 2019. https://ieeexplore.ieee.org/abstract/document/8862604/

Sai Kumar DV, Chaurasia R, Misra A, Misra PK, Khang A. "Heart Disease and Liver Disease Prediction using Machine Learning," *Data-Centric AI Solutions and Emerging Technologies in the Healthcare Ecosystem* (1st Ed., p. 4). CRC Press, 2023. https://doi.org/10.1201/9781003356189-13

Saxena K, Sharma R. "Efficient Heart Disease Prediction System," *Procedia Computer Science*, 85, 962–969, 2016. www.sciencedirect.com/science/article/pii/S187705091630638X

Shanta K, Kumaraswamy YS, Patil B. "Predictive Data Mining for Medical Diagnosis of Heart Disease Prediction," *Indian Journal of Computer Science and Engineering*, 17, 2011.

Shashi KG, Khang A, Somani P, Dixit CK, Pathak A. "Data Mining Processes and Decision-Making Models in Personnel Management System," *Designing Workforce Management Systems for Industry 4.0: Data-Centric and AI-Enabled Approaches* (1st Ed., pp. 89–112). CRC Press, 2023. https://doi.org/10.1201/9781003357070-6

Tailor RK, Pareek R, Khang A (Eds.). "Robot Process Automation in Blockchain," *The Data-Driven Blockchain Ecosystem: Fundamentals, Applications, and Emerging Technologies* (1st Ed., pp. 149–164). CRC Press, 2022. https://doi.org/10.1201/9781003269281-8

12 Cloud-Integrated Industrial Internet of Things and Its Applications

Fatima M. and Jyoti Jain

12.1 INTRODUCTION

The Industrial Internet of Things (IIoT) refers to the integration of internet-connected devices, machines, and technologies in the manufacturing and production industries. This innovative approach to industrial processes and systems offers numerous benefits and has the potential to revolutionize the way we approach production and management.

The IIoT enables industrial machines and devices to communicate with each other, as well as with other systems, in real time. This communication enables companies to gather data from multiple sources and analyze it to improve processes and increase efficiency.

With the IIoT, manufacturers can gain real-time insights into the performance and status of their machinery, allowing for predictive maintenance and improved overall reliability (Rani et al., 2021).

The IIoT is a methodology, a practice, and an implementation sweeping through businesses and industries worldwide. On a basic level, the IIoT is a way of congregating data that was previously inaccessible and locked within inflexible data streams. This provides all stakeholders with a more complete and comprehensive view of operations.

Imagine smart TVs and watches or security cameras—devices that were historically lacking in internet connection but now have that capability. This is the IoT, or the Internet of Things. The IIoT is used to refer to industrial equipment and plant assets that are now integrated.

IoT generally refers to consumer items like home thermostats and smart speakers. These devices have tiny sensors in them that collect data. They then relay that data to another device (like a phone) over a network (Bluetooth, Wi-Fi, or cellular).

The IIoT takes those same principles and applies them to a business setting. It connects machines that are mission-critical to the way a business runs. The original Industrial Revolution brought steam-powered automation to manufacturing. The IIoT is changing traditional thinking and models, optimizing plant processes previously not thought possible, and even helping to build smart cities. It will redefine the markets for key players who are willing to digitally transform their business.

DOI: 10.1201/9781003392224-12

For businesses, IIoT implementations amplify the power and usefulness of data. The IIoT affects every industry and organization; realize it has developed beyond the 'trend', now actively restructuring global markets and economic models. As businesses stand to gain the ability to optimize and reduce operational costs and energy consumption, as well as manage maintenance (Pritiprada et al., 2024).

Market analysts proposed that IoT will connect 46.1 billion devices by 2025, generating 79.4 zettabytes of data. Within interconnected marketplaces, a lack of access to integration is a death knell for business growth.

Without digitization, there can be no transformation. The IIoT is not just the means for industry to improve and optimize its processes. It also works to empower your workforce by providing the right data and the right visibility—all at the right time. It is a way of delivering the Fourth Industrial Revolution for all stakeholders, aligning people and processes into proactive, coherent systems.

An organization relies on its workforce. So if you neglect to empower that workforce with the right applications, you could restrict the capabilities of a critical element in business productivity and success.

The modern workforce expects a higher level of interaction with plant equipment and other company assets, even extending toward the need for a cloud-based ecosystem detailing everything from overall efficiency to the granular data. This means access to the system anywhere, whether an employee finds themselves on the shop floor or even at home.

The IIoT's biggest benefit is data collection. This technology provides a huge amount of accurate, real-time data. Business owners can use it to make smarter, faster decisions. The IIoT connects machines, sensors, hardware tools, software tools, and users, among others, to collect the data. These data will be analyzed and give alerts and connectivity as Figure 12.1.

By monitoring machinery in real time, manufacturers can detect and diagnose issues before they become serious problems, allowing for quick and efficient repairs. Additionally, the IIoT allows for remote monitoring and control, enabling manufacturers to remotely access and manage machinery, reducing the need for on-site visits, and minimizing downtime.

Another important benefit of the IIoT is its ability to enhance safety. With real-time monitoring and control, manufacturers can quickly identify and respond to potential safety hazards, reducing the risk of accidents and injuries in the workplace. This is especially important in industries where machinery is used extensively, such as construction, manufacturing, and mining.

In addition to reducing downtime and enhancing safety, the IIoT also provides manufacturers with valuable data and insights that can be used to improve processes and increase efficiency. By analyzing data from various sources, manufacturers can identify bottlenecks, inefficiencies, and areas for improvement, allowing them to make data-driven decisions that drive their operations forward. IIoT can improve the following:

- Process automation
- Quality control
- Asset performance
- Supply-chain traceability

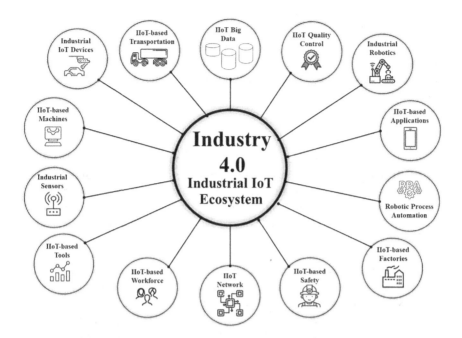

FIGURE 12.1   Industrial Internet of Things.

Source: Khang (2021).

- Energy management
- Predictive maintenance
- Field service
- Overall efficiency
- Accuracy
- Product and process optimization
- Predictive maintenance and analysis
- Remote accessibility and monitoring
- Security
- Scalability of network
- On time for machines and process
- Power savings
- Cost-effectiveness

Despite its many benefits, the IIoT also presents challenges and risks, including cybersecurity threats, data privacy concerns, and the need for specialized skills and expertise. Companies must take steps to secure their networks and data, ensuring that sensitive information remains protected.

Additionally, manufacturers must invest in training and development programs to equip employees with the necessary skills and knowledge to effectively implement and manage IIoT systems.

12.2 IIOT COMPONENTS

The IIoT helps in making industries smart. This is possible through IIoT components like Arduino, Raspberry Pi, sensors, open-source cloud platforms, and others. These components are explained in the following sections. Afterward, a case study is presented.

12.2.1 ARDUINO

Arduino is an open-source electronics platform based on easy-to-use hardware and software. It is intended for anyone making interactive projects. It can connect different sensors, such as temperature and others. Arduino is a microcontroller. It is an open-source platform with input/output (I/O) pins (digital in/out, analog in, and pulse-width modulation (PWM)). The Starter Kit is approximately $40–80 as shown in Figure 12.2.

Arduinos contain a number of different parts and interfaces together on a single circuit board. The design has changed over the years, and some variations include other parts as well. On a basic board, it is likely to find the following pieces:

A number of pins, which are used to connect with various components, might be used with the Arduino. These pins come in two varieties: Digital pins, which can read and write a single state, on or off. Most Arduinos have 14 digital I/O pins.

Analog pins can read a range of values and are useful for more fine-grained control. Most Arduinos have six of these analog pins. These pins are arranged in a specific pattern so that if you buy an add-on board designed to fit into them, typically called a "shield," it should fit into most Arduino-compatible devices easily.

A power connector, which provides power to both the device itself, and provides a low voltage to components like light emitting diode (LED) and various sensors,

FIGURE 12.2 Arduino UNO Hardware.

Source: ElectronicsComp.com.

FIGURE 12.3 A screenshot of the Arduino integrated development environment for programming hardware.

Source: javatpoint.com.

provided their power needs are reasonably low. The power connector can connect to either an AC adapter or a small battery (Khang et al., 2022c).

A microcontroller, the primary chip, allows the programming of the Arduino in order for it to be able to execute commands and make decisions based on various inputs. The exact chip varies depending on what type of Arduino, but they are generally Atmel controllers, usually an ATmega8, ATmega168, ATmega328, ATmega1280, or ATmega2560.

The differences between these chips are subtle, but the biggest difference a beginner will notice is the different amounts of onboard memory. A serial connector, on most new boards, is implemented through a standard USB port. This connector allows communication to the board from a computer, as well as loading new programs onto the device.

Arduinos can also be powered through the USB port, removing the need for a separate power connection. A variety of other small components, like an oscillator and/or a voltage regulator, provide important capabilities to the board.

The official integrated development environment (IDE) is used for Arduino programming. The Arduino IDE is open-source software that is written in Java and will work on a variety of platforms: Windows, Mac, and Linux. The IDE enables to write code in a special environment with syntax highlighting and other features, which will make coding easier as shown in Figure 12.3. Then the code is loaded onto the device with a simple click of a button.

12.2.2 RASPBERRY PI

Raspberry Pi is a series of small, low-cost, single-board computers designed by the Raspberry Pi Foundation. They are popular for use in educational, hobbyist, and industrial projects and can run various operating systems such as Linux, Windows 10 IoT, and others. They have a variety of input and output options, making them versatile for a wide range of applications.

The Raspberry Pi was first introduced in 2012 and has since undergone several hardware revisions and software updates.

The most recent iteration is the Raspberry Pi 4, released in 2019, which features a faster processor, more RAM, and improved connectivity options compared to previous models. It is often used for a variety of projects, including media centers, home automation systems, game consoles, and even as the basis for small servers. It can run a variety of operating systems, including Linux-based distributions, as well as Windows 10 IoT Core and other custom-built software. It has a variety of input and output options, including HDMI, USB, Ethernet, and GPIO (general purpose input/output) pins as shown in Figure 12.4.

The GPIO pins allow for connection to a variety of sensors, motors, and other electronic components, making the Raspberry Pi a versatile platform for a wide range of projects.

One of the key features of the Raspberry Pi is its low cost. The Raspberry Pi 4 is available for as little as $35, making it an affordable option for educators, students, and hobbyists. Its affordability has helped to make it a popular platform for educational initiatives, including coding camps, robotics workshops, and other science, technology, engineering, and mathematics–related programs.

Its foundation also provides a variety of resources for learning and working with the Raspberry Pi, including online forums, tutorials, and project ideas. These

FIGURE 12.4   Raspberry Pi.

Source: Wikipedia.

resources have helped build a vibrant community of users and developers around the platform and have contributed to its popularity and continued development. While it has many benefits, it also has some limitations. Its size and form factor may limit its use in certain projects where space is a concern.

Overall, the Raspberry Pi is a powerful and affordable platform for learning and experimentation in computer science and electronics. Its versatility and wide range of capabilities have made it a popular choice for a variety of projects and initiatives, and its continued development and support by the Raspberry Pi Foundation ensures that it will remain a valuable resource for years to come as shown in Figure 12.4.

12.2.3 SENSORS

IoT sensors are a critical component of the IoT technology. They are small, wireless devices that can detect and transmit data about the environment around them.

IoT sensors are used in a wide range of applications, from monitoring temperature and humidity in homes to tracking the location of vehicles and measuring the performance of industrial machinery as shown in Figure 12.5.

The following are some of the most common types of IoT sensors:

- **Temperature Sensors:** These sensors measure the temperature of an environment and are used in a variety of applications, including climate control systems, food and beverage manufacturing, and healthcare (Khang et al., 2022a).
- **Humidity Sensors:** These sensors measure the amount of moisture in the air and are used in applications such as agriculture, climate control systems, and data centers (Hajimahmud et al., 2023a).
- **Light Sensors:** These sensors measure the amount of light in an environment and are used in applications such as lighting control systems and energy management.

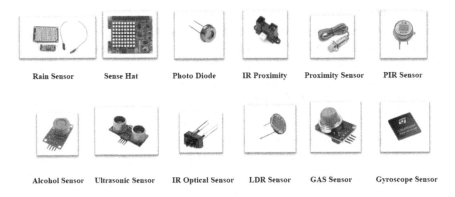

FIGURE 12.5   Types of sensors.

Source: Electricaltechnology.org.

- **Motion Sensors:** These sensors detect movement and are used in applications such as security systems, home automation systems, and vehicle tracking systems.
- **Pressure Sensors:** These sensors measure the pressure of liquids and gases and are used in applications such as fluid control systems and process monitoring in manufacturing.
- **Proximity Sensors:** These sensors detect the presence of objects and are used in applications such as security systems and industrial automation systems.
- **Accelerometer Sensors:** These sensors measure acceleration and are used in applications such as sports equipment, wearable devices, and automotive safety systems.
- **Magnetic Sensors:** These sensors detect magnetic fields and are used in applications such as security systems, navigation systems, and industrial automation systems.

IoT sensors are connected to a network, such as the internet, allowing the data they collect to be transmitted and analyzed in real time. This enables real-time monitoring and control of industrial systems, machines, and equipment, improving efficiency and reducing downtime.

In conclusion, IoT sensors are a critical component of IoT technology and are used in a wide range of applications. They enable the collection of data from the environment and transmit it over a network, allowing for real-time monitoring and control of industrial systems, machines, and equipment.

12.2.4 OPEN-SOURCE CLOUD PLATFORM

OpenStack is an example of an open-source cloud platform. It provides infrastructure as a service (IaaS) and allows users to manage and deploy virtual machines, storage, and networking resources in a cloud environment.

Other examples of open-source cloud platforms include Apache CloudStack and OpenNebula. Different types of cloud services are as follows.

12.2.4.1 Platform as a Service

Examples: Microsoft Azure, Amazon Web Services (AWS). Most commonly used for database management, development tools, and operating systems. Platform as a service (PaaS) is important for developers. A PaaS business rents out its technology library. You can use it to develop, test, deliver, and manage software applications. You do not need to buy underlying hardware or software. Instead, you can focus on developing and managing apps.

12.2.4.2 Infrastructure as a Service

Examples: Microsoft Azure, AWS, Google Cloud. Most commonly used for servers, networking, storage, and more. Infrastructure-as-a-service companies rent out an entire information technology infrastructure.

You can pay to use theirs rather than hosting everything yourself. Rent servers, storage, networks, and operating systems. All are available on a pay-as-you-go basis. Because this model is so broad, it also allows for the most flexibility and customization.

12.2.4.3   Software as a Service

Examples: Salesforce, Microsoft Office 365 apps, MailChimp, Kinetic (a new name for Epicor ERP), ThingSpeak. Most commonly used for out-of-the-box solutions. The most common type of cloud service, software as a service (SaaS) is software that users can access over the internet.

SaaS may require a subscription. Access to the app is available from any compatible device over the internet. Sometimes, you can download the software directly to your computer.

12.2.4.4   ThingSpeak

ThingSpeak is a cloud-based platform for the IoT that allows devices to send and receive data to and from the cloud. It provides a simple, accessible platform for developers and users to connect, collect, analyze, and visualize data from a variety of IoT devices.

Sensors connected with Arduino can send data to ThingSpeak through Wi-Fi for real-time analysis as shown in Figure 12.6.

Some of features of ThingSpeak are:

- **Data Collection:** ThingSpeak enables devices to send data to the cloud, where it can be stored and analyzed. These data can then be used to monitor and control devices, track changes in the environment, and perform analytics.
- **Real-Time Visualization:** ThingSpeak provides real-time visualization of data collected from IoT devices, allowing users to see changes in real time and respond quickly to any issues.

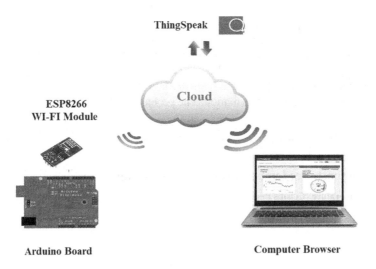

FIGURE 12.6   Sensors data on Thingspeak.

- **Application Development:** ThingSpeak provides a range of tools and application programming interfaces (API) that allow developers to create applications and services that use IoT data.
- **Scalability:** ThingSpeak can handle large amounts of data and can be scaled up or down as needed, making it suitable for a wide range of applications.
- **Integration:** ThingSpeak can be integrated with a variety of other services and platforms, such as MATLAB, AWS, and Microsoft Power BI.

ThingSpeak is widely used in a variety of industries, including agriculture, energy, manufacturing, and transportation. It is particularly useful for monitoring and controlling systems and devices in real time, such as climate control systems, machinery, and vehicles.

In conclusion, ThingSpeak is a powerful and accessible platform for the IoT, providing features for data collection, real-time visualization, application development, scalability, and integration. It is widely used in a variety of industries, providing a simple and effective way to connect and monitor devices and systems. It can send data from any internet-connected device directly to ThingSpeak using a Rest API or Message Queuing Telemetry Transport (MQTT).

In addition, cloud-to-cloud integrations with The Things Network, Senet, the Libelium Meshlium gateway, and Particle.io enable sensor data to reach ThingSpeak over LoRaWAN (low-power WAN (wide area networking protocol)) and 4G/3G cellular connections. It can store and analyze data in the cloud without configuring web servers and can create sophisticated event-based email alerts that trigger based on data coming in from your connected devices. Its service is operated by MathWorks.

In order to sign up for ThingSpeak, need to create a new MathWorks account or log in to your existing MathWorks account. It is free for small non-commercial projects. This includes a web service (REST API) that allows the collection and storage of sensor data in the cloud and develops Internet of Things applications. It works with Arduino, Raspberry Pi, and MATLAB (premade libraries and APIs exist).

12.3 INDUSTRIAL PROJECT

A project was implemented in the fabrication industry in Madhya Pradesh, India. Details are given in Table 12.1 and Figure 12.7.

Long electrodes (12 ft) are stored in the fabrication industry. These long electrodes are used in ship stator formation. The problem is that if the electrodes are kept

TABLE 12.1
Demonstrates Temperature Value and Heater Status

Temperature (Celsius degrees)	Heater status
>31	OFF
<31	ON

in a humid atmosphere, that is, in the winter and the rainy season, these electrodes will be damaged and cannot be used.

The storage of these electrodes without damage is not an easy task in humid seasons. To cope with this issue, a heater is used at the storage place to maintain the temperature and humidity.

This heater is on for 24×7 in humid season. This wastes lots of electricity. Therefore, we decided to set the heater to auto on/off as required to maintain the temperature and humidity and named it an auto heater. Details are given in Section 12.3.1.

12.3.1 PROJECT DESCRIPTION

The heater is connected to the contactor, and the Arduino is also connected to the contactor. Programming is done on the Arduino. The output of the Arduino is given to the contactor if the temperature is lower than 31 Celsius degrees. Then the contactor will connect the power supply to the heater.

The heater will start as soon as it receives the supply. Otherwise, the heater will be cut off from supply and so on. The heater will be off if the temperature is higher than 31 Celsius degrees as shown in Table 12.1.

This heater had an 11KW power rating and working on a three-phase power supply. Because of the heater's high-power rating, instead of a relay, the contactor uses a connect/break for the power supply to this heater. A block diagram of auto heater setup is shown in Figure 12.7.

12.3.2 CIRCUIT

A circuit is made for control of 11KW heater as shown in Figure 12.8. Arduino has analog pins from A0 to A5 and digital pins from 2 to 13. Pin 0 is for reception, and pin 1 is for transmission. Pin 14 is the ground. The circuit is demonstrated in Figure 12.8.

An LCD display is connected at digital pins 6, 9, and 10. Input is given at pin 8, and the output is taken from pin 7 as shown in Figure 12.8.

The data availability can be made per second, minute, or hour. We called for data after 10 min. ThingSpeak made a graph of these data as shown in Figure 12.9.

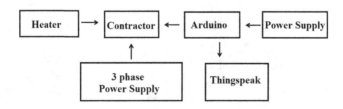

FIGURE 12.7 Block diagram of the auto heater.

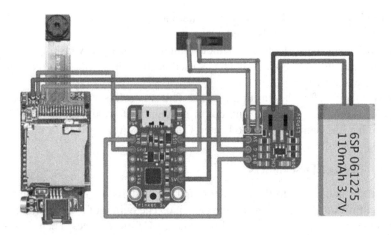

FIGURE 12.8 Circuit diagram.

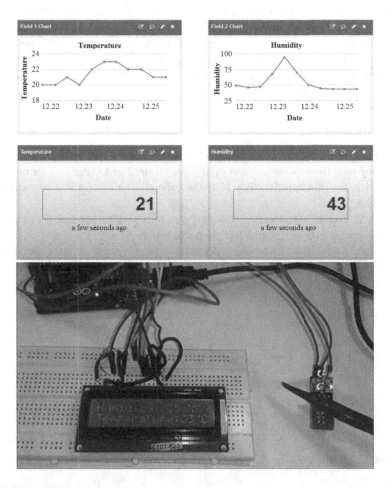

FIGURE 12.9 Data on ThingSpeak.

12.3.3 PROGRAM

The following is the Python programming of the auto heater.

```
#include <LiquidCrystal.h>
#include<DHT.h>
DHT dht(8, DHT22);
// initialize the library with the numbers of the interface pins
LiquidCrystal lcd(12, 11, 5, 4, 3, 2);
int red_light_pin = 6;
int green_light_pin = 9;
int blue_light_pin = 10;
void setup() {
   // set up the LCD's number of columns and rows:
   lcd.begin(16, 2);
     Serial.begin(115200);
     delay(100);
     Serial.println();
     Serial.println();
     dht.begin();
     // Print a message to the LCD.
     lcd.print("Temp: Humidity:");
     pinMode(red_light_pin, OUTPUT);
     pinMode(green_light_pin, OUTPUT);
     pinMode(blue_light_pin, OUTPUT);
     pinMode(8, INPUT);
     pinMode(7, OUTPUT);
}
void loop() {
   delay(2000);
   lcd.setCursor(0, 1);
   // Reading temperature or humidity takes about 250
   float h = dht.readHumidity(); milliseconds!
   float t = dht.readTemperature();
   if (isnan(h) || isnan(t)) {
     lcd.print("ERROR");
     return;
   }
```

```
lcd.print(t);
lcd.setCursor(7,1);
lcd.print(h);
Serial.print("Temperature");
Serial.println(t);
Serial.print("Humidity");
Serial.print(h);
/*Serial.print(val);*/
if (t < 31.0) {
        digitalWrite(7,LOW);
        Serial.println(digitalRead(7));
        Serial.print("Heater is ON");
        Serial.print("FAN is ON");
        digitalWrite(green_light_pin,LOW); // Green
        delay(2000);
        digitalWrite(blue_light_pin,HIGH); // Blue
        delay(2000);
        digitalWrite(red_light_pin,HIGH); // Red
        delay(2000);
}
else {
        digitalWrite(7,HIGH);
        Serial.println(digitalRead(7));
        Serial.print("Heater is OFF");
        digitalWrite(green_light_pin,HIGH); // Green
        delay(2000);
        digitalWrite(blue_light_pin,HIGH); // Blue
        delay(2000);
        digitalWrite(red_light_pin,LOW); // Red
        delay(2000);
}
delay (2000);
}
```

With this arrangement, long electrodes are kept safe without any damage to the storage area of the fabrication industry. In this manner, the industrial issue is solved through cloud-integrated IIoT. Similarly, other issues of industries can be solved. There are many sectors in which such IIoT arrangements can be made. Also, there are many advantages to the IIoT.

12.4   ADVANTAGES OF THE IIOT

The IIoT is a rapidly growing technology that is transforming the way industrial companies operate. It involves the integration of advanced technologies, such as sensors, network connectivity, cloud computing, and artificial intelligence, into traditional industrial processes to increase efficiency, reduce downtime, and improve product quality. The following are some of the main advantages of the IIoT in the industrial sector (Khang et al., 2024a).

12.4.1   IMPROVED EFFICIENCY

One of the main advantages of the IIoT is the improvement of efficiency in industrial processes. With the help of sensors and network connectivity, the IIoT enables real-time monitoring and control of industrial systems, machines, and equipment. This allows for quick adjustments to be made to improve efficiency and reduce downtime.

For example, in a manufacturing process, real-time monitoring of machinery can help detect early-warning signs of potential breakdowns. With this information, maintenance teams can quickly schedule repairs before the machine fails, reducing downtime and increasing machine uptime.

Additionally, the IIoT enables the automation of many industrial processes, reducing the need for manual labor and increasing productivity. Automated processes can run 24/7 without the need for human intervention, reducing the risk of human error and increasing efficiency (Khang et al., 2023a).

12.4.2   PREDICTIVE MAINTENANCE

The IIoT enables the use of predictive maintenance, which involves using data and machine learning algorithms to predict when a machine is likely to fail. This helps reduce downtime and increase machine uptime, resulting in higher production and lower maintenance costs (Hajimahmud et al., 2023b).

Predictive maintenance is based on the collection of data from machines, such as vibration, temperature, and other performance metrics. The data are analyzed and used to create predictive models that can predict when a machine is likely to fail. The models are then used to send notifications to maintenance teams, who can schedule repairs before the machine fails.

By addressing problems before they become critical, predictive maintenance helps reduce the need for emergency repairs and maintenance costs. It also helps ensure that machines are functioning properly, reducing the risk of product defects and improving product quality.

12.4.3   INCREASED PRODUCTIVITY

The IIoT enables the automation of many industrial processes, reducing the need for manual labor and increasing productivity. Automated processes can run 24/7 without the need for human intervention, reducing the risk of human error and increasing efficiency.

In addition, the IIoT enables real-time monitoring and control of industrial systems, machines, and equipment. This allows for quick adjustments to be made to improve efficiency and reduce downtime, increasing productivity and reducing the need for manual labor.

12.4.4   IMPROVED PRODUCT QUALITY

With the help of real-time monitoring and control, the IIoT helps ensure that machines are functioning properly, reducing the risk of product defects and improving product quality. In addition, predictive maintenance helps to ensure that machines are functioning properly, reducing the risk of product defects, and improving product quality.

For example, in a food and beverage manufacturing process, the real-time monitoring of equipment can help detect early-warning signs of contamination. With this information, the production process can be quickly adjusted to prevent contamination and improve product quality (Khang et al., 2024b).

12.4.5   REMOTE ACCESS AND CONTROL

The IIoT enables remote access and control of industrial systems, machines, and equipment. With the help of network connectivity and cloud computing, industrial processes can be monitored and controlled from anywhere in the world (Rani et al., 2023).

Remote access and control are particularly useful for companies with multiple locations or those with operations in remote areas. With the IIoT, industrial processes can be monitored and controlled from anywhere in the world, allowing for quick adjustments to be made to improve efficiency and reduce downtime. The benefits of remote access and control include the following:

- Increased efficiency: By monitoring and controlling industrial processes from anywhere in the world, quick adjustments can be made to improve efficiency and reduce downtime.
- Reduced downtime: Remote access and control allows for quick identification and resolution of problems, reducing downtime and increasing machine uptime.

12.4.6   FASTER IMPLEMENTATION OF IMPROVEMENTS

The IIoT generates valuable information so that those in charge of improving processes in an industrial business model (process, quality, or manufacturing engineers) can access data and analyze it faster and automatically and remotely perform the necessary process adjustments.

This also increases the speed at which changes and improvements are applied in operational intelligence and business intelligence—changes that are already offering competitive advantages to a myriad of industrial businesses.

12.4.7 PINPOINT INVENTORIES

The use of IIoT system allows for the automated monitoring of inventory, certifying whether plans are followed and issuing an alert in case of deviations. It is yet another essential IIoT application to maintain a constant and efficient workflow.

12.4.8 SUPPLY CHAIN OPTIMIZATION

Among the IIoT applications aimed at achieving higher efficiency, we can find the ability to have real-time in-transit information regarding the status of a company's supply chain. This allows for the detection of various hidden opportunities for improvement or pinpointing the issues that are hindering processes, making them inefficient or unprofitable.

12.4.9 PLANT SAFETY IMPROVEMENT

Machines that are part of the IIoT can generate real-time data regarding the situation on the plant. Through the monitoring of equipment for damage, plant air quality, and the frequency of illnesses in a company, among other indicators, it is possible to avoid hazardous scenarios that imply a threat to the workers. This boosts not only safety in the facility but also productivity and employee motivation. In addition, economic and reputation costs that result from poor management of company safety are minimized.

12.4.10 IMPROVED DATA CONTEXT

When real-time data are collected in the cloud for enhanced visibility and analytics functions, this is a benefit from a single source. These data are accessible company-wide, meaning teams can easily find the data and context they need.

12.4.11 IMPROVED DECISION-MAKING

When the previous value points of the IIoT are combined, an environment that actively promotes smarter decision-making can be gained. The mixture of artificial intelligence (AI), machine learning (ML), and cloud analytics and data accessibility is the foundation for business decisions that can be made with confidence and assuredness (Jaiswal et al., 2023).

12.4.12 REAL-TIME MONITORING AND CONTROL

The IIoT enables real-time monitoring and control of industrial systems, machines, and equipment. With the help of sensors and network connectivity, industrial processes can be monitored in real time, allowing for quick adjustments to be made to improve efficiency and reduce downtime.

Real-time monitoring and control are particularly important in the manufacturing and industrial industries, where quick decisions and adjustments are necessary to ensure efficient operations (Khang et al., 2022b).

With the IIoT, industrial processes can be monitored in real time, allowing for quick adjustments to be made to improve efficiency and reduce downtime. The benefits of real-time monitoring and control include the following:

- Increased efficiency: By monitoring industrial processes in real-time, quick adjustments can be made to improve efficiency and reduce downtime.
- Reduced downtime: Real-time monitoring and control allows for quick identification and resolution of problems, reducing downtime and increasing machine uptime.
- Improved product quality: Real-time monitoring and control helps to ensure that machines are functioning properly, reducing the risk of product defects and improving product quality.

12.5 PRACTICAL APPLICATION AREAS OF THE IIOT

12.5.1 INDUSTRIAL AUTOMATION

Industrial automation is one of the most significant and common applications of the IIoT. Automation of machines and tools enables companies to operate in an efficient way with sophisticated software tools to monitor and make improvements for the next process iterations as shown in Figure 12.10.

The accuracy of the process stages can be improved to a greater level using machine automation. Automation tools like PLC (programmable logic control) and PAC (programmable automation control) are used with smart sensor networks connected to a central cloud system to collect huge amounts of data. Specially designed software and applications are used to analyze the data and its behavior for improvements.

Industrial automation improves accuracy and efficiency, reduces errors, is easy to control, and is remotely accessible via applications. Machines can operate in harsh environments than humans; the automation of machines and tools reduces manpower requirements for specific tasks (Tailor et al., 2022).

FIGURE 12.10 Machine automation.

The connected-factory concept is an effective solution for improvements in all areas of operation. Major components, such as machines, tools, and sensors, will be connected to a network for easier management and access.

An overview of the process flow, the monitoring down time, status checking inventory and shipments, scheduling maintenance, and stopping/pausing a particular process for further analysis, among other processes, can be done remotely using IIoT solutions (Khang et al., 2023b).

12.5.2 Smart Robotics

Many companies are developing intelligent robotics systems for IoT-enabled factories. Smart robotics ensures the smooth handling of tools and materials in the manufacturing line with precise accuracy and efficiency (Khang et al., 2024c).

Predefined specifications can be set for maximum precision (up to a few nanometers in scale for some applications) using intelligent robotic arms as shown in Figure 12.11.

Robots can be programmed to perform complex tasks with high-end embedded sensors for real-time analysis as shown in Figure 12.11. These robotics networks are connected to a secure cloud for monitoring and controlling. Engineering teams can access and analyze these data to take quick actions for product improvements or prevent an unexpected failure due to machine fault.

12.5.3 Integration of Smart Tools/Wearables

The integration of smart sensors into tools and machines enables the workforce to perform the task with improved accuracy and efficiency. Specially designed wearables and smart glasses help employees reduce errors and improve safety in work environments (Khang et al., 2023c).

Smart wearables can trigger instant warning messages to employees during emergency situations like gas leaks or fires. Wearables can monitor the health conditions of individuals continuously and feedback if not fit for a particular task (Khang et al., 2023d).

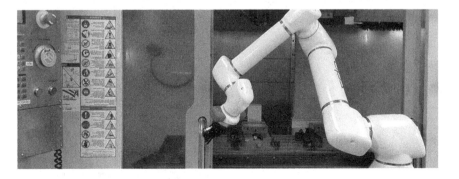

FIGURE 12.11   Smart robots.

Source: ABB.

12.5.4   SMART LOGISTICS MANAGEMENT

Logistics is one of the important areas in many industries, which needs continuous improvements to support increasing demands. Smart sensor technology is a perfect fit to solve many of the complex logistics operations and manage goods efficiently. Retail giants like Amazon are using drones to deliver goods to their customers.

Advanced technologies like drones offer better efficiency, accessibility, and speed, and they require less manpower as shown in Figure 12.12.

Airline is another major industry, which uses IoT for its daily operations at the production and predictive maintenance of airplanes in service. At the manufacturing plant, airline companies use IoT solutions to track thousands of components required every day at work.

A centralized management of inventories helps manage its supplies effortlessly. Suppliers will be automatically informed if any items are required to top up. Without much human action, inventory management can be effectively implemented using IoT. Smart sensors continuously monitor the airplane's machinery, and the data are collected in real time and sent to the airplane manufacturer.

Maintenance of any part of an airplane will be triggered, the concerned team will be informed, and maintenance will be carried out once the plane has landed without any delay. Manufacturers can plan and deliver spare parts efficiently based on the data shared by the system.

12.5.5   AUTONOMOUS VEHICLES

Automotive industries are using IoT enable self-driving vehicles to supply goods and logistics management within their company premises (Hajimahmud et al., 2022).

FIGURE 12.12   Smart logistics (delivery by drone).

FIGURE 12.13   Autonomous vehicles.

Source: www.dhl.com.

Smart vehicles can detect traffic congestion along their path and make deviations to reach their destination in the shortest time as shown in Figure 12.13.

12.5.6   Software Integration for Product Optimization

Smart analytics solutions are one of the most important components of any IoT system that further enhances the possibilities of the system for improvement and optimization.

Major companies are implementing customized software for deep analyses of the huge amounts of data collected from large sensor networks and machines. Detailed analysis of data and understanding behavior over time give a much better overview of process improvement strategies for product optimization.

12.5.7   Smart Package Management

Package management using IoT technology gives a lot of convenience and efficiency for manufacturing units. Smart sensors can monitor each stage of packing and update statuses in a real-time manner.

Embedded sensors can detect vibrations and atmospheric conditions like temperature and humidity, among others, and provide feedback if something goes wrong during transit or storage.

12.5.8   Enhanced Quality and Security

The introduction of IoT technology into manufacturing offers enhanced product quality. Continuous monitoring and analysis of each stage ensures better quality by improving process steps for optimum quality.

The integration of smart tools and software-assisted procedures offers a higher level of security. Software-controlled automation and data collection from huge sensor networks are connected to a highly secure gateway and cloud server platform. Complex encryption techniques are used in the IIoT platform for enhanced security (Khang et al., 2022c).

12.5.9   POWER MANAGEMENT

The IoT can offer better solutions for power management in industries. Specific sensors can detect environmental changes and trigger on/off control of lights, air conditioners, humidity controls, and liquid flow, among others, for efficient power management.

12.5.10   ENVIRONMENTAL MONITORING

Environmental data are not useful at the weather station or field sensor. Putting data in the cloud means that the data can be monitored and acted on it immediately. Environmental sensors can measure the presence of pollution as shown in Figure 12.14.

FIGURE 12.14   Environmental sensors.

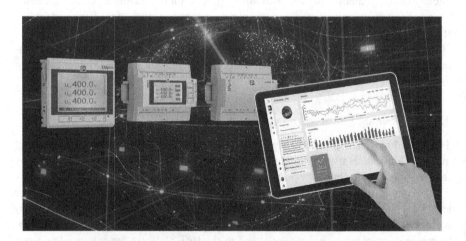

FIGURE 12.15   Energy monitoring.

IoT systems can be built without setting up servers or developing web software. ThingSpeak provides a hosted solution that you can use in production.

12.5.11 ENERGY MONITORING

Energy use and consumption can be monitored to increase efficiency and investigate consumption patterns. The use IoT sensors to monitor and understand energy use is given in Figure 12.15.

12.6 CONCLUSION

In this chapter, we present components, advantages, and application areas of the IIoT. Also, an industrial project was discussed. We concluded that the IIoT has the potential to revolutionize the way we approach production and management in the manufacturing and production industries, with its ability to reduce downtime, enhance safety, and provide valuable data and insights (Rana et al., 2021).

IIoT represents a powerful tool for manufacturers looking to improve their operations and stay ahead of the curve. Existing industries can adapt to an IIoT ecosystem for process improvements, better management, cost-effectiveness, and overall efficiency (Khang et al., 2023a). Companies that invest in IIoT will be well positioned to take advantage of its many benefits, as well as address the challenges and risks that come with implementing this innovative technology (Bhambri et al., 2022).

Future industries can utilize the power of IIoT infrastructure for product optimization by analyzing big data from thousands of tiny sensors. The IIoT is a fast-growing technology with limitless possibilities for future industries and manufacturing units (Khanh and Khang, 2021).

REFERENCES

Bhambri P, Rani S, Gupta G, Khang A. *Cloud and Fog Computing Platforms for Internet of Things*. CRC Press, 2022. ISBN: 978-1-032-101507. https://doi.org/10.1201/9781003213888

Hajimahmud VA, Khang A, Gupta SK, Babasaheb J, Morris G. *AI-Centric Modelling and Analytics: Concepts, Designs, Technologies, and Applications* (1st Ed.). CRC Press, 2023a. https://doi.org/10.1201/9781003400110

Hajimahmud VA, Khang A, Hahanov V, Litvinova E, Chumachenko S, Alyar AV. "Autonomous Robots for Smart City: Closer to Augmented Humanity," *AI-Centric Smart City Ecosystems: Technologies, Design and Implementation* (1st Ed.). CRC Press, 2022. https://doi.org/10.1201/9781003252542-7

Hajimahmud VA. et al. (Eds.). "The Role of Data in Business and Production," *AI-Aided IoT Technologies and Applications in the Smart Business and Production*. CRC Press, 2023b. https://doi.org/10.1201/9781003392224-2

Jaiswal N, Misra A, Misra PK, Khang A (Eds.). "Role of the Internet of Things (IoT) Technologies in Business and Production," *AI-Aided IoT Technologies and Applications in the Smart Business and Production*. CRC Press, 2023. https://doi.org/10.1201/9781003392224-1

Khang A. "Material4Studies," *Material of Computer Science, Artificial Intelligence, Data Science, IoT, Blockchain, Cloud, Metaverse, Cybersecurity for Studies*, 2021. www.researchgate.net/publication/370156102_Material4Studies

Khang A (Ed.). *AI-Oriented Competency Framework for Talent Management in the Digital Economy: Models, Technologies, Applications, and Implementation*. CRC Press, 2023b. https://doi.org/10.1201/9781003440901

Khang A. *Advanced Technologies and AI-Equipped IoT Applications in High-Tech Agriculture* (1st Ed.). IGI Global Press, 2024b. https://doi.org/10.4018/9781668492314

Khang A, Abdullayev V, Hahanov V, Shah V. *Advanced IoT Technologies and Applications in the Industry 4.0 Digital Economy* (1st Ed.). CRC Press, 2024c. https://doi.org/10.1201/9781003434269

Khang A, Gupta SK, Rani S, Karras DA. *Smart Cities: IoT Technologies, Big Data Solutions, Cloud Platforms, and Cybersecurity Techniques* (1st Ed.). CRC Press, 2023c. https://doi.org/10.1201/9781003376064

Khang A, Gupta SK, Shah V, Misra A (Eds.). *AI-Aided IoT Technologies and Applications in the Smart Business and Production*. CRC Press, 2023a. https://doi.org/10.1201/9781003392224

Khang A, Hahanov V, Abbas GL, Hajimahmud VA. "Cyber-Physical-Social System and İncident Management," *AI-Centric Smart City Ecosystems: Technologies, Design and Implementation* (1st Ed.). CRC Press, 2022c. https://doi.org/10.1201/9781003252542-2

Khang A, Hahanov V, Litvinova E, Chumachenko S, Hajimahmud VA, Alyar AV. "The Key Assistant of Smart City—Sensors and Tools," *AI-Centric Smart City Ecosystems: Technologies, Design and Implementation* (1st Ed.). CRC Press, 2022a. https://doi.org/10.1201/9781003252542-17

Khang A, Misra A, Abdullayev V, Eugenia L. *Machine Vision and Industrial Robotics in Manufacturing: Approaches, Technologies, and Applications* (1st Ed.). CRC Press, 2024a. https://doi.org/10.1201/9781003438137

Khang A, Ragimova NA, Hajimahmud VA, Alyar AV. "Advanced Technologies and Data Management in the Smart Healthcare System," *AI-Centric Smart City Ecosystems: Technologies, Design and Implementation* (1st Ed.). CRC Press, 2022b. https://doi.org/10.1201/9781003252542-16

Khang A, Rani S, Gujrati R, Uygun H, Gupta SK (Eds.). *Designing Workforce Management Systems for Industry 4.0: Data-Centric and AI-Enabled Approaches*. CRC Press, 2023d. https://doi.org/10.1201/9781003357070

Khang A, Rani S, Sivaraman AK. *AI-Centric Smart City Ecosystems: Technologies, Design and Implementation* (1st Ed.). CRC Press, 2022a. https://doi.org/10.1201/9781003252542

Khang A, Shah V, Rani S. *AI-Based Technologies and Applications in the Era of the Metaverse* (1st Ed.). IGI Global Press, 2023e. https://doi.org/10.4018/9781668488515

Khanh HH, Khang A. "The Role of Artificial Intelligence in Blockchain Applications," *Reinventing Manufacturing and Business Processes Through Artificial Intelligence* (pp. 20–40). CRC Press, 2021. https://doi.org/10.1201/9781003145011-2

Misra A, Khang A, Gupta SK, Shah V. *AI-Aided IoT Technologies and Applications in the Smart Business and Production* (1st Ed.). CRC Press, 2023. https://doi.org/10.1201/9781003392224

Pritiprada P, Satpathy I, Patnaik BCM, Patnaik A, Khang A (Eds.). "Role of the Internet of Things (IoT) in Enhancing the Effectiveness of the Self-Help Groups (SHG) in Smart City," *Smart Cities: IoT Technologies, Big Data Solutions, Cloud Platforms, and Cybersecurity Techniques*. CRC Press, 2024. https://doi.org/10.1201/9781003376064-14

Rana G, Khang A, Sharma R, Goel AK, Dubey AK (Eds.). *Reinventing Manufacturing and Business Processes through Artificial Intelligence*. CRC Press, 2021. https://doi.org/10.1201/9781003145011

Rani S, Bhambri P, Kataria A, Khang A, Sivaraman AK. *Big Data, Cloud Computing and IoT: Tools and Applications* (1st Ed.). Chapman and Hall/CRC Press, 2023. https://doi.org/10.1201/9781003298335

Rani S, Chauhan M, Kataria A, Khang A (Eds.). "IoT Equipped Intelligent Distributed Framework for Smart Healthcare Systems," *Networking and Internet Architecture*. CRC Press, 2021. https://doi.org/10.48550/arXiv.2110.04997

Tailor RK, Pareek R, Khang A (Eds.). "Robot Process Automation in Blockchain," *The Data-Driven Blockchain Ecosystem: Fundamentals, Applications, and Emerging Technologies* (1st Ed., pp. 149–164). CRC Press, 2022. https://doi.org/10.1201/9781003269281-8

13 Smart Notice Board Using the Internet of Things–Based NODEMCU ESP8266

*Suman Turpati, Richi Sumith Raj P.,
Mohammed Taj S., Naveen Kumar S.,
and Ranga Reddy S. V.*

13.1 INTRODUCTION

Erecting an Internet of Things (IoT)–grounded system gives the fast metamorphosis of data and the stoner can pierce the data from anywhere in the world (Abeer et al., 2019). In this design, we have developed an IoT-grounded smart notice board (Gaurav Bhardwaj et al., 2020). The main idea of this design is to develop an automatic, tone-enabled, and largely dependable electronic notice board (Normanyo et al., 2020).

A display connected with the pall will continuously stay for the communication from the stoner, if the stoner uploads the data in the ThingSpeak pall, it will automatically upload to the fleck matrix liquid-crystal display (LED) control LED display module (Valerie Popp et al., 2020). By using NodeMCU ESP8266, the stoner can upload the communication to the fleck matrix LED control LED display module by penetrating the web server (Pawan et al., 2018).

The stoner can write the data from anywhere in the world to the fleck matrix LED control LED display module (Fraser et al., 2021). This will reduce the time to modernize the data as well as it will efficiently transfer the data to the end stoner. This smart notice board is available in numerous educational institutions, banks, and public places (Al-Balas et al., 2021). Global System for Mobile Communication (GSM) and a microcontroller are used for this, by transferring a communication, the stoner can get the data in the LED display. So, the system uses GSM-grounded notice boards (Elavarashi et al., 2022).

The notice board is extensively used at the moment in some of the places that need critical notices like the council, and road shares of stations—requests, we require a real-time notification since this notice should be in real time (Agarwal et al., 2022). With this design, we are attempting to usher in a new era of real-time notification via SMS and the internet. Operations (Bahadur Sinha and Dhanalakshmi, 2021) design is about writing the communication on a web runner or in simple SMS and

DOI: 10.1201/9781003392224-13

sending it to a remote garcon (Kam et al., 2021). This communication is brought into a microcontroller, where it is shown on a DOT-matrix LED display (Donnelly et al., 2021).

With the arrival of digital technology, it is effective to represent the information on digital bias (Caravaggio et al., 2021). Nowadays, the internet is the primary mode of communication far and wide (Robert et al., 2020). Notice boards play a vital part in conveying communication in any association. To achieve green information technology, it is obligatory to use digital media rather than earlier conventional media like paper printing (Ankit et al., 2020).

This chapter details the enforcement of an Android-controlled smart notice board using IoT (Internet of Effects), which uses the NodeMCU ESP8266, and the IoT, which is controlled by a web cyber surfer (Asin et al., 2020). With the help of this design, an authenticated person can convey the communication/notice, indeed, from a remote place on the digital bias like a dot-matrix LED screen (Gaurav et al., 2020). The proposed system reduces resources like force and time. In this design, IoT technology has been espoused in which the internet is used as a tool to connect and change the data among the different biases (Sharma et al., 2020).

In earlier days, people used paper as the medium to convey any dispatches, and the paper material, which is made up of tree pulp, could often be scarce (Kariofyllis et al., 2020). It is pivotal to save trees by espousing digital technology rather than conventional methods, which are veritably expensive (Sharma et al., 2020). A desire for this is the operation of digital technology in daily life (Coehoorn et al., 2020).

Notice boards are used in institutions, associations, hospitals, railway stations, shopping centers, boardwalks, and numerous other public places (Rani et al., 2021). The conventional notice boards are not so accessible for the variations to be dispatched to be conveyed, and quire-ready coffers and force (Mala et al., 2020). Due to the fashionability of the internet, we choose the internet as a medium for transferring information.

The World Wide Web affects the network of physical things (IoT) bias, automobiles, home appliances, and others, including implanted electronic devices (Cavdir, 2020). The operating system enables these objects to connect and change data. To give security, we add a username and word type authentication system (Khang et al., 2022a). So only separate authorities can shoot information, jeer Pi is the heart of our system (Sandeep et al., 2020).

An examiner connives with it, so information in the form of textbooks and PDFs can be displayed on the large screen. In this design, we have developed IoT grounded smart noticeboard (Alisha et al., 2020). Then, we are using NodeMCU ESP8266 and fleck matrix LED control LED display module display to show the data. A cheap open-source platform is NodeMCU. We may connect to the WIFI network using NodeMCU (Muhammad et al., 2020).

The stoner can write the data in the ThingSpeak cloud which they want to show on the notice board. After writing the data, the end stoner can get this data in fleck matrix LED control LED display module displays. These displays are more helpful in educational institutions, banks, and public transport similar to railroads, airways, and others (Nellya et al., 2022).

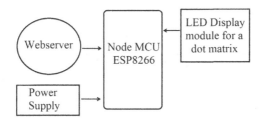

FIGURE 13.1   Block diagram.

13.2   BLOCK DIAGRAM

In this block diagram, NodeMCU ESP8266 is used as a main controller. The web server cloud is used to write the data and display them (Bhambri et al., 2022). The dot-matrix LED controls LED display module is connected as the output device, which is used to show the data to the end users.

In this project, we have developed an IoT-based smart noticeboard. Here, we are using NodeMCU ESP8266 and dot matrix LED control LED display module display to show the data. A cheap open-source platform is NodeMCU (Khang et al., 2024a).

By using NodeMCU, we can connect through the Wi-Fi network. The user can write the data that they want to show on the notice board in the ThingSpeak cloud. After writing the data, the end user can get these data in dot-matrix LED control LED display module displays (Rani et al., 2023).

13.3   FLOW CHART

Figure 13.2 shows the inflow map for the digital notice board using the IOT. As the display is turned on, the first communication that it displays is the dereliction communication set up in the programming.

The dereliction of communication is decided by the stoner. There is a DHT11 detector that is used for seeing temperature and moisture. Temperature and moisture are also displayed on the P10 LED scrolling display.

The regulator used is the AT mega 328 Microcontroller. The communication is transferred to the microcontroller through a personal Android mobile phone. The communication that is to be shown is entered by the AT mega 328 Microcontroller, which further decodes the communication and sends it to the display unit.

The communication that is transferred to the microcontroller through the GSM is defended with a word. If the communication transferred has the correct word also GSM receives the communication as shown in Figure 13.2.

As the message sent has a correct password and is sent to the GSM, then the display unit, the P10 LED display, displays the sent message. If the password for the sent message is incorrect, then the GSM does not receive the message, and the display unit displays the default message. If the password for the GSM is correct, it is further displayed on the P10 LED scrolling display.

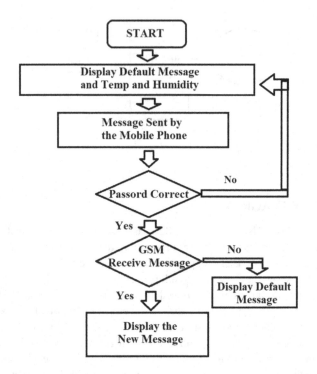

FIGURE 13.2   Digital notice board flow chart.

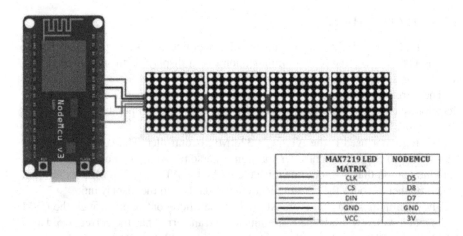

	MAX7219 LED MATRIX	NODEMCU
	CLK	D5
	CS	D8
	DIN	D7
	GND	GND
	VCC	3V

FIGURE 13.3   NodeMCU description.

13.4   DESCRIPTION OF COMPONENTS

The ESP Wi-Fi allows the NodeMCU to produce a web server using a Wi-Fi network and then to control the LED matrix, using MD MAX72xx.h, as shown in Figure 13.3.

Where it is law, the URL may be verified after the file is uploaded to the NodeMCU. Periodical examiner and used to pierce a website runner for regulating the textbook displayed on the MAX7219 LED display.

Using this web runner, we could open a communication; plus, that communication will be shown on the grid as shown in Figure 13.4.

A large and brilliant 512 LED matrix panel with onboard regulator straight from NodeMCU board, easy-to-use circuitry is used. Its brilliance position is over 3500nits to 4500nits; one plate requires 3.5 a current and 5V.

13.4.1   GSM MODULE

It acts like a communication device. The serial text format data from a mobile are given to the microcontroller through the GSM module as shown in Figure 13.5.

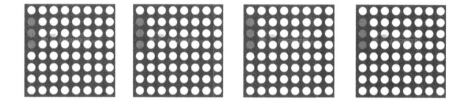

FIGURE 13.4   LED P10 Display.

FIGURE 13.5   GSM modem.

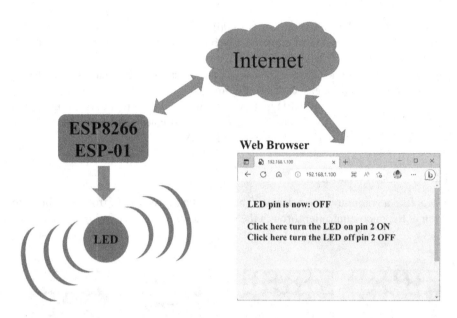

FIGURE 13.6   Data processing.

13.4.2 SOFTWARE INSTALLATION

Using a USB, a bootloader must be flashed to a periodical, including ST-Link (SWD). Flashing the bootloader is shown in Figure 13.6. Be aware that you might need to position the board after flashing the bootloader in "perpetual bootloader".

Before uploading a design, connect a resistor between leg PC14 and 3.3 V and the board will reset. Now you ought to be able to remove the resistor to cause the sketch to flash and renew the board, which should function normally afterward when fresh drawings are submitted. However, Device Firmware Update (DFU) claims that there is no DFU bias if you are there, you might need to change the maple-upload script in the apparatus brochure if you find that the IDE successfully resets your board. Increase the value that is given to the line that calls upload-reset.

13.5 HARDWARE DESCRIPTION

13.5.1 NODEMCU FIRMWARE

An open-source IoT platform is NodeMCU. The scripting language Lua is employed. It is built using the ESP8266 SDK 0.9.5 and is based on the Lua design. It makes use of several open-source programs like Lua-Jason and spiffs. It contains firmware that is powered by the ESP8266 Wi-Fi SoC and controls that are anchored by the ESP-12 module.

A short while after NodeMCU was created, the ESP8266 was released. On December 30, 2013, Espressif Systems launched the ESP8266 device. The ESP8266 is a Wi-Fi SoC integrated with a Tensilica Xtensa LX106 core, extensively used in

IoT operations. NodeMCU started on October 13, 2014, when Hong committed the first train of NodeMCU firmware to GitHub.

Two months later, the design expanded to include an open-tackle platform when inventor Huang R committed the Gerber train of an ESP8266 board, named devkit1.0. At that month's end, Tuan PM ported the MQTT customer library from Contiki, which was committed to NodeMCU and the ESP8266 SoC platform. Also, designing in NodeMCU was suitable for utilizing Lua to support the MQTT IoT protocol pierce the MQTT IoT protocol, using Lua to pierce the broker for MQTT. On January 30, 2015, Devsaurus released a significant upgrade by porting the u8glib to the NodeMCU platform design, allowing the NodeMCU to fluently operate a TV, display, An organic light-emitting diode (OLED), and indeed Video Graphics Array (VGA) monitors.

13.5.2 NODE MCU ESP8266

Shortly after the ESP8266 was released, NodeMCU was developed. On December 30, 2013, the Tensilica Xtensa LX106 core is included in the Wi-Fi system-on-a-chip (SoC) known as the ESP8266, which is extensively utilized in IoT operations as shown in Figure 13.7.

NodeMCU launched on October 13, 2014, when Hong made the first train firmware for the NodeMCU to GitHub. Two months later, the concept was developed to incorporate an open tackle and elevated when inventor Huang R broke the Gerber law (NodeMCU, 2023). On January 30, 2015, Devsaurus released a significant upgrade by porting the u8glib to the NodeMCU platform design, allowing the NodeMCU to fluently drive TV, display, OLED, and indeed VGA monitors. In the summer of 2015,

FIGURE 13.7 ESP8266 unit.

the generators gave up on the firmware design and a set of autonomous but devoted Participants took control.

When Arduino C/C started creating new MCU boards based on non-automatic voltage regulator (AVR) processors, such as the ARM/SAM MCU used in the Arduino Due, they demanded that the Arduino IDE be modified to make it relatively simple to support different tool chains that would enable Arduino C/C to be compiled down to these new processors. With the help of the SAM core and the board director's preface, they accomplished this. A "core" is the group of software components required by the board director and the Arduino IDE to translate an Arduino C/C source train into the machine language of the destination MCU (Khang et al., 2024b).

On the GitHub ESP8266 Core webpage, some inventive ESP8266 suckers have created an Arduino core for the ESP8266 WIFI SoC. It has become one of the top software development platforms for the vibrant ESP8266 grounded modules and development boards, including NodeMCUs. This is what is commonly referred to as the "ESP8266 Core for the Arduino IDE". Peter R. Jennings created the Button, a Wi-Fi-connected driving button. The Button is made to do solitary, internet-enabled tasks. When the button is pressed, a connection to a web service provider that will carry out the requested action is established. Operations feature a panic button or doorbell.

The open IoT platform NodeUSB is a typical USB stick size. It was made to interact with NodeMCU (Lua) for simple programming and features a backup USB port. It is perfect for plug-and-play outcomes, making it simple for inventors to prototype. With an OLED screen and NodeMCU firmware, the watch is a free and open-source Wi-Fi smartwatch. It might be the first smartwatch.

Electronic outfits and transmission cables for signals and power subordinated voltage spikes evident in radio-frequency broadcasts, lightning, electrostatic discharge, switching beats (harpoons), and disquiet in the control force. Remote lightning strikes can cause surges of up to 10 kV, furthering the voltage limitations of numerous voltaic factors. High voltages can be intentionally incorporated into a circuit, in which case it has to be a safe, dependable method of uniting its voltage in high factors with low voltages.

An opto-primary person's job is to obstruct similar so that a high voltage and voltage transients swell in a certain system component won't sabotage or obliterate the other corridor. In our system, the print coupler is PC817 design. It is related to the two legs of the ARM processor PV3, PC4, or PC5 at VCC it is leg number 1 and number 2. The affair of ARM has transferred a relay PCB. The input voltage at leg 4 is being produced by the SMPS. A VCC with 5 V and Rel 1 (the first relay) also has 5 V, and no current overflows between these two coupler locations, there is an LED; therefore, there will not be a link between the third and fourth legs of the couple. Therefore, voltages at 5 V would be an entry at leg 4, that is, R1.

On the other side, there will not be any if Rel 1 receives 0 V, an implicit distinction between leg 1 and 2. Therefore, the LED will illuminate as a result of electricity flowing through it. The secondary connection's timber. It will provide leg 4 earth eventuality, and therefore, R1 will not admit the SMPS's 5-V output voltages. To put it another way, the relay card receives power if Rel1's voltage is at just 5 V, and the relay PCB will not receive current if the voltage is higher or lower. The Rel1 leg is at ground voltage. This 5-V, 2-channel relay board is compact and simple to use. This relay board enables you to operate two 240-V electrical appliances straight from low-voltage circuits or 5-V ARM processors.

13.5.3 Dot-Matrix 4-in-1 Display Module MAX7219

Four-in-one display fleck MAX7219, the matrix module is a combined periodic input/affair common-cathode display driver that combines the microprocessor's seven-member digital LED display with eight integers as well as a bar graph display or 64 independent LEDs. A multichannel scanning circle member word motorist, a B-type binary-coded decimal (BCD) encoder chip, and an 8 × 8 static RAM are all included on the device. To adjust the member current of each LED, just one external register is used.

13.5.3.1 Features and Specifications
The guidelines the input on the left side of the module harborage, the legal right to affair harborage.

- Confines 128 × 32 × 1.5 mm (L × W × H; all 4-in-1)
- Dimensions 32 × 32 × 1.5 mm (L × W × H; single module)
- An 8 × 8 fleck matrix common cathode may be driven by a single module.
- 55 grams in weight
- Fixing 64-hole screws with a periphery of 3 mm
- Module with input and affair interfaces

Simply input when a single module is under control harborage linked to a CPU when several protruded inputs, modules, and affair CPU termination, an input outstation of the alternate affair the first two modules of the input, after the first module outstation and so forth, of the three terminal modules (Khang et al., 2023a).

13.5.3.2 Description
VCC 5V, GND, DIN P 2.2, CS P 2.1, and CLK P 2.0 as shown in Figure 13.8.

13.6 RESULT AND DISCUSSION

Using mobile hotspot Wi-Fi, we can connect to the NodeMCU, a firmware as shown in Figure 13.9. It can take us to Chrome, where the IP address of the firmware displaying in dot-matrix LED has to be typed.

Now we have to type the message that has to be displayed. After typing the message, we want, can be displayed on the dot-matrix LED as Figure 13.10.

This default message will continue to be displayed on the LED display until the user sends a message for the new message that is to be displayed as shown in Figure 13.12.

When the user sends a message from their mobile phone that is to be displayed on the notice board it uses the normal messaging app on the Android mobile phone of the user (Rani et al., 2022).

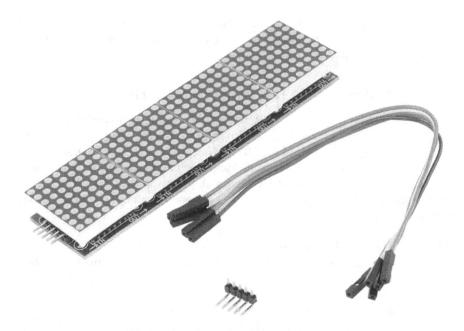

FIGURE 13.8   Dot-matrix LED and jumper wires.

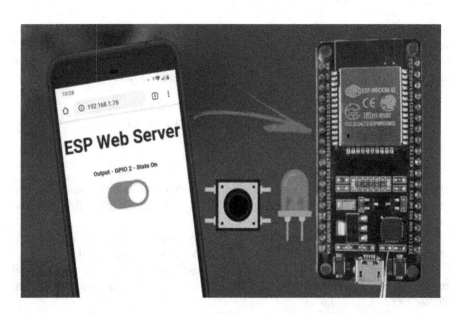

FIGURE 13.9   ESP web server and NodeMCU.

FIGURE 13.10 Message on ESP Web Server.

FIGURE 13.11 Display of message on NodeMCU.

FIGURE 13.12 Display of the message sent by the user.

13.7 CONCLUSION

This project focuses on a cutting-edge, IoT-based notice board that overcomes the benefits of the current noticing system. To the best of our knowledge, the prototype that was produced can be utilized to replace the system and has been successfully tested with troubleshooting. Each block within has been rationalized and supported (Khang et al., 2022b).

The project is incredibly marketable and cost-effective, and the parts utilized are fairly straightforward and readily available. We think that this system can be made commercially viable and employed at institutions like colleges, banks, and railroad stations (Khang et al., 2023b) (Aggarwal and Khang, 2023).

Last but not least, we conclude that the idea is built on the widely adopted Wi-Fi technology. It can be altered by its uses and provides more potential for development and study in the future (Luke et al., 2024).

REFERENCES

Abeer A. *Metamorphosis of Mosque Semiotics from Sacred to Secular Power Metaphors—the Case of State Mosques* (vol. 13, pp. 204–217), March 22, 2019. https://ieeexplore.ieee.org/abstract/document/8528677/

Agarwal A, Ray K, Pradhan BK, Kumari V. "GSM Based Smart Digital Wireless Electronic Notice Board," *Journal of Information Technology and Digital World*, 4, September 2022. https://doi.org/10.36548/jitdw

Aggarwal P, Khang A. "A Study on the Impact of the Industry 4.0 on the Employees Performance in Banking Sector," *Designing Workforce Management Systems for Industry 4.0: Data-Centric and AI-Enabled Approaches* (1st Ed., pp. 384–400). CRC Press, 2023. https://doi.org/10.1201/9781003357070-20

Al-Balas M, Al-Balas HI, Jaber HM, Obeidat K, Al-Balas H, Aborajooh EA, Al-Balas RATB. "Distance Learning in Clinical Medical Education Amid COVID-19 Pandemic in Jordan: Current Situation, Challenges, and Perspectives," *BMC Medical Education*, 20, 1–7, 2021. https://doi.org/10.1186/s12909-020-02257-4

Alisha P, Jelen B, Siek KA, Chan J, Lazar A. *Understanding Older Adults' Participation in Design Workshops*, April 23, 2020. https://dl.acm.org/doi/abs/10.1145/3313831.3376299

Ankit K, Dwivedi A, Dutta MK. *A Zero watermarking Approach for Biometric Image Security*, April 27, 2020. https://ieeexplore.ieee.org/abstract/document/9077148/

Asin Ali M. Mia MS, Hasan MM. *Design and Construction of Voice-Controlled Smart Electronic Notice Board* (vol. 7, p. 11), November 2020. www.academia.edu/download/65352491/IRJET_V7I11280.pdf

Bahadur Sinha B, Dhanalakshmi R. "Recent Advancements and Challenges of the Internet of Things in Smart Agriculture: A Survey," *Future Generation Computer Systems*, August 8, 2021. https://doi.org/10.1016/j.future.2021.08.006

Bhambri P, Rani S, Gupta G, Khang A. *Cloud and Fog Computing Platforms for Internet of Things*. CRC Press, 2022. https://doi.org/10.1201/9781003213888

Caravaggio A, Olin Kirsty AB, Franklin A, Dudley SP. "Twitter Conferences as a Low-Carbon, Far-Reaching and Inclusive Way of Communicating Research in Ornithology and Ecology," *Ibis*, May 19, 2021. https://doi.org/10.1111/ibi.12959

Cavdir D. *Embedding Electronics in Additive Manufacturing for Digital Musical Instrument Design*, June 26, 2020. www.researchgate.net/profile/Doga-Cavdir/publication/342854392_Embedding_Electronics_in_Additive_Manufacturing_for_Digital_Musical_Instrument_Design/links/5f09116592851c52d628cae9/Embedding-Electronics-in-Additive-Manufacturing-for-Digital-Musical-Instrument-Design.pdf

Coehoorn CJ, Stuart-Hill LA, Abimbola W, Neary JP, Krigolson OE. "Firefighter Neural Function and Decision-Making Following Rapid Heat Stress," *Fire Safety Journal*, October 29, 2020. https://doi.org/10.1016/j.firesaf.2020.103240

Donnelly N, Stapleton L. *Digital Enterprise Technologies: Do Enterprise Control and Automation Technologies Reinforce Gender Biases and Marginalisation?* September 14–17, 2021. www.sciencedirect.com/science/article/pii/S2405896321019443

Elavarashi M, Shifana M, Gayathri K. "Ultra Protection for Future Generation Women Safety," *Physical Review*, 47(6), 777–780, 2022. https://giirj.com/index.php/giirj/article/view/3626

Fraser RT, Surawski NC, Fleck R, Irga PJ. *Effective Reduction of Roadside Air Pollution with Botanical Biofiltration*, February 26, 2021. www.sciencedirect.com/science/article/pii/S030438942100529X

Gaurav B, Sahu G, Mishra RK. *IoT Based Smart Notice Board* (vol. 9, p. 6), June 2020. www.academia.edu/download/63936953/iot-based-smart-notice-board-IJERTV9IS06048520200716-12734-1u2ar5x.pdf

Gaurav SK, Dhamange PV, Duratkar PS, Vaidya SS, Zilpe AR, Gujar SS. *Bluetooth Based Digital Notice Board with Solar* (vol. 7), March 3, 2020. https://ieeexplore.ieee.org/abstract/document/8863483/

Kam KA, Kumar V, Lamport ZA, Kymissis I. "A Laboratory Course on Information Display Technologies for Remote Learning," *Journal of the Society for Information Display*, March 30, 2021. https://doi.org/10.1002/jsid.1000

Kariofyllis P, Nicopoulos C, Sirakoulis GC, Dimitrakopoulos G. *RISC-V2: A Scalable RISC-V Vector Processor*, September 28, 2020. https://ieeexplore.ieee.org/abstract/document/9181071/

Khang A, Abdullayev V, Hahanov V, Shah V. *Advanced IoT Technologies and Applications in the Industry 4.0 Digital Economy* (1st Ed.). CRC Press, 2024a. https://doi.org/10.1201/9781003434269

Khang A, Gupta SK, Rani S, Karras DA (Eds.). *Smart Cities: IoT Technologies, Big Data Solutions, Cloud Platforms, and Cybersecurity Techniques*. CRC Press, 2023a. https://doi.org/10.1201/9781003376064

Khang A, Gupta SK, Shah V, Misra A (Eds.). *AI-Aided IoT Technologies and Applications in the Smart Business and Production*. CRC Press, 2023b. https://doi.org/10.1201/9781003392224

Khang A, Hahanov V, Abbas GL, Hajimahmud VA. "Cyber-Physical-Social System and İncident Management," *AI-Centric Smart City Ecosystems: Technologies, Design and Implementation* (1st Ed.). CRC Press, 2022b. https://doi.org/10.1201/9781003252542-2

Khang A, Misra A, Abdullayev V, Eugenia L, *Machine Vision and Industrial Robotics in Manufacturing: Approaches, Technologies, and Applications* (1st Ed.). CRC Press, 2024b. https://doi.org/10.1201/9781003438137

Khang A, Rani S, Sivaraman AK. *AI-Centric Smart City Ecosystems: Technologies, Design and Implementation* (1st Ed.). CRC Press, 2022a. https://doi.org/10.1201/9781003252542

Luke J, Khang A, Chandrasekar V, Pravin AR, Sriram K (Eds.). "Smart City Concepts, Models, Technologies and Applications," *Smart Cities: IoT Technologies, Big Data Solutions, Cloud Platforms, and Cybersecurity Techniques*. CRC Press, 2024. https://doi.org/10.1201/9781003376064-1

Mala R, Saini RP. *Dispatch Strategies-Based Performance Analysis of a Hybrid Renewable Energy System for a Remote Rural Area in India*, February 19, 2020. https://search.proquest.com/openview/f104820244c626f65779b51c1a2360bd/1?pq-origsite=gscholar&cbl=2030013

Muhammad SN, Amin M. *Monitoring System for Temperature and Humidity Measurement with DHT11 Sensor Using NodeMCU* (vol. 5, p. 10), October 2020. www.ijisrt.com/assets/upload/files/IJISRT20OCT142.pdf

Nellya A, Lecocq T, Fourier C, Nivelle R, Fleck C, Fontaine P, Pasquet A, Thomas M. *A Multi-Trait Evaluation Framework to Assess the Consequences of Polyculture in Fish Production: An Application for Pikeperch in Recirculated Aquaculture Systems*, September 24, 2022. www.sciencedirect.com/science/article/pii/S2352513422003453

NodeMCU. *NodeMCU Is a Low-Cost Open Source IoT Platform*, 2023. https://en.wikipedia.org/wiki/NodeMCU

Normanyo E, Danquah IA. *Application of Solar Powered Electronic Notice Board to Blasting Schedules in Mining Operations* (vol. 4, p. 2), March 2020. http://www2.umat.edu.gh/gjt/index.php/gjt/article/view/24

Pawan Sai R, Sunil MP. *Non-Invasive Heart Rate Measurement on Wrist Using IR LED with IoT Sync to Web Server* (vol. 104), October 2, 2018. https://link.springer.com/chapter/10.1007/978-981-13-1921-1_7

Rani S, Bhambri P, Kataria A, Khang A, Sivaraman AK. *Big Data, Cloud Computing and IoT: Tools and Applications* (1st Ed.). Chapman and Hall/CRC Press, 2023. https://doi.org/10.1201/9781003298335

Rani S, Bhambri P, Kataria A, Khang A. "Smart City Ecosystem: Concept, Sustainability, Design Principles and Technologies," *AI-Centric Smart City Ecosystems: Technologies, Design and Implementation* (1st Ed.). CRC Press, 2022. https://doi.org/10.1201/9781003252542-1

Rani S, Chauhan M, Kataria A, Khang A (Eds.). "IoT Equipped Intelligent Distributed Framework for Smart Healthcare Systems," *Networking and Internet Architecture*. CRC Press, 2021. https://doi.org/10.48550/arXiv.2110.04997

Robert H, Kikut A, Jesch E, Woko C, Siegel L, Kim K. "Association of COVID-19 Misinformation with Face Mask Wearing and Social Distancing in a Nationally Representative US Sample," *Health Communication*, November 22, 2020. https://doi.org/10.1080/10410236.2020.1847437

Sandeep SS, Hariharan U, Rajkumar K. *Multimodal Biometric Authentication System Using Deep Learning Method*, March 12–14, 2020. https://ieeexplore.ieee.org/abstract/document/9167512/

Sharma R, Singh G, Sharma S. "Modelling Internet Banking Adoption in Fiji: A Developing Country Perspective," *International Journal of Information Management*, March 12, 2020. https://doi.org/10.1016/j.ijinfomgt.2020.102116

Sharma S, Singh G, Sharma R, Jones P, Kraus S, Dwivedi YK. *Digital Health Innovation: Exploring Adoption of COVID-19 Digital Contact Tracing Apps*, September 15, 2020. https://ieeexplore.ieee.org/abstract/document/9198147/

Valerie P, Fleck B, Neumann C. *Temporal Coherence Properties of Laser Modules Used in Headlamps Determined by a Michelson Interferometer*, November 23, 2020. www.degruyter.com/document/doi/10.1515/aot-2020-0039/html

14 Automotive Internet of Things

Accelerating the Automobile Industry's Long-Term Sustainability in a Smart City Development Strategy

Vrushank Shah, Suketu Jani, and Alex Khang

14.1 INTRODUCTION

The Internet of Things (IoT) enables an object to connect to the internet so that it can be tracked and observed. The IoT is a network that enables the connection of numerous devices and can be used to receive, collect, and exchange data. In order to make running the tool simple we only require internet connectivity. As a result, the system needs a connection between users and hardware, one of which is the IoT platform.

Automotive IoT describes the incorporation of sensors, devices, clouds, and applications into cars, as well as their use as a sophisticated system for fleet management, predictive maintenance, and other functions. Cars now have "near-artificial intelligence" thanks to IoT solutions (Rana et al., 2021). Automotive industry leaders can use IoT solutions to improve vehicle efficiency and reduce the environmental impact of their products. Incorporating IoT monitoring systems increases vehicle efficiencies and enhances precision around maintenance, driver behavior, and navigation (Subhashini et al., 2024).

Today's cars are already connected and have been connected for some time, since they can link to smartphones, offer emergency roadside assistance, register real-time traffic alerts, and more (Khang et al., 2023a). The automobile industry is on the brink of a revolution, moving to the self-driving automobile industry, and the driving force behind this is the fast-developing technology, the IoT.

The IoT will transform the automobile industry, and at the same time, the automobile industry will provide a big boost to the IoT. The potential and prospects of this technology are astonishing. A fleet management system is a scientific approach that enables businesses that rely on logistics business to eliminate or reduce the risks associated with vehicle investment, improving productivity and efficiency and lowering the overall logistics cost and staff cost.

Fleet management system includes a wide range of functions, including supply chain management, driver management, fuel management, driver behavior, vehicle maintenance, and vehicle telematics, among others. With the help of sensors and a variety of technologies, including the IoT, cloud computing, computer vision, machine learning, deep learning, and embedded system, the proposed fleet management system can be set up (Hahanov et al., 2022).

Vehicle-to-vehicle communication technology—commonly described as V2V—is a smart technology that enables vehicle data to be exchanged from one vehicle to another. Communication for V2V technology is based on dedicated short-range communications (Demba et al., 2018).

V2V communication enables motor vehicles to access information about the speed and position of other V2V-enabled vehicles surrounding it using a wireless communication protocol similar to that of Wi-Fi. Those data are then used to alert drivers of potential dangers, helping reduce accidents and traffic congestion. The wireless transfer of data between vehicles and the road infrastructure is known as vehicle-to-infrastructure (V2I) communication. V2I communication is typically wireless and bidirectional, made possible by a system of hardware, software, and firmware. Infrastructure elements like lane markings, traffic lights, and road signs can wirelessly provide information to the vehicle and vice versa.

Timely information can be used to enable a variety of safety, mobility, and environmental benefits due to the volume of data being collected and shared (Bhambri et al., 2022). Vehicle-to-everything, or V2X, is a term that refers to both V2V and V2I technology. Every vehicle on the road is made smarter and safer by V2X technology because it gives them the ability to "communicate" with the infrastructure and other vehicles in the traffic system.

Drivers can receive alerts from V2X about hazardous weather conditions, nearby accidents, traffic jams, and other dangerous behaviors. In order to speed up the driver's response time, V2X transmits a lot of the information that humans can access directly to the car or truck (Jayashree et al., 2024). By automating toll and parking payments, V2X also simplifies the act of driving. Based on the aforementioned areas, the long-term sustainability of the automotive industry is discussed in this chapter.

14.2   CAR ECOSYSTEM

The IoT has a big impact on telecom service providers, software companies, and automakers. To construct the connected car, they are all working together. A connected car is one that can communicate with other cars and other gadgets using internet access while driving. A connected car is one that uses its onboard sensors and internet connectivity to improve its passengers' experiences while driving (Rani et al., 2021).

To provide online music and navigation, the IoT enables in-car entertainment systems to interface with other devices both within and outside the vehicle (Menon et al. 2022; Kumar et al., 2022). A change in the industry will result from Google-supported Android infotainment systems. The Play Store, Google Maps, and Google Assistant apps may all be accessed through in-car infotainment systems.

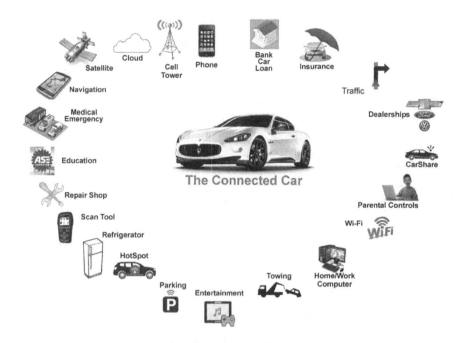

FIGURE 14.1 Connected car features and applications.

Currently, only a small percentage of cars are internet-ready, but the market for connected cars as shown in Figure 14.1 will undoubtedly experience exponential growth due to the demands of the current generation and their exposure to smartphones, and it is expected that the percentage will more than double within the next few years (Bajaj et al., 2018).

The IoT is experiencing a significant change in thinking due to the introduction of brand-new technologies that have the power to significantly improve car features and applications (Ghosh et al., 2022). The IoT offers high-speed connection and control over longer distances with increased reliability and security with the introduction of 5G and 6G (Khang et al., 2022a). The automobile industry has seen the most astounding developments, even though IoT has already demonstrated its utility in the current environment (Luke et al., 2024).

Self-driving vehicles are already on the road, and federated learning, artificial intelligence (AI), and machine learning are helping improve the navigation system over time (Khang et al., 2024). The purpose of this chapter is to provide a clear and brief overview of the Intelligent IoT and its uses in the automotive sector (Khanh and Khang, 2021).

14.3 INFOTAINMENT

Infotainment refers to a system in automobiles that provides both information and entertainment. An infotainment system's typical features include providing navigation while driving, manipulating audio/visual entertainment content, delivering rear-seat entertainment, and connecting with smartphones for a hands-free experience via voice commands. Infotainment options should assist drivers in keeping their eyes on the road and their hands on the wheel, resulting in a safer car experience (Pančík et al., 2019).

To reduce driver distraction, infotainment options such as applications and menus must be neatly structured, and accessing features must be very straightforward. The in-car application platform must also be supported by speech and audio commands, and drivers must be able to use them.

To navigate menus or compose messages using voice commands users' vocal commands should be the main method of acceptance. With the rapid development of smartphones and cloud technology, consumers are demanding more (Khang et al., 2022b). For internet radio and live-streaming music, concepts for cutting-edge infotainment features combine users with the newest generation of cloud-based infotainment systems. There are numerous smartphone integration devices, such as Car Play, Google Projected Mode, Mirror Link, and others (Rani et al., 2022).

14.4 AUTOMOTIVE COMMUNICATION

A completely new paradigm of vehicle communications is about to take shape as more and more connected cars appear on the road and in-vehicle integrated connection becomes standard.

14.4.1 V2V

To prevent accidents and provide the chance to dramatically increase safety, V2V communication uses a wireless network in which the vehicles broadcast speed and location data to the vehicles nearby (Weiß et al., 2011; Iskandarani et al., 2021).

As seen in Figure 14.2, each vehicle continuously broadcasts a message to the neighboring vehicles through an ad hoc mesh network with the speed and position data. Cars are able to communicate with other vehicles using dedicated short-range communication (DSRC) technology. A vehicular ad hoc network (VANET), a wireless ad hoc network for cars, is used by V2V as a means of data exchange.

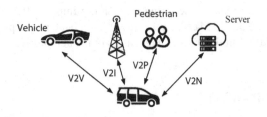

FIGURE 14.2 Automotive communication.

14.4.2 V2I

V2I communication will enhance autonomous vehicles' ability to communicate online in a 5G environment. In order to increase road capacity, V2I communication is utilized to control the platooning of cars on metropolitan roadways (Liu et al., 2021). Quality of service (QoS) factors like delay and interference are used to measure how well V2I communication occurs. The complexity of the environment increases the difficulty of the interference analysis of wireless communications (Geraets, 2016).

14.4.3 V2X

A wide range of further safety, mobility, and environmental advantages are provided by V2X communication, which is the wireless transmission of information between cars and roadside infrastructure (Montanaro, 2019). Through intelligent safety applications, the vehicles can communicate with the roadways, digital signage, traffic signals, safety, and control systems to prevent collisions and traffic jams (Kotte et al., 2017).

Although the idea of connected automobiles, talking cars, and driverless cars has been around for a while, research and development are currently ongoing. The advancement in V2V, V2I, and V2X enhanced road safety and the comfortability of driving, maximized fuel efficiency, and reduced traffic congestion and accidents, which are among the crucial performance issues (Abou et al., 2022).

14.5 FLEET MANAGEMENT SYSTEM

A fleet management system is a scientific approach that enables the companies that rely on logistics business to eliminate or reduce the risks associated with vehicle investment, improving productivity and efficiency, and lowering the overall logistics cost and staff cost (Khang et al., 2023b).

While some goods and services can be used alone, the best performance comes from a comprehensive system that combines and gathers data from many functions. Vehicle tracking systems offer a variety of information about driver behaviors, geolocation, fuel usage, and engine problems (Khang et al., 2023c).

Fleet data include information on upkeep, fuel use, transactions, individual vehicle documents like registration certificates and travel permits, supply chain information like vehicle identification numbers and warranties, vehicle identifying information, driver-centric information like safety training completion rates and acceptance of company policies, as well as demographic information on job types. The more systems and data points a fleet management system integrates, the more specialized functions it can perform (Jaiswal et al., 2023).

14.6 CONCLUSION

The IoT advances in the automotive industry are discussed in this chapter, including connected car services and applications, vehicle communications, and fleet management in intelligent transportation (Rani et al., 2023).

As vehicles become more intelligent and interconnected with other vehicles, smartphones, and other objects, appropriate analytical processing can be applied to operational parameters to give original equipment manufacturers (OEM), drivers, and road safety authorities better insights into the performance of the vehicle as well as the overall traffic situation, enabling them to take timely action (Khang et al., 2022c).

The previously mentioned IoT-based function was required for the next-generation vehicle and would increase the sustainability of the automotive industry. The aforementioned IoT-based function will improve the capabilities of next-generation vehicles and increase the automotive industry's long-term viability (Hajimahmud et al., 2023).

REFERENCES

Abou El Hassan A, Kerrakchou I, El Mehdi A, Saber M. "Road Safety Enhancement of Intelligent Transportation Systems: From Cellular LTE-V2X toward 5G-V2X," in *Digital Technologies and Applications* (pp. 745–754), 2022. https://link.springer.com/chapter/10.1007/978-3-031-02447-4_77

Bajaj RK, Rao M, Agrawal H. "Internet of Things (IoT) In the Smart Automotive Sector: A Review," *Internet Things*, 9, 2018. https://ieeexplore.ieee.org/abstract/document/10047604/

Bhambri P, Rani S, Gupta G, Khang A. *Cloud and Fog Computing Platforms for Internet of Things*. CRC Press, 2022. https://doi.org/10.1201/9781003213888

Demba A, Moller DPF. "Vehicle-to-Vehicle Communication Technology," *2018 IEEE International Conference on Electro/Information Technology (EIT), Rochester, MI*, 0459–0464, May 2018. https://doi.org/10.1109/EIT.2018.8500189

Geraets M. "V2V and V2I Communications—From Vision to Reality," *Energy Consumption and Autonomous Driving* (pp. 33–35), 2016. https://link.springer.com/chapter/10.1007/978-3-319-19818-7_4

Ghosh RK, Banerjee A, Aich P, Basu D, Ghosh U. "Intelligent IoT for Automotive Industry 4.0: Challenges, Opportunities, and Future Trends," Ghosh U, Chakraborty C, Garg L, Srivastava G (Eds.) *Intelligent Internet of Things for Healthcare and Industry* (pp. 327–352). Springer International Publishing, 2022. https://doi.org/10.1007/978-3-030-81473-1_16

Hahanov V, Khang A, Litvinova E, Chumachenko S, Hajimahmud VA, Alyar AV. "The Key Assistant of Smart City—Sensors and Tools," *AI-Centric Smart City Ecosystems: Technologies, Design and Implementation* (1st Ed.). CRC Press, 2022. https://doi.org/10.1201/9781003252542-17

Hajimahmud VA. et al. (Eds.). "The Role of Data in Business and Production," *AI-Aided IoT Technologies and Applications in the Smart Business and Production*. CRC Press, 2023. https://doi.org/10.1201/9781003392224-2

Iskandarani MZ. "Prediction of Road Congestion Through Application of Neural Networks and Correlative Algorithm to V2V Communication (NN-CA-V2V)," *Intelligent Computing* (pp. 1203–1221), 2021. https://link.springer.com/chapter/10.1007/978-3-030-80126-7_84

Jaiswal N, Misra A, Misra PK, Khang A (Eds.). "Role of the Internet of Things (IoT) Technologies in Business and Production," *AI-aided IoT Technologies and Applications in the Smart Business and Production*. CRC Press, 2023. https://doi.org/10.1201/9781003392224-1

Jayashree M. et al. (Eds.). "Vehicle and Passenger Identification in Public Transportation to Fortify Smart City Indices," *Smart Cities: IoT Technologies, Big Data Solutions, Cloud Platforms, and Cybersecurity Techniques*. CRC Press, 2024. https://doi.org/10.1201/9781003376064-13

Khang A, Abdullayev V, Hahanov V, Shah V. *Advanced IoT Technologies and Applications in the Industry 4.0 Digital Economy* (1st Ed.). CRC Press, 2024. https://doi.org/10.1201/9781003434269

Khang A, Gupta SK, Rani S, Karras DA. *Smart Cities: IoT Technologies, Big Data Solutions, Cloud Platforms, and Cybersecurity Techniques* (1st Ed.). CRC Press, 2023a. https://doi.org/10.1201/9781003376064

Khang A, Gupta SK, Shah V, Misra A (Eds.). *AI-Aided IoT Technologies and Applications in the Smart Business and Production.* CRC Press, 2023c. https://doi.org/10.1201/9781003392224

Khang A, Hahanov V, Abbas GL, Hajimahmud VA. "Cyber-Physical-Social System and Incident Management," *AI-Centric Smart City Ecosystems: Technologies, Design and Implementation* (1st Ed.). CRC Press, 2022c. https://doi.org/10.1201/9781003252542-2

Khang A, Ragimova NA, Hajimahmud VA, Alyar AV. "Advanced Technologies and Data Management in the Smart Healthcare System," *AI-Centric Smart City Ecosystems: Technologies, Design and Implementation* (1st Ed.). CRC Press, 2022b. https://doi.org/10.1201/9781003252542-16

Khang A, Rani S, Sivaraman AK. *AI-Centric Smart City Ecosystems: Technologies, Design and Implementation* (1st Ed.). CRC Press, 2022a. https://doi.org/10.1201/9781003252542

Khang A, Shah V, Rani S. *AI-Based Technologies and Applications in the Era of the Metaverse* (1st Ed.) IGI Global Press, 2023b. https://doi.org/10.4018/9781668488515

Khanh HH, Khang A. "The Role of Artificial Intelligence in Blockchain Applications," *Reinventing Manufacturing and Business Processes Through Artificial Intelligence* (pp. 20–40), CRC Press, 2021. https://doi.org/10.1201/9781003145011-2

Kotte J, Schmeichel C, Zlocki A, Gathmann H, Eckstein L. "Concept of an Enhanced V2X Pedestrian Collision Avoidance System with a Cost Function–Based Pedestrian Model," *Traffic Injury Prevention*, 18(sup1), S37–S43, May 2017. https://doi.org/10.1080/15389588.2017.1310380

Kumar A, Akhtar MAK, Pandey A. "Design of Internet of Things (IoT) System Based Smart City Model on Raspberry Pi," *IETE Journal of Research*, 1–8, July 2022. https://doi.org/10.1080/03772063.2022.2088629

Liu P, Fan WD. "Exploring the Impact of Connected and Autonomous Vehicles on Mobility and Environment at Signalized Intersections Through Vehicle-To-Infrastructure (V2I) and Infrastructure-To-Vehicle (I2V) Communications," *Transportation Planning and Technology*, 44(2), 129–138, February 2021. https://doi.org/10.1080/03081060.2020.1868088

Luke J, Khang A. "Smart City Concepts, Models, Technologies and Applications," Chandrasekar V, Pravin AR, Sriram K (Eds.) *Smart Cities: IoT Technologies, Big Data Solutions, Cloud Platforms, and Cybersecurity Techniques* (1st Ed.). CRC Press, 2024. https://doi.org/10.1201/9781003376064-1

Menon VG, Jacob S, Joseph S, Sehdev P, Khosravi MR, Al-Turjman F. "An IoT-Enabled Intelligent Automobile System for Smart Cities," *Internet Things*, 18, 100213, May 2022. https://doi.org/10.1016/j.IoT.2020.100213

Montanaro U. et al. "Towards Connected Autonomous Driving: Review of Use-Cases," *Vehicle System Dynamics*, 57(6), 779–814, June 2019. https://doi.org/10.1080/00423114.2018.1492142

Pančík J, Beneš V. "IoT Challenge: Older Test Machines Modernization in an Automotive Plant," Cagáňová D, Balog M, Knapčíková L, Soviar J, Mezarcıöz S (Eds.) *Smart Technology Trends in Industrial and Business Management* (pp. 85–100). Springer International Publishing, 2019. https://doi.org/10.1007/978-3-319-76998-1_7

Rana G, Khang A, Sharma R, Goel AK, Dubey AK (Eds.). *Reinventing Manufacturing and Business Processes Through Artificial Intelligence.* CRC Press, 2021. https://doi.org/10.1201/9781003145011

Rani S, Bhambri P, Kataria A, Khang A. "Smart City Ecosystem: Concept, Sustainability, Design Principles and Technologies," *AI-Centric Smart City Ecosystems: Technologies, Design and Implementation* (1st Ed.). CRC Press, 2022. https://doi.org/10.1201/9781003252542-1

Rani S, Bhambri P, Kataria A, Khang A, Sivaraman AK. *Big Data, Cloud Computing and IoT: Tools and Applications* (1st Ed.). Chapman and Hall/CRC Press, 2023. https://doi.org/10.1201/9781003298335

Rani S, Chauhan M, Kataria A, Khang A (Eds.). "IoT Equipped Intelligent Distributed Framework for Smart Healthcare Systems," *Networking and Internet Architecture*. CRC Press, 2021. https://doi.org/10.48550/arXiv.2110.04997

Subhashini R, Khang, A. "The Role of Internet of Things (IoT) in Smart City Framework," *Smart Cities: IoT Technologies, Big Data Solutions, Cloud Platforms, and Cybersecurity Techniques* (1st Ed.). CRC Press, 2024. https://doi.org/10.1201/9781003376064-3

Weiß C. "V2X Communication in Europe—From Research Projects Towards Standardization and Field Testing of Vehicle Communication Technology," *Computer Networks*, 55(14), 3103–3119, October 2011. https://doi.org/10.1016/j.comnet.2011.03.016

15 A Bibliometric Analysis of the Internet of Things and Insurance Technology

Durga Prasad Singh Samanta, B. C. M. Patnaik, and Ipseeta Satpathy

15.1 INTRODUCTION

The Internet of Things (IoT) and insurance technology (InsurTech) are two emerging technologies that have the potential to revolutionize industries such as insurance. The IoT refers to the network of physical objects, devices, vehicles, and other items that are embedded with sensors (Hahanov et al., 2022), software, and other technologies to enable them to connect and exchange data with each other and with the internet (Subhashini et al., 2024). InsurTech, by comparison, refers to the use of technology to innovate and improve insurance products and services, ranging from underwriting and claims processing to customer engagement and risk management.

In this bibliometric analysis, we aim to understand the current state of research on these topics and identify key trends and areas for future research. Using a combination of online databases and manual searches, we identified and analyzed a sample of relevant research articles published in the past decade (Hussain et al., 2022). Our analysis revealed that there has been a steady increase in the number of publications on IoT and InsurTech over the years, indicating the growing interest and importance of these technologies in the insurance industry.

Specifically, we found that there has been a particular focus on topics such as risk assessment, customer engagement, and data analytics. One key trend that emerged from our analysis is the integration of IoT and InsurTech with other technologies such as blockchain and artificial intelligence (AI) (Khanh and Khang, 2021).

For example, some studies have explored the use of blockchain for secure data sharing and storage, while others have investigated the use of AI for fraud detection and claims processing. These integrations have the potential to enhance the efficiency and effectiveness of insurance operations and improve customer satisfaction (Rana et al., 2021). However, our analysis also identified several challenges and areas for future research. One key challenge is the ethical and regulatory considerations surrounding the use of the IoT and InsurTech in the insurance industry.

For example, the use of personal data collected from IoT devices raises concerns about privacy and security, while the use of AI for decision-making raises questions about fairness and bias (Rani et al., 2021). Another challenge is the need for insurers to adapt to the changing landscape of IoT and InsurTech. Insurers must be able to leverage these technologies to remain competitive in the market and provide value to their customers. This requires investment in new technologies and training of personnel, as well as collaboration with other stakeholders such as technology providers and regulators (Shah et al., 2024).

15.2 PROBLEM STATEMENT

In recent years, the advancement of technology has rapidly transformed various industries and the insurance sector is no exception.

With the emergence of the IoT and InsurTech, developed nations across the world have been exploring innovative ways to improve insurance operations and enhance customer experience (Shah et al., 2023).

This literature review aims to examine the latest studies conducted globally on IoT and InsurTech in the insurance industry, shedding light on the current trends, challenges, and opportunities in this rapidly evolving field (Khang et al., 2022a).

The following tabulation provides an overview of the various developed countries, year of the publication, authors, methodology used, and findings of the studies included in this review (Khang et al., 2023a) as shown in Table 15.1 for the United Kingdom, Table 15.2 for the United States of America, Table 15.3 for Israel, Table 15.4 for Japan, Table 15.5 for China, Table 15.6 for India, and Table 15.7 for Germany.

TABLE 15.1

Research on the Internet of Things (IoT) and Insurance Technology (InsurTech) in the United Kingdom.

Authors	Methodology	Findings
Ayeni et al. (2020)	A survey was conducted among insurance professionals in the United Kingdom to collect data on their perception of the impact of IoT on the insurance industry.	IoT has a significant impact on the insurance industry in the United Kingdom, particularly in terms of risk management and pricing
Roulstone et al. (2017)	A literature review was conducted to explore the concept of InsurTech and its potential impact on the insurance industry.	InsurTech has the potential to disrupt the traditional insurance industry in the United Kingdom and globally by offering new products, services, and business models.
Meijer et al. (2020).	A case study approach was used to explore the potential of InsurTech to enhance customer engagement in the insurance industry in the United Kingdom.	InsurTech has the potential to improve customer engagement in the insurance industry by offering personalized and real-time insurance solutions.
Ojo et al. (2018)	A legal and regulatory analysis was conducted to explore the regulatory challenges and opportunities of IoT in the insurance sector.	The regulatory framework is not yet fully equipped to deal with the regulatory challenges of IoT in the insurance sector and proposed several recommendations to address this issue.

TABLE 15.2

Research on the Internet of Things (IoT) and Insurance Technology (InsurTech) in the United States

Authors	Methodology	Findings
Hu et al. (2019)	A systematic review of the literature was conducted to analyze the impact of IoT on the insurance industry in the United States and globally.	The study found that IoT has a significant impact on the insurance industry, particularly in terms of risk assessment, personalized pricing, and claims management.
Fleischman et al. (2018)	A literature review was conducted to explore the concept of InsurTech and its impact on the insurance industry in the United States.	The study found that InsurTech has the potential to disrupt the traditional insurance industry in the United States by offering new products, services, and business models.
Wang et al. (2020)	A quantitative analysis was conducted to explore the impact of IoT on insurance risk assessment and management in the United States.	The study found that IoT has a significant impact on insurance risk assessment and management in the United States, particularly in terms of data collection and analysis and risk prediction.
Lee et al. (2021)	A survey was conducted among insurance customers in the United States to explore the role of InsurTech in improving customer satisfaction.	The study found that InsurTech has a significant role in improving customer satisfaction in the insurance industry in the United States by offering personalized and real-time insurance solutions.

TABLE 15.3

Research on the Internet of Things (IoT) and Insurance Technology (InsurTech) in Israel

Authors	Methodology	Findings
Karkalakos et al. (2020)	A systematic literature review was conducted to explore the applications of the IoT and InsurTech in the insurance industry in Israel and globally	The study found that the IoT and InsurTech have a significant impact on the insurance industry, particularly in terms of risk assessment, personalized pricing, and claims management.
Chen et al. (2019)	A case study approach was used to explore the applications of blockchain and the IoT in the insurance industry in Israel and globally.	The study found that blockchain and the IoT have the potential to transform the insurance industry in Israel and globally by offering new products, services, and business models.
Gertner et al. (2018).	A qualitative analysis was conducted to explore the future of InsurTech in Israel.	The study found that Israel has a vibrant InsurTech ecosystem with a growing number of start-ups and innovation centers and has the potential to become a global hub for InsurTech innovation.
Lavie et al. (2019)	A survey was conducted among insurance professionals in Israel to explore the opportunities and challenges of InsurTech and the IoT in the insurance industry.	The study found that InsurTech and IoT have the potential to transform the insurance industry in Israel by offering new products, services, and business models, but it also faces challenges related to data privacy and cybersecurity.

TABLE 15.4
Research on the Internet of Things (IoT) and Insurance Technology (InsurTech) in Japan

Authors	Methodology	Findings
Li et al. (2019)	A systematic review of the literature was conducted to analyze the impact of the IoT on the insurance industry in Japan and globally.	The study found that the IoT has a significant impact on the insurance industry in Japan, particularly in terms of risk assessment, personalized pricing, and claims management.
Iino et al. (2018)	A literature review was conducted to explore the concept of InsurTech and its impact on the insurance industry in Japan.	The study found that InsurTech has the potential to not only transform the traditional insurance industry in Japan by offering new products, services, and business models, but it also faces challenges related to regulations and customer trust.
Inoue et al. (2020)	A case study approach was used to explore the role of the IoT in the transformation of the insurance industry in Japan.	The study found that the IoT has the potential to transform the insurance industry in Japan by offering new data sources for risk assessment, improving claims management, and enabling new insurance products and services.
Yamada et al. (2018)	A qualitative analysis was conducted to explore the InsurTech ecosystem in Japan and its opportunities for insurers and start-ups.	The study found that Japan has a growing InsurTech ecosystem with a high level of collaboration between insurers and start-ups and offers new opportunities for innovation in the insurance industry.

TABLE 15.5
Research on the Internet of Things (IoT) and Insurance Technology (InsurTech) in China

Authors	Methodology	Findings
Wang et al. (2020)	A survey was conducted among insurance professionals in China to explore the impact of the IoT on the insurance industry.	The study found that the IoT has a significant impact on the insurance industry in China, particularly in terms of risk assessment, personalized pricing, and claims management.
Ding et al. (2019)	A survey was conducted among insurance professionals in China to explore the innovation of the IoT and InsurTech in the insurance industry.	The study found that the IoT and InsurTech have the potential to not only transform the insurance industry in China by offering new products, services, and business models but also face challenges related to regulations and data privacy.
Xiang et al. (2020)	A literature review was conducted to analyze the impact of the IoT on the insurance industry in China and globally.	The study found that the IoT has a significant impact on the insurance industry in China, particularly in terms of risk assessment, personalized pricing, and claims management.
Zhang et al. (2021)	A literature review was conducted to provide an overview of the current state and future prospects of InsurTech in China.	The study found that China has a vibrant InsurTech ecosystem with a growing number of start-ups and innovation centers and has the potential to become a global leader in InsurTech innovation.

TABLE 15.6

Research on the Internet of Things (IoT) and Insurance Technology (InsurTech) in India

Authors	Methodology	Findings
Bhandari et al. (2020)	A literature review was conducted to explore the role of the IoT in the Indian insurance industry.	The study found that the IoT has the potential to transform the Indian insurance industry by offering new data sources for risk assessment, improving claims management, and enabling new insurance products and services.
Gupta et al. (2019)	A literature review was conducted to explore the concept of InsurTech and its impact on the insurance industry in India.	The study found that InsurTech has the potential to not only transform the traditional insurance industry in India by offering new products, services, and business models, but it also faces challenges related to regulations and customer trust.
Shashikiran et al. (2019)	A survey was conducted among insurance professionals in India to explore the current scenario and future prospects of the IoT in the insurance industry.	The study found that the IoT is still in its early stages of adoption in the Indian insurance industry, but has the potential to improve risk assessment, claims management, and customer engagement.
Verma et al. (2019)	A survey was conducted among insurance professionals in India to explore the role of InsurTech in the growth of the insurance industry.	The study found that InsurTech has the potential to enhance the growth of the Indian insurance industry by improving operational efficiency, customer engagement, and innovation. However, challenges related to data privacy and regulations need to be addressed.

15.3 RESEARCH METHODOLOGY

In March 2022, searches were conducted in the Scopus database using the search terms "(TITLE (internet of things* OR IoT OR insutech OR insurance technology))" and "(LIMIT-TO (DOCTYPE,"article" OR "research paper"))," with no other restrictions applied except for limiting the search to publication types "article" and "review".

The search aimed to gather information on the volume and annual trends of publications, authors, journals, open access, document types, language, and country of publication, author affiliations, funding sources, highly cited publications, and top publishing authors. The collected papers' trends were identified and noted, and bibliometric networks were constructed using VOSviewer to analyze the data. To ensure accuracy, the results were exported on the same day of the search to avoid any errors from database updates as shown in Figure 15.1.

The utilization of bibliometric analysis will enhance the understanding of the limitations of the three search terms, specifically in the context of the IoT and InsurTech. Through the examination of a relevant subset of articles, a subsequent analysis of the bibliographic network revealed valuable insights on the citation network, co-occurrence network, and fundamental data such as nations and documents.

While the first qualitative evaluation relies on researchers' opinions and explanatory methods, the bibliometric study offers objective insights using quantitative and

TABLE 15.7

Research on the Internet of Things (IoT) and Insurance Technology (InsurTech) in Germany

Authors	Methodology	Findings
Kimberly et al. (2022)	A conceptual analysis was conducted to explore the potential impact of the IoT on the insurance industry in Germany.	The study found that the IoT has the potential to transform the insurance industry in Germany by offering new data sources for risk assessment, improving claims management, and enabling new insurance products and services.
Canhoto and Arp (2017)	A case study approach was used to explore the opportunities and challenges of the IoT for the insurance industry in Germany.	The study found that the IoT not only offers new opportunities for the insurance industry in Germany, such as improving risk assessment and customer engagement, but it also faces challenges related to data privacy and cybersecurity.
Radwan (2019)	A survey was conducted among insurance professionals in Germany to explore the impact of digitalization on the insurance industry.	The study found that digitalization, including the IoT and InsurTech, not only has the potential to transform the insurance industry in Germany by offering new products, services, and business models, but it also faces challenges related to regulations and customer trust.
Shetty et al. (2022)	A comparative analysis was conducted to explore the adoption and impact of the IoT in the insurance industry in Japan and Germany.	The study found that while IoT is still in its early stages of adoption in the insurance industry in both Japan and Germany, there are differences in the perception and utilization of the IoT between the two countries.

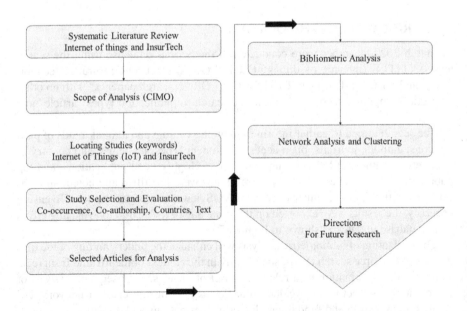

FIGURE 15.1   Flow of the study is represented in a graphical format.

statistical data. Alyev et al. (2019) suggest that bibliometric approaches examine notable author names, journal names, article titles, article keywords, and publication years. Per Block and Fisch (2020), bibliometric data analysis includes the examination of these elements.

15.4 OBJECTIVES OF THE STUDY

The study mentioned in this text aims to conduct a systematic literature network analysis to analyze the development of the field of the IoT and InsurTech over time.

The analysis involves determining the number of published documents on the selected keywords, recognizing the authors, organizations, and countries that have contributed to the field, and emphasizing interdisciplinary connections to track information movement within and between fields. The use of specific keywords and exclusion/inclusion criteria, along with stringent search protocols, ensures a systematic approach to reduce the risk of researchers' selection biases (Pallavi et al., 2023). Additionally, quantitative bibliographic investigations are used to avoid biases and allow for repeatable analyses (Jaiswal et al., 2023).

The study offers a broader and more up-to-date scope, leading to evidence-based conclusions and minimizing the likelihood of authors presenting a subjective argument. Figure 15.1 provides a detailed description of each phase of the process, and the subsequent section presents a scientific literature review of IoT and InsurTech topics.

15.5 DATA ANALYSIS

This study utilized VOSviewer to analyze citation maps created from Scopus database files initially in CSV format. The analysis allowed for the identification of co-authorship, keyword co-occurrence, citation bibliography coupling, and citation map based on the bibliography data (Hajimahmud et al., 2023).

The systematic literature network analysis (SLNA) was performed using the search terms "IoT" and "INSURTECH" in the Scopus database to determine the number of papers and authors discussing artificial intelligence in this field. The primary objective of this study was to investigate the contributions related to the IoT in InsurTech, including the authors, organizations, and countries involved.

15.5.1 AUTHORS' NETWORK ANALYSIS (VOS CLUSTERING)

The analysis employed in this study was co-authorship, where the authors were fully counted, and a maximum of one author per document was considered. A minimum document count of 1 was applied to identify authors who had significant contributions as Figure 15.2.

15.5.2 TEXT ANALYSIS (VOS CLUSTERING)

A threshold of a minimum of 1 keyword occurrence was set, resulting in only 129 keywords meeting the criteria out of the initial 446 keywords. To investigate the

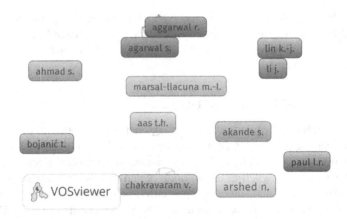

FIGURE 15.2 Visualization of clusters that represent the link strength between the co-authors regarding co-occurrence and relativity in research.

Source: Authors' own analysis from VOSviewer software version 1.6.18 and Scopus, www.scopus.com.

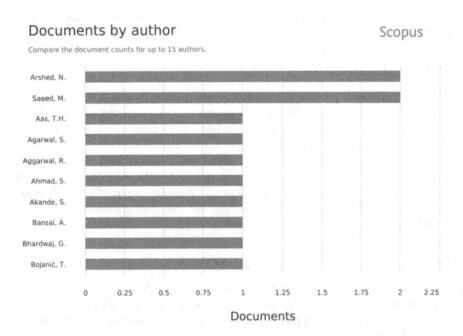

FIGURE 15.3 Documents by top ten authors having more than or equal to two publications represented in a radar chart.

Source: Scopus, www.scopus.com.

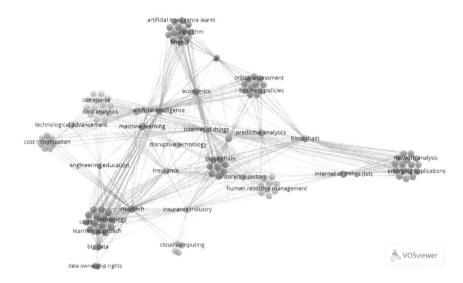

FIGURE 15.4 Co-occurrence author keywords network.

Source: Authors' own analysis from VOSviewer software version 1.6.18 and Scopus, www.scopus.com.

most common terms in the literature and to identify potential research directions within each cluster, the following aspects were explored using the VOSviewer visualization results as Figure 15.4.

15.5.3 COUNTRIES NETWORK ANALYSIS (VOS CLUSTERING)

The abundance of literature on IoT and its use in the fields of InsurTech and insurance is evident (Bhambri et al., 2022) as shown in Figure 15.5.

There are several academic scholars who have three publications, such as India and the United Kingdom; who have two publications, such as China, the United Arab Emirates, and the United States; and who have one publication, such as Norway, Pakistan, Serbia, and Spain; these countries have made significant contributions to the study of the IoT in insurance, particularly in the InsurTech as Figure 15.6.

India is ranked first, followed by the United Kingdom in second place and China in third place in terms of the number of documents related to research on IoT in InsurTech as shown in Table 15.8.

In this regard, developed countries continue to have a significant influence on how researchers from developing countries view the IoT and InsurTech in general, how they are supported and funded by their governments, or how they can work toward achieving their goals of contributing to the insurance industries as a whole (Khang et al., 2023c).

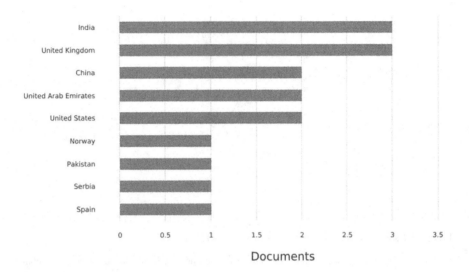

FIGURE 15.5 Documents by countries.

Source: Scopus, www.scopus.com.

FIGURE 15.6 Documents by countries grouped by link strength considering the co-occurrence and co-authorship and clusters have been colored on the basis of publication year.

Source: Authors' own analysis from Google Sheets.

TABLE 15.8
Documents by Country

Country	Documents
India	3
United Kingdom	3
China	2
United Arab Emirates	2
United States	2
Norway	1
Pakistan	1
Serbia	1

15.6 LIMITATIONS OF THE STUDY

The study focused only on two keywords, namely, "IoT" or "Internet of Things" and "InsurTech", which were pertinent to the research subject and only 11 publications were found.

However, the authors did not include any other keywords, which could have made the study more exhaustive. To further enhance the study, additional information from physical libraries that are not published online could be included if they were accessible. It is worth noting that the study only relied on Scopus and that other literature databases like Web of Science and PubMed were not included in the study's scope.

15.7 CONCLUSION

In conclusion, our analysis highlights the need for ongoing research and discussion on the potential impact of the IoT and InsurTech on the insurance industry and society at large. As these technologies continue to evolve, it is important to stay abreast of the latest developments and identify strategies for addressing the challenges and opportunities that they present (Khang et al., 2023d).

The integration of the IoT and InsurTech with other emerging technologies, along with ethical and regulatory considerations, will be key areas for future research in this field (Pritiprada et al., 2024). The IoT has revolutionized various industries, including the insurance sector. InsurTech, which refers to the use of technology to enhance insurance processes, has benefited greatly from IoT advancements. One of the most significant contributions of the IoT to InsurTech is the availability of real-time data.

Connected devices, such as telematics sensors in vehicles or smart home devices, collect data that insurers can use to assess risks accurately (Khang et al., 2022b). These data allow insurers to offer personalized policies and premiums, which can help attract and retain customers. Moreover, IoT devices can help insurers prevent or reduce losses by alerting policyholders to potential risks or incidents (Khang et al., 2022c).

For instance, smart sensors in homes can detect water leaks or fires and notify homeowners and insurers in real time. This timely notification can help minimize damages and claims, resulting in lower costs for insurers. Furthermore, the IoT can help streamline insurance processes, including claims management. By automating claim processes and leveraging real-time data, insurers can expedite the claims process, reduce fraud, and provide better customer service (Tailor et al., 2022).

The IoT is of utmost importance in the field of InsurTech. By leveraging the IoT's capabilities, insurers can provide more accurate, personalized policies and reduce costs by preventing or minimizing losses. As the IoT continues to evolve, it will undoubtedly play an even more significant role in the insurance industry (Khang et al., 2023e).

The IoT has revolutionized various industries, including the insurance sector. InsurTech, which refers to the use of technology to enhance insurance processes, has benefited greatly from IoT advancements. One of the most significant contributions of the IoT to InsurTech is the availability of real-time data (Hajimahmud et al., 2022).

Connected devices, such as telematics sensors in vehicles or smart home devices, collect data that insurers can use to assess risks accurately. These data allow insurers to offer personalized policies and premiums, which can help attract and retain customers (Rani et al., 2023).

Moreover, IoT devices can help insurers prevent or reduce losses by alerting policyholders to potential risks or incidents. For instance, smart sensors in homes can detect water leaks or fires and notify homeowners and insurers in real time.

This timely notification can help minimize damages and claims, resulting in lower costs for insurers. Furthermore, the IoT can help streamline insurance processes, including claims management (Shah et al., 2023).

By automating claim processes and leveraging real-time data, insurers can expedite the claims process, reduce fraud, and provide better customer service. In conclusion, the IoT is of utmost importance in the field of InsurTech.

By leveraging the IoT's capabilities, insurers can provide more accurate, personalized policies and reduce costs by preventing or minimizing losses. As the IoT continues to evolve, it will undoubtedly play an even more significant role in the insurance industry.

REFERENCES

Ayeni A, Adedoyin F. "The Impact of IoT on the Insurance Industry in the United Kingdom," *Journal of Insurance and Financial Management*, 6(23), 123–133, 2020. https://ieeexplore.ieee.org/abstract/document/8528677/

Bhambri P, Rani S, Gupta G, Khang A. *Cloud and Fog Computing Platforms for Internet of Things*. CRC Press, 2022. https://doi.org/10.1201/9781003213888

Bhandari R, Kaur P. "Role of the Internet of Things in the Indian Insurance Industry," *Journal of Insurance Issues*, 43(2), 1–24, 2020. https://doi.org/10.1201/9781003145011-2

Block JH, Fisch C. "Eight Tips and Questions for Your Bibliographic Study in Business and Management Research," *Management Review Quarterly*, 70, 307–312, 2020. https://link.springer.com/article/10.1007/s11301-020-00188-4

Canhoto AI, Arp S. "Exploring the Factors That Support Adoption and Sustained Use of Health and Fitness Wearables," *Journal of Marketing Management*, 33(1–2), 32–60, 2017. https://www.tandfonline.com/doi/abs/10.1080/0267257X.2016.1234505

Chen W, Tang J. "InsurTech Ecosystem: Applications of Blockchain and IoT in Insurance," *Technological Forecasting and Social Change*, 138, 221–234, 2019. https://doi.org/10.1201/9781003145011-2

Ding Y, Gao F, Huang D, Li X. "Empirical Study of Internet of Things and Insurance Innovation in China," *Journal of International Technology and Information Management*, 28(1), 15–34, 2019. https://doi.org/10.1201/9781003145011

Fleischman L, Hanson W. "InsurTech and Its Impact on the Insurance Industry," *Journal of Insurance Issues*, 41(1), 1–17, 2018. https://doi.org/10.1201/9781003145011-2

Gertner D, Levin A. "The Future of InsurTech in Israel," *Journal of Insurance Issues*, 41(1), 18–33, 2018. https://doi.org/10.48550/arXiv.2110.04997

Gupta R, Anand A. "InsurTech in India: A Review and Research Agenda," *Journal of Insurance Issues*, 42(1), 1–26, 2019. www.sciencedirect.com/science/article/pii/S2667096822000374

Hahanov V, Khang A, Litvinova E, Chumachenko S, Hajimahmud VA, Alyar AV. "The Key Assistant of Smart City—Sensors and Tools," *AI-Centric Smart City Ecosystems: Technologies, Design and Implementation* (1st Ed.). CRC Press, 2022. https://doi.org/10.1201/9781003252542-17

Hajimahmud VA, Khang A, Hahanov V, Litvinova E, Chumachenko S, Alyar AV. "Autonomous Robots for Smart City: Closer to Augmented Humanity," *AI-Centric Smart City Ecosystems: Technologies, Design and Implementation* (1st Ed.). CRC Press, 2022. https://doi.org/10.1201/9781003252542-7

Hajimahmud VA. et al. (Eds.). "The Role of Data in Business and Production," *AI-Aided IoT Technologies and Applications in the Smart Business and Production*. CRC Press, 2023. https://doi.org/10.1201/9781003392224-2

Hu Y, Chen H. "The Impact of the Internet of Things (IoT) on the Insurance Industry: A Systematic Review," *Technological Forecasting and Social Change*, 140, 341–348, 2019. https://link.springer.com/article/10.1007/s10796-012-9374-9

Hussain SH, Sivakumar TB, Khang A (Eds.). "Cryptocurrency Methodologies and Techniques," *The Data-Driven Blockchain Ecosystem: Fundamentals, Applications, and Emerging Technologies* (1st Ed., pp. 149–164). CRC Press, 2022. https://doi.org/10.1201/9781003269281-2

Iino T, Yabushita M. "InsurTech and Its Impact on the Insurance Industry in Japan," *Journal of Insurance Issues*, 41(1), 34–49, 2018. https://doi.org/10.48550/arXiv.2110.04997

Inoue T, Tamura H. "The Role of IoT in the Transformation of the Japanese Insurance Industry," *Technological Forecasting and Social Change*, 151, 119774, 2020. www.google.com/books?hl=en&lr=&id=peCVEAAAQBAJ&oi=fnd&pg=PA117&d-q=Inoue+The+role+of+IoT+in+the+transformation+of+the+Japanese+insurance+industry.+Technological+Forecasting+and+Social+Change&ots=8lq9sz-k7Ru&sig=FX-ayqsmCaRQKxKrU1_CE-1jGEk

Jaiswal N, Misra A, Misra PK, Khang A (Eds.). "Role of the Internet of Things (IoT) Technologies in Business and Production," *AI-Aided IoT Technologies and Applications in the Smart Business and Production*. CRC Press, 2023. https://doi.org/10.1201/9781003392224-1

Karkalakos S, Katsabekis S, Kameas A, Menychtas A. "Internet of Things and InsurTech Applications: A Systematic Literature Review," *IEEE Access*, 8, 188730–188754, 2020. https://doi.org/10.1201/9781003213888

Khang A (Ed.). *AI-Oriented Competency Framework for Talent Management in the Digital Economy: Models, Technologies, Applications, and Implementation*. CRC Press, 2023d. https://doi.org/10.1201/9781003440901

Khang A, Gupta SK, Rani S, Karras DA (Eds.). *Smart Cities: IoT Technologies, Big Data Solutions, Cloud Platforms, and Cybersecurity Techniques*. CRC Press, 2023a. https://doi.org/10.1201/9781003376064

Khang A, Gupta SK, Shah V, Misra A (Eds.). *AI-aided IoT Technologies and Applications in the Smart Business and Production*. CRC Press, 2023c. https://doi.org/10.1201/9781003392224

Khang A, Hahanov V, Abbas GL, Hajimahmud VA. "Cyber-Physical-Social System and İncident Management," *AI-Centric Smart City Ecosystems: Technologies, Design and Implementation* (1st Ed.). CRC Press, 2022c. https://doi.org/10.1201/9781003252542-2

Khang A, Ragimova NA, Hajimahmud VA, Alyar AV. "Advanced Technologies and Data Management in the Smart Healthcare System," *AI-Centric Smart City Ecosystems: Technologies, Design and Implementation* (1st Ed.). CRC Press, 2022b. https://doi.org/10.1201/9781003252542-16

Khang A, Rana G, Tailor RK, Hajimahmud VA (Eds.). *Data-Centric AI Solutions and Emerging Technologies in the Healthcare Ecosystem*. CRC Press, 2023e. https://doi.org/10.1201/9781003356189

Khang A, Rani S, Sivaraman AK. *AI-Centric Smart City Ecosystems: Technologies, Design and Implementation* (1st Ed.). CRC Press, 2022. https://doi.org/10.1201/9781003252542

Khang A, Shah V, Rani S. *AI-Based Technologies and Applications in the Era of the Metaverse*. (1st Ed.). IGI Global Press, 2023b. https://doi.org/10.4018/9781668488515

Khanh HH, Khang A. "The Role of Artificial Intelligence in Blockchain Applications," *Reinventing Manufacturing and Business Processes Through Artificial Intelligence* (pp. 20–40). CRC Press, 2021. https://doi.org/10.1201/9781003145011-2

Kimberly P, Grima S, Özen E. "Perceived Effectiveness of Digital Transformation and InsurTech Use in Malta: A Study in the Context of the European Union's Green Deal." *Big Data: A Game Changer for Insurance Industry* (pp. 239–263). Emerald Publishing Limited, 2022. https://ieeexplore.ieee.org/abstract/document/8528677/

Lavie D, Rosenbaum S. "InsurTech and IoT: Opportunities and Challenges for the Insurance Industry in Israel," *Journal of Financial Services Marketing*, 24(3), 129–141, 2019. https://doi.org/10.1201/9781003392224

Lee J, Lee H, Lee J. "The Role of InsurTech in Improving Customer Satisfaction in the United States," *Journal of Risk and Insurance*, 88(1), 157–179, 2021. https://doi.org/10.1201/9781003298335

Li X, Goto Y. "Internet of Things and Insurance: A Review and Research Agenda," *Technological Forecasting and Social Change*, 146, 73–81, 2019.

Meijer A, Faber E. "The Potential of InsurTech to Enhance Customer Engagement in the Insurance Industry," *Journal of Financial Services Marketing*, 25(3), 156–168, 2020. https://doi.org/10.1201/9781003252542

Ojo A, Janowski T. "Towards an IoT-Enabled Insurance Sector in the United Kingdom: A Regulatory Perspective," *Journal of Internet Law*, 21(11), 1–16, 2018. www.tandfonline.com/doi/abs/10.1080/16258312.2022.2135972

Pallavi J, Tripathi V, Malladi R, Khang A. "Data-driven AI Models in the Workforce Development Planning," *Designing Workforce Management Systems for Industry 4.0: Data-Centric and AI-Enabled Approaches* (1st Ed., pp. 179–198). CRC Press, 2023. https://doi.org/10.1201/9781003357070-10

Pritiprada P, Satpathy I, Patnaik BCM, Patnaik A, Khang A (Eds.). "Role of the Internet of Things (IoT) in Enhancing the Effectiveness of the Self-Help Groups (SHG) in Smart City," *Smart Cities: IoT Technologies, Big Data Solutions, Cloud Platforms, and Cybersecurity Techniques*. CRC Press, 2024. https://doi.org/10.1201/9781003376064-14

Radwan SM. *The Impact of Digital Technologies on Insurance Industry in Light of Digital Transformation*. Blom Egypt Investments and Insurance Brokerage & Consultancy, 2019. https://aqabaconf.com/images/uploads/The_Impact_of_digital_Technologies_on_Insurance_Industry_.pdf

Rana G, Khang A., Sharma R, Goel AK, Dubey AK (Eds.). *Reinventing Manufacturing and Business Processes Through Artificial Intelligence*. CRC Press, 2021. https://doi. org/10.1201/9781003145011

Rani S, Bhambri P, Kataria A, Khang A, Sivaraman AK. *Big Data, Cloud Computing and IoT: Tools and Applications* (1st Ed.). Chapman and Hall/CRC Press, 2023. https://doi. org/10.1201/9781003298335

Rani S, Chauhan M, Kataria A, Khang A. "IoT Equipped Intelligent Distributed Framework for Smart Healthcare Systems," *Networking and Internet Architecture*, 2, 30, 2021. https://doi.org/10.48550/arXiv.2110.04997

Roulstone A, Marriott J. "InsurTech: Innovation in the Insurance Sector," *Journal of Financial Perspectives*, 5(3), 21–39, 2017. https://diglib.uibk.ac.at/ulbtirolhs/content/ titleinfo/7801620/full.pdf

Shah V, Jani S, Khang A (Eds.). "Automotive IoT: Accelerating the Automobile Industry's Long-Term Sustainability in Smart City Development Strategy," *Smart Cities: IoT Technologies, Big Data Solutions, Cloud Platforms, and Cybersecurity Techniques*. CRC Press, 2024. https://doi.org/10.1201/9781003376064-9

Shah V, Vidhi T, Khang A. "Electronic Health Records Security and Privacy Enhancement Using Blockchain Technology," *Data-Centric AI Solutions and Emerging Technologies in the Healthcare Ecosystem* (1st Ed., p. 1). CRC Press, 2023. https://doi. org/10.1201/9781003356189-1

Shashikiran K, Shanthi S. "Internet of Things (IoT) in Indian Insurance Industry: Current Scenario and Future Prospects," *Journal of Advanced Research in Dynamical and Control Systems*, 11(14), 1338–1345, 2019. https://doi.org/10.1201/9781003269281-2

Shetty A. et al. "Block Chain Application in Insurance Services: A Systematic Review of the Evidence." *SAGE Open*, 12(1). https://journals.sagepub.com/doi/abs/10.1177/ 21582440221079877

Subhashini R, Khang A (Eds.). "The Role of Internet of Things (IoT) in Smart City Framework," *Smart Cities: IoT Technologies, Big Data Solutions, Cloud Platforms, and Cybersecurity Techniques*. CRC Press, 2024. https://doi.org/10.1201/9781003376064-3

Tailor RK, Pareek R, Khang A (Eds.). "Robot Process Automation in Blockchain," *The Data-Driven Blockchain Ecosystem: Fundamentals, Applications, and Emerging Technologies* (1st Ed., pp. 149–164). CRC Press, 2022. https://doi.org/10.1201/9781003269281-8

Verma M, Sharma N. "Role of InsurTech in the Growth of Indian Insurance Industry: An Empirical Study," *International Journal of Management, Technology, and Social Sciences*, 4(2), 90–108, 2019. www.inderscienceonline.com/doi/abs/10.1504/IJMCP.2020.111023

Wang L, Sun Y, Zhang X, Yang Y. "The Impact of IoT on Insurance Risk Assessment and Management: Evidence from the United States," *Technological Forecasting and Social Change*, 156, 120018, 2020. https://doi.org/10.1201/9781003252542-2

Wang M, Wang S. "The Impact of the Internet of Things on the Insurance Industry: Evidence from China," *International Journal of Information Management*, 52, 102070, 2020. www.hindawi.com/journals/misy/2022/6922337/

Xiang X, Du R. "The Internet of Things and Insurance: A Literature Review and Future Research Directions," *Journal of Risk and Insurance*, 87(2), 297–326, 2020. www. sciencedirect.com/science/article/pii/S2090447922000843

Yamada T, Nakagawa H, Nakajima H. "InsurTech Ecosystem in Japan: New Opportunities for Insurers and Startups," *Journal of Risk and Insurance*, 85(3), 753–775, 2018. www. mdpi.com/2078-2489/14/2/122

Zhang X, Wu D. "InsurTech in China: An Overview of the Current State and Future Prospects," *Journal of Insurance Issues*, 44(1), 1–24, 2021. https://pubsonline.informs.org/ doi/abs/10.1287/isre.2021.0997

16 Cybersecurity

Techniques and Applications to Combat Vicious Threats in Modern-Era Indices

Mauparna Nandan and Sharmistha Dey

16.1 INTRODUCTION

In this present era of digital technologies, the world changes very rapidly. We have become digital, but such dominance of technology in the digital era has also led to cyber-related crimes. As per a recently published security audit, every year we have to face 800,000 security attacks, and every 39 seconds an intruder tries to attack (Kaur et al., 2021).

Per the statistics, the rate of cybercrime has increased during lockdown phases. Today, most of the work has been dependent on technology irrespective of their disciplines, such as the business sector, the education sector, the banking sector, and many more.

At present, society urges an essential web-like connection among different sections like financial institutions, businesses, governments, and citizens, crossing all the political, social, and cultural boundaries. With the advent of digitalization and the modernization of technology, a quantum jump has been taken by the entire world, proving the advancements in technology to be an undeniable boon.

This chapter discusses about basic concepts of cybersecurity in Section 16.2, where different types of attacks and threats have been included. Section 16.3 discusses the importance of cybersecurity; Section 16.4 includes a literature study; Section 16.5 gives an idea about cybersecurity architecture; different cybersecurity tools, techniques, and various cybersecurity applications are explained in Sections 16.6 and 16.7, respectively; and the impact of AI and other emerging technologies on cybersecurity are elucidated in Sections 16.8 and 16.9.

16.2 OVERVIEW OF THE WEB

Overall, what we see or browse on the internet is only a limited percentage of the overall internet we can see as a normal user is called surface web, it is just 4–5% of the total internet. Mainly different URLs, search engines, and social media sites are included in this (Khang et al., 2022a).

DOI: 10.1201/9781003392224-16

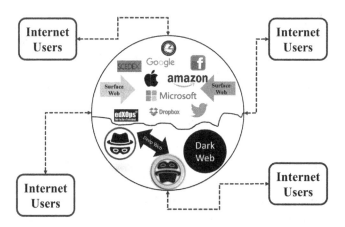

FIGURE 16.1 The surface web, the deep web, and the dark web.

Source: Khang (2021).

The major portion of the internet consists of the deep web, which is almost 95%, covers the information not indexed by normal search engines. The term *deep net* refers to online content that search engines cannot access due to password protection or being stored behind internet services.

As a result, search engine crawlers, or "spiders," are not visible. The dark web, which is a part of the deep web, comprises hidden sites, accessible with an onion browser. The dark net is the home to websites that are associated with illicit activities and underground marketplaces like human trafficking, illegal weapon exports, and the like. Figure 16.1 shows the surface, deep, and dark webs.

16.2.1 Types of Cybersecurity Attacks

Cybercrimes are crimes that use a computer or electronic system in an illegal way. They are considered a major risk as they may lead to sensitive data being lost, defamation, financial loss, system failure, and more. The concept of cybercrime is not a new one.

In 1962, Allen et al. (1962) initiated what is believed to be the first instance of cybercrime by launching a cyberattack on the computer networks at the Massachusetts Institute of Technology (Douligeris et al., 2004), Douligeris used a punch card to steal passwords from their database. In 1971, Bob Thomas created the world's first computer virus as part of his research at BBN Technologies.

In the present era, cybercrimes are increasing rapidly, and cyber attackers are exploring innovative ways day by day. Cyberattacks mainly can be categorized two ways: an active attack and a passive attack. In the case of active attacks, the intruder accesses the system un-authentically for the purpose of tampering with the data or defamation and data loss are almost obvious. But in the case of a passive attack, they silently monitor a system and work like a sniffer. Some common cybersecurity attacks follow.

16.2.1.1 Denial of Service

Any type of attack that causes a network to slow down or that disables a server from serving clients is known as a denial-of-service (DoS) attack. These types of attacks are accomplished by flooding the targeted machine or sending unnecessary requests to slow down the server, which prevents the fulfillment of important requests.

As many companies use similar technology, this becomes an advantage for the perpetrators to carry out DoS attacks. Figure 16.2 gives an idea how a distributed DOS (DDOS) attack works.

Types of DDoS attacks:

a. **Flooding**—The perpetrators use useless packets to consume available bandwidth by flooding the network.
b. **Protocol Violation Attack**—In this attack the server resources or intermediate communication equipment is consumed and is measured in packets per second (PPS).
c. **Central Processing Unit (CPU) Power and Service**—The main motive of the attackers behind this attack is to have control of the victim's computer totally. They generate useless processes on the victim's machine, which occupy the memory of the computer and result in a complete and full breakdown. The attackers then get access to the victim's services before anyone else.

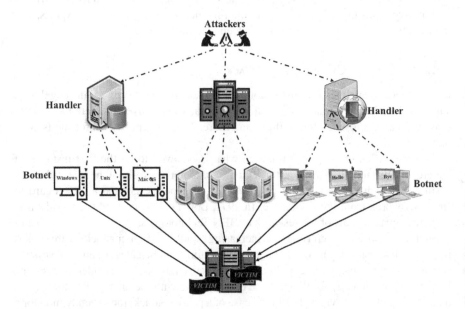

FIGURE 16.2 Distributed denial-of-service attack.

Source: Khang (2021).

16.2.1.2 Malware

Malware is pernicious software that the attackers design to damage a computer without the user's consent. Cyber threats are increasing daily, especially during some international or national events or when the world is going through a hard time like the pandemic situation. Some of the following incidents are where the attackers took advantage of the time.

Cybercrime increased rapidly in 2020 when the world was fighting COVID-19. People around the world were downloading applications to know the symptoms of COVID-19, and the attackers used it to their advantage. They designed and launched a ransomware attack called CovidLock, which, once installed by a user, will immediately lock accessibility until the amount demanded by the attackers is paid (Li et al., 2021).

16.2.1.3 Phishing

The information about the victim is collected using social engineering and technology and various methods of communication such as email, messages, and so on.

In this attack, the criminals pose to be a legitimate institution to attract the victim and then convince them to share their personal details, bank details, passwords, and so on, which then are used for identity theft and financial loss as shown in Figure 16.3.

16.2.1.4 Man-in-the-Middle Attack or Session Hijacking

It is a type of cyberattack in which the intruder accesses the system un-authentically and accesses the information and/or tampers with data without the knowledge of the original user. Figure 16.4 shows how a man-in-the-middle attack takes place.

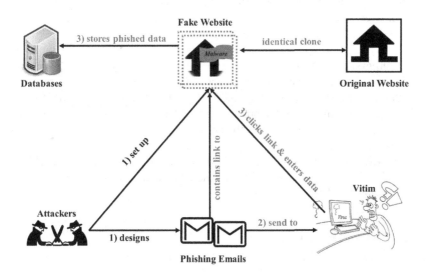

FIGURE 16.3 How a phishing attack happens.

Source: Khang (2021).

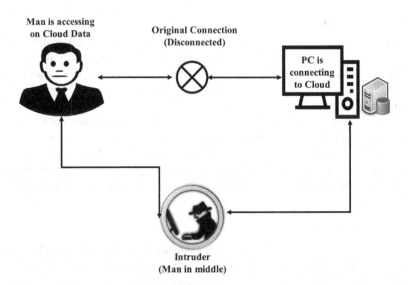

FIGURE 16.4   Man-in-the-middle attack.

Source: Khang (2021).

16.2.1.5 Cross-Site Scripting

A cross-site scripting (XSS) attack involves injecting a malicious script into a web application. An attacker can exploit a vulnerability in the application by sending harmful code. A JavaScript code is run by the attacker in the machine corresponding to the targeted victim, thereby stealing information or data.

Mainly, this type of attack can be categorized into three groups:

- **Nonpersistent attacks**—These attacks are the most common web vulnerability, where the target is accomplished in just a single response or request.
- **Persistent attacks**—These attacks are carried out in two steps and are more dangerous compared to the previous type. Step 1 is when the malicious code is injected and stored in the web application, and Step 2 is when the victim loads the malicious page.
- **Document object model (DOM)-based attacks**—The vulnerability appears in a document-like model, where the attackers use documents such as write () and other functions.

16.2.1.6 Zero-Day Exploits

These attacks are flaws in software or hardware that are unknown to the user.

16.2.1.7 Data Breaches

Data theft by cybercriminals. The motive is crime or embrace an institution or espionage.

16.2.1.8 Attacks on Internet of Things Devices

Hackers use Internet of Things (IoT) devices to launch some DDoS attacks and collect all the information present in the device prior to their access (Subhashini et al., 2024).

16.2.2 RECENT STATISTICS ABOUT CYBERSECURITY ATTACKS

In accordance to a recent study by AV-Test and the web root threat report from March 2021 to February 2022, there were new 153 million malware programs, which, in turn, was a nearly 5% increase over the previous year's data (Carley et al., 2018) as shown in Figure 16.5.

Figure 16.5 shows the year-wise cybercrime numbers in popular industries from 2014 to 2021. It shows a trend of cybercrime worldwide.

According to a recent survey done by a Naval Criminal Investigative Service (NCSI) project team, all over the world, a national cybersecurity index and a digital development index have been published for all the countries, along with their ranking.

India ranks 51, with a 69.72 national cybersecurity index and a 40.02 digital development index as shown in Figure 16.6. Figure 16.6 shows the top 10 countries regarding the national cybersecurity index (Easttom et al., 2019).

The core contributions of this chapter are as follows:

1. To explore the issues and challenges of cybersecurity
2. To elaborate present trend of cybersecurity research and discuss cybersecurity tools and techniques available in the market
3. To indicate the future direction of cybersecurity techniques
4. To analyze the overall scenario of cybersecurity techniques adapted nowadays to combat cybercrimes

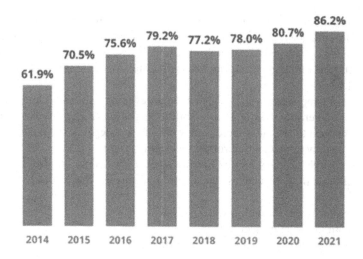

FIGURE 16.5   Year-wise cybercrime numbers in industries.

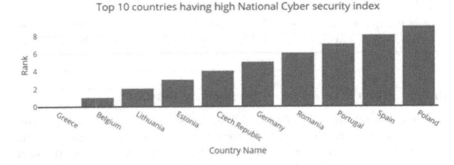

FIGURE 16.6   Top 10 countries in the national cyber security index.

Source: Carley et al. (2018).

In the next section, the importance of cybersecurity is discussed, which will help with understanding why research on cybersecurity tools and techniques should be highlighted.

16.3   IMPORTANCE OF CYBERSECURITY

Cybersecurity is a good practice aimed at safeguarding devices from hackers to ensure that they are more secure. Cybersecurity professionals implement security measures and operations to safeguard our data and devices (Jayashree et al., 2024).

   In essence, cybersecurity aims to prevent unlawful access to our network, devices, and data. Hackers and cybercriminals exploit the internet to gain access to other people's devices by using spyware and malware and conducting cyberattacks (Sakthivel et al., 2022).

   Cybersecurity aims primarily to shield internet users from digital threats such as infected files, malware, and cyberattacks that could expose sensitive personal information or lead to ransom demands by hackers. It also helps prevent disruption of critical infrastructure like power supplies and military operations that can arise from such attacks.

   The role of cybersecurity is to address the preexisting vulnerabilities in software applications and ensure their ongoing stability. As the number of internet-connected devices continues to rise, it is increasingly critical to safeguard them against unauthorized access by implementing robust security measures.

   Cybersecurity's primary goal is to prevent data loss, compromise, or attack. At least one of these three objectives can be used to evaluate cybersecurity:

- Keep sensitive information secret.
- Keep all information secure.
- Increase access to information for legitimate users.

The availability, integrity, and confidentiality (AIC) triangle, which comprises of these principles, is the foundation of any security plan. Security practices for protecting sensitive data within a business or organization can be informed by the AIC trinity.

To avoid any association with the Central Intelligence Agency (CIA), this paradigm is alternatively known as the AIC triad (Easttom et al., 2019). The triad comprises the three most important facets of security.

1. **Confidentiality**—Confidentiality is the way of keeping sensitive information private without informing the original user. It assures the secrecy of the message.
2. **Integrity**—Data integrity refers to the processes that guarantee information is genuine, correct, and protected from tampering by unauthorized parties. There has been no unauthorized tampering with the data, and they come from a reliable source.
3. **Availability**—Information has high availability if it can be accessed and updated quickly by those who are authorized to do so. It ensures that only those with proper authorization can access our private information at all times.

The AIC triad is represented in Figure 16.7.

16.4 LITERATURE REVIEW

Research on cybersecurity is not new. Li and Liu (2021) and Shieh et al. (2021) in their studies have given the overall evolution of cybersecurity research. Through their survey, they showed the emerging trends and recent developments in this sector. They also discussed cybersecurity policies.

This chapter helps with finding a road map for improving cybersecurity research. Cascavilla et al. (2021) performed another survey on cybersecurity threat intelligence.

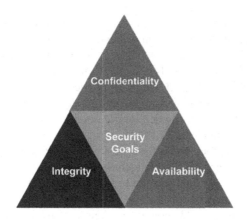

FIGURE 16.7 The availability, integrity, and confidentiality triangle.

They have given a proper taxonomy related to all cyber threats and a vision statement (Rahman et al., 2020).

Rahman et al. (2020; Mbaziira et al., 2016) have developed a model for cybercrime detection using a big data approach. They used an Apache Hadoop platform. They did a performance analysis before using Hadoop, at the beginning, in the middle, and when the request returned from Hadoop. Their proposed model helps in cybercrime prediction and in reducing the rate of cybercrime (Rani et al., 2021).

Mbaziira et al. (2016; Rupa et al., 2020) have worked on online data theft and identity theft, and they have applied classification algorithms and traditional machine learning (ML) algorithms like support vector machine (SVM) or naïve Bayes. They used a Waikato Environment for Knowledge Analysis (WEKA) data analytics tool. k-nearest neighbors (KNN) and SVM predict fraud with an accuracy of 40% and 20%, respectively. With naïve Bayes, the receiver operating characteristic curve (ROC) is 0.976, and with SVM, it is 0.971.

Rupa et al. (2020) worked on copyright attacks and identity theft using classification algorithms. The main plus of this work was generating an analysis report. They collected data sources from Kaggle and CERT IN. It consisted of more than 2,000 records and had a 99% accuracy rate (Lim et al., 2020).

There is a lot of research not only on ML algorithms but also on deep learning algorithms. Xin et al. (2021) did a comparative study on several traditional, as well as deep learning, algorithms, and through this study, a clear picture of the research has been elaborated. They used the Knowledge Discovery in Databases (KDD) Cup 99 database (Mbaziira et al., 2017). KDD Cup 99 is the data set used for the Third International Knowledge Discovery and Data Mining Tools Competition, which was held in conjunction with KDD Cup 99 (the Fifth International Conference on Knowledge Discovery and Data Mining).

Another survey work on deep learning models for cybersecurity technique detection was performed by Samtani et al. (2023). They designed a four-phase framework that contains raw data collection, data preprocessing, representation learning, and classification learning. Through their work, the challenges of using deep learning models were revealed (Xin et al., 2021).

Lim et al. (2020; Choi et al., 2020) have worked on detecting criminal networks using a deep reinforcement learning model and fusion matrix. They used the Cavier drug data set from UCINET (UCINET is a comprehensive package for the analysis of social network data and other 1-mode and 2-mode data, developed by Analytic Technologies) and proposed FDRL (Fluid Dynamics Research Laboratory) link prediction model based on a deep reinforcement model and fusion matrix to predict criminal networks. They have divided their dataset into a 3:1 ratio for training and testing, respectively. (Ullah et al., 2019)

Patel et al. (2023), in their work, have shown the implementation of feature modeling and natural language processing techniques to perform text-based cybercrime detection. They tested two data sets, and overall performance with those data sets were 60% and 80%, respectively.

Eunaicy and Suguna (2022) conducted their research on software piracy and malware attacks on the IoT, which is considered a very high-priority attack today. They used a data set collected from Google Code Jam and checked software piracy attacks on that.

TABLE 16.1

Comparative Analysis of Relative Works in Cybersecurity Techniques

References	Cybercrime/attack covered	Approach to solve	Performance
Using Traditional Machine Learning Algorithms			
(Rahaman et al., 2020; Cascavilla et al., 2021)	Online fraud	K-means algorithm, big data analytics	82.3% accuracy
(Mbaziira and Jones 2016; Rupa et al., 2020)	Online data theft, identity theft	Naive Bayes, support vector machine, k-nearest neighbor algorithm	60% predictive accuracy
(Rupa et al., 2020; Lim M et al., 2020)	Copyright attack, identity theft	Clustering, classification	99% accuracy
Using Deep Learning Algorithms			
References	Cybercrime/attack covered	Approach to solve	Performance
(Lim et al., 2019; Mbaziira et al., 2017)	To detect criminal network	Deep reinforcement learning, fusion-oriented network analysis model	Area under the curve is 0.59, for 1,500 iterations
(Choi et al., 2020)	Software piracy, malware attacks (on IoT)	Deep convolutional neural network and feature extraction	F-measure score 97%, classification Accuracy 97.47%
(Eunaicy and Suguna, 2022; Ullah et al., 2019)	Web-based attack	Artificial neural network (ANN), convolutional neural network, recurrent neural network	Highest with recurrent neural network (RNN), 94% accuracy
(Dey et al., 2020; Patel et al., 2023)	DDoS, social engineering	Hybrid meta-heuristic-based feature selection, reinforcement learning-based, genetic algorithm	99.48% accuracy, with 13 selected features

Samtani et al. (2023) have worked on IoT-enabled devices, and they used deep reinforcement learning to detect cybersecurity techniques. They have used ToN-IoT device. They proposed a meta-heuristic techniques based on feature selection technique and the accuracy shows 99.48%, with 13 selected features (Pritiprada et al., 2024).

Table 16.1 represents a comparative study of different research works performed in this area.

Table 16.1 shows different traditional ML models as well as deep learning models in cases of cybersecurity. Most of the attacks have been covered as software piracy, web-based attacks, and phishing, among others. The use of AI and ML in cybersecurity has been beautifully explained by Patel (2023; Dey et al., 2023).

16.5 CYBERSECURITY ARCHITECTURE

The term cyber security architecture refers to a structure that outlines the proper alignment and interaction between a computer network's security protocols and operational procedures.

A cybersecurity framework is a fundamental component of this architecture, serving as the basis for constructing and developing a comprehensive product or system (Maleh et al., 2020).

The concept of security architecture is a structured approach that outlines how a company's security measures and countermeasures integrate with the overall system structure. The primary objective of these measures is to ensure the quality attributes of critical systems, such as confidentiality, integrity, and availability. Furthermore, achieving an effective security architecture requires a combination of expertise in hardware and software comprehension, programming skills, research abilities, and policy development.

To safeguard one's business against business threats, it is crucial to have antivirus software, firewalls, and intrusion detection systems. The organizations should create a comprehensive security architecture that incorporates all these components to maintain and enhance one's network security technologies, as well as any existing rules and processes that are currently in place.

16.5.1 NIST CYBERSECURITY FRAMEWORK

The NIST cybersecurity framework (Krumay et al., 2018) is a valuable resource for streamlining and enhancing an organization's cybersecurity program. It comprises a collection of principles and optimal approaches that can assist organizations in strengthening their cybersecurity position.

The framework presents a set of protocols and benchmarks that empower organizations to detect and recognize cyber threats more efficiently and offers guidance on how to prevent, respond to, and recover from cybersecurity incidents.

The NIST developed the NIST Cybersecurity Framework (NIST CSF) to establish a reliable set of regulations, recommendations, and benchmarks that organizations across various industries can follow. The framework has the potential to be valuable as a security management tool at a broad level, assessing cybersecurity risks across the organization, regardless of whether the cybersecurity program is new or already established.

16.5.2 CORE FUNCTIONS OF THE NIST CSF

16.5.2.1 Identify
The Identify function aims to establish a solid foundation for an efficient cybersecurity program. Its primary objective is to assist in creating an organizational comprehension of how to manage cybersecurity risks to systems, people, assets, data, and capabilities.

16.5.2.2 Protect
The Protect function defines the necessary measures to guarantee the provision of essential infrastructure services and enable controlling or minimizing the significance of a potential cybersecurity incident.

16.5.2.3 Detect
The Detect function is useful in identifying possible cybersecurity breaches, which is extremely crucial, and this function specifies the suitable actions to promptly recognize the presence of a cybersecurity occurrence.

16.5.2.4 Respond

The Respond function mainly concentrates on taking suitable measures when a cybersecurity incident is detected and aids in limiting the consequences of a possible cybersecurity event.

16.5.2.5 Recover

The Recover feature identifies the appropriate measures to revise and maintain strategies for withstanding adverse events and restoring any functionalities or services that were impacted by a cybersecurity event. The emphasis is on the timely restoration of regular operations to minimize the effects of a cybersecurity event.

The framework classifies all cybersecurity abilities, initiatives, procedures, and daily tasks into five primary functions, which are summarized in Figure 16.8.

16.6 SOME POPULAR CYBERSECURITY TOOLS

Safeguarding our information technology (IT) infrastructure is a matter of utmost importance, and every organization must accord cybersecurity the necessary attention. Hackers, malware, viruses, and other security threats pose a tangible danger to businesses of all sizes. It is crucial for companies to acknowledge these hazards and take appropriate measures to protect themselves.

Various aspects of cyber defense may warrant attention, and six critical tools and services that companies must consider to bolster their cybersecurity are outlined in Figure 16.9 (Matari et al., 2018).

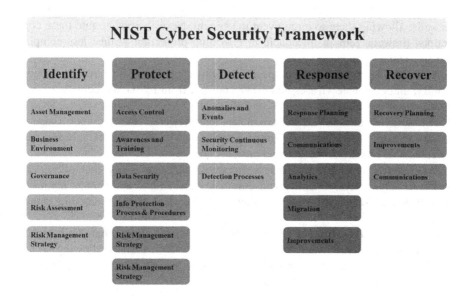

FIGURE 16.8 NIST Cybersecurity Framework.

Source: Khang (2021).

FIGURE 16.9 Cybersecurity tools.

Firewalls—Firewalls are widely recognized as a critical component of security tools and are among the most essential instruments to ensure safety. They function to prevent unauthorized access to and from a private network and can be implemented either as software, hardware, or a combination of both.

Every incoming or outgoing message is evaluated by the firewall, and those that do not satisfy predetermined security criteria are blocked. While firewalls are highly beneficial, they do have some limitations (Khang, 2021).

Experienced hackers can create data and programs that closely resemble trusted firewalls, allowing them to circumvent the system. Nevertheless, despite these constraints, firewalls remain invaluable in providing protection against less complex malicious attacks on our systems.

Antivirus Software—It is a specialized program that is intended to impede, identify, and eliminate viruses and other malware from individual computers, networks, and IT systems.

This software also shields our computers and networks from a diverse range of viruses and threats like worms, Trojan horses, rootkits, key loggers, adware, spyware, browser hijackers, botnets, and ransomware.

Many antivirus programs possess an auto-update capability that allows the system to automatically identify and eliminate new viruses and threats regularly. Additionally, these programs offer supplementary services, such as screening emails and others.

Public Key Infrastructure (PKI) Services—PKI is a system that facilitates the distribution and identification of public encryption keys, thereby allowing secure data exchanges between users and computer systems over the internet while ensuring the identity of the communicating parties.

While it is possible to exchange sensitive information without using PKI, the lack of authenticating other parties makes it risky.

Managed Detection and Response (MDR) Service—MDR is an advanced security service that offers threat hunting, threat intelligence, security monitoring, incident analysis, and incident response.

This service has emerged from the need of organizations, which may lack the necessary resources, to be more aware of potential risks and enhance their ability to detect and respond to threats.

MDR also leverages the power of AI and ML to automatically detect and investigate threats, thereby enabling faster response times.

Penetration Testing—Penetration testing, or pen-testing, is a crucial method for accessing the security of a business's IT infrastructure and security systems by attempting to exploit vulnerabilities in a safe manner.

Penetration testing refers to the process of imitating a genuine attack on a network or application, with the purpose of assessing the types of attacks that a business may encounter from malicious hackers.

Staff Training—Although staff training is not typically considered a cybersecurity tool, it is still one of the most effective ways to defend against cyberattacks.

Educating employees and enhancing their knowledge of cybersecurity is crucial in protecting a business. Nowadays, there exist numerous training tools that can teach a company's staff about the best cybersecurity practices.

Every business has the opportunity to organize and implement these training tools to educate their employees and help them understand their role in cybersecurity.

16.7   APPLICATIONS AND TYPES OF CYBERSECURITY

The major applications of cybersecurity (Jun et al., 2021) follow:

DDoS Security—A DDoS attack is a type of digital assault that stands for Distributed Denial of Service. The attacker deploys several devices to flood the web server with numerous requests, thereby causing it to become overwhelmed and unable to serve legitimate traffic.

This results in fake website traffic being generated on the server. To combat this issue, cybersecurity offers a DDoS mitigation service that redirects the traffic to other cloud-based servers, which ultimately resolves the situation (Bhambri et al., 2022).

Web Firewall—A firewall based on web application servers is installed across a wide area network that monitors all incoming and outgoing server traffic and possesses the ability to automatically identify and eliminate fraudulent and harmful website traffic.

This cybersecurity technique aids in detecting and implementing automatic traffic monitoring and reducing the risk of attacks.

Bots—Currently, a lot of hackers and attackers utilize bots to generate high levels of traffic from multiple devices, which can cause the server to crash.

To address this issue, cybersecurity assists in detecting bogus users (i.e., bots) and terminating their sessions to prevent them from negatively impacting the experience of regular users.

Antivirus and Antimalware—The purpose of cybersecurity is to create antivirus and antimalware software to safeguard computers and prevent them from being subjected to digital attacks.

This includes protecting devices against data breaches, unauthorized attacks, and other types of digital threats initiated by hackers. Additionally, cybersecurity is involved in maintaining the network security and firewall systems for all connected devices on the network.

Threat management systems—Cybersecurity is involved in addressing digital threats and attacks on computer systems. It identifies potential vulnerabilities and system flaws that hackers and attackers could exploit and automatically corrects these defects to improve performance issues.

Moreover, it enhances the capacity to swiftly respond to digital attacks and grants users greater control over vulnerability issues.

Critical systems—Cybersecurity is essential for addressing critical attacks that occur on large servers connected to wide-area networks. It upholds strict safety protocols that users must follow to ensure device protection.

Additionally, it continuously monitors all applications in real time and regularly checks the safety of the servers, network, and users to maintain high levels of cybersecurity.

Rules and regulations—The implementation of cybersecurity measures leads to the establishment of guidelines and standards that must be adhered to by the network users, including both malicious and non-malicious individuals.

This empowers the authorities to identify and address security concerns, ultimately improving the overall functionality of the network.

Managing user privileges—When users are provided more system privileges or access to data than they actually require, it can enhance the likelihood of their misuse or compromise. To mitigate this risk, it is recommended that all users be given a moderate (yet sufficient) level of system privileges and that data access should be based on their specific job responsibilities.

The allocation of extremely high-level system privileges must be closely monitored and regulated. This concept is often known as the 'least privilege' principle.

Risk management regime—An effective risk management program within the organization must be implemented to evaluate the potential risks to the security of information and computer systems. The Board and senior management should endorse this program.

Additionally, it is essential to ensure that all employees, contractors, and suppliers are familiar with the approach and any relevant risk limitations.

Secure configuration—Establishing a plan to identify standard technology setups and procedures for managing system configurations can significantly enhance security. It is recommended to create a plan to eliminate or deactivate any redundant system functions and swiftly address known security vulnerabilities, often by applying software updates.

Neglecting to undertake these measures may raise the likelihood of system and information breaches.

16.8 THE IMPACT OF AI ON CYBERSECURITY: FUTURE DIRECTIVE

ML is a crucial component of AI that utilizes algorithms to enable computers to analyze data, learn from past experiences, and make decisions in a manner that mimics human behavior (Rana et al., 2021).

Although AI has the potential to enhance security, it can also provide cybercriminals with unmonitored access to systems. The following points illustrate the positive impacts that AI has on cybersecurity (Ansari et al., 2022).

Vulnerability management—Businesses are facing challenges in managing and prioritizing the significant volume of new vulnerabilities they encounter on a daily basis. Traditional vulnerability management practices typically react to incidents only after cybercriminals have already taken advantage of the vulnerability.

However, AI and ML technologies can enhance the vulnerability management capabilities of databases that track these vulnerabilities. Furthermore, User and Entity Behavior Analytics (UEBA) tools can analyze user behavior on servers and endpoints, and identify anomalies that may indicate a previously unknown attack when powered by AI (Khang et al., 2022b).

Threat hunting—Conventional security tools rely on signatures or attack indicators to recognize threats, but they are only able to identify threats that have been previously discovered. These signature-based tools have limitations in detecting new or unknown threats and can only identify about 90% of threats.

The use of AI can enhance the detection rate of traditional techniques, potentially up to 95%. Furthermore, AI can enhance threat hunting by integrating behavior analysis (Rani et al., 2022).

Network security—Traditional approaches to secure traditional networks have two primary focuses: establishing security policies and comprehending the network environment. The following points should be kept in mind:

Policies—Security policies can aid in distinguishing between legitimate and malicious network connections and may enforce a zero-trust paradigm.

Environment—By learning the patterns of network traffic, AI has the potential to improve network security by suggesting both security policies and functional workload grouping.

Data centers—AI can oversee and fine-tune crucial data center procedures such as power usage, backup power, internal temperatures, bandwidth utilization, and cooling filters. AI can provide insights into which values can enhance the security and effectiveness of data center infrastructure.

Using AI can help decrease maintenance costs. AI alerts can inform when a hardware failure occurs and need attention. This enables us to repair equipment before any additional damage takes place (Khang et al., 2023a).

16.8.1 AI Applications in Cybersecurity: Real-Life Examples

The process of ML involves analyzing vast quantities of data at a rapid pace through statistical analysis. Given the enormous amount of data generated by contemporary organizations, it is plausible why ML has become an inevitable resource today (Taddeo et al., 2019).

Security screening—Immigration and customs officers use security screening methods to identify individuals who are not truthful about their intentions. To address these challenges, the United States Department of Homeland Security has developed a system called AVATAR.

This technology implements AI and big data to analyze subtle changes in facial expressions and body gestures that may indicate deceptive behavior (Luke et al., 2024).

Security and crime prevention—The New York Police Department has been using the Computer Statistics (CompStat) AI system since 1995. CompStat combines organizational management and philosophy with various software tools, and it was the first technology to be used for 'predictive policing'. As a result, numerous police stations across the US have adopted CompStat for criminal investigations.

More recent AI-based crime analysis tools like Armorway, based in California, use AI and game theory to predict terrorist threats.

Analyze mobile endpoints—Google is utilizing AI to examine potential security risks for mobile devices. This analysis can be utilized by organizations to safeguard the increasing quantity of personal mobile devices. Zimperium and MobileIron have teamed up to assist organizations in implementing anti-malware mobile solutions that incorporate AI.

AI-powered threat detection—ED&F Man Holdings, a commodities trader, encountered a security issue a few years ago and to address this, the company turned to Vectra's Cognito, a threat detection and response platform powered by AI.

Cognito acquires and preserves network metadata and complements it with unique security insights. It utilizes ML techniques to identify and prioritize attacks in real time.

Detection of sophisticated cyberattacks—Darktrace's Enterprise Immune System is a platform that utilizes machine learning technology to model the conduct of every network, device, and user and recognize distinctive patterns. Darktrace instantly identifies any unusual behavior and notifies the company in real time.

Reducing threat response time—To enhance the bank's threat detection and response, its security team implemented Paladon's AI-powered MDR service. Paladon's threat-hunting service relies on data science and ML capabilities.

16.8.2 Limitations of Implementing AI for Cybersecurity

Resources—The employment of AI technology is a widespread security tool that is hindered by certain constraints. One of these limitations is the requirement for significant resources, such as computing power, memory, and data, that an organization must possess.

Data sets—In order to train an AI system, security firms must utilize a multitude of distinct data sets composed of anomalies and malware codes. Acquiring precise and reliable datasets can be a costly and time-consuming process, which may not be feasible for certain companies due to resource constraints.

Hackers in AI—The utilization of AI by cybercriminals to refine and augment their malware can pose a significant threat, as it possesses the capability to generate increasingly sophisticated attacks by assimilating knowledge from preexisting AI-based models.

Neural fuzzing—The process of identifying software vulnerabilities involves examining extensive volumes of arbitrary input data through a technique known as 'Neural Fuzzing'. An attacker may employ a combination of neural fuzzing and neural networks to obtain insights into the target system or software and uncover its weaknesses.

16.9 CYBERSECURITY CHALLENGES

Nowadays, cybersecurity has proved to be a vital element of a nation's comprehensive security policies that include both economic and national security strategies.

In India, there are numerous cybersecurity challenges that need to be addressed due to the rising number of cyberattacks. Therefore, it has become inevitable for every organization to have security analysts who can ensure the safety of their systems.

However, these security analysts encounter various cybersecurity challenges, such as safeguarding confidential information of government organizations and protecting private organization servers as shown in Figure 16.10 (Yampolskiy et al., 2016).

Ransomware evolution—Ransomware refers to malicious software that encrypts the data on a victim's computer, making it inaccessible until a ransom is paid. Once the payment is done, the victim is granted access to their data again.

This type of malware has become a significant threat to cybersecurity and is causing concern among data professionals, IT experts, and executives. A proper recovery plan must include strategies for restoring corporate and customer data and applications, as well as reporting any breaches under the Notifiable Data Breaches scheme.

To defend against ransomware attacks, disaster recovery as a service (DRaaS) solutions are highly recommended. With DRaaS, automatic backups of files are created, and it is easy to identify which backup is clean. In the event of a ransomware attack, a fail-over can be launched quickly with just the press of a button, ensuring data are not lost or compromised.

Blockchain revolution—The invention of blockchain technology is considered the most significant milestone in the history of computing. This is because, for the first time, we possess a truly digital means of exchanging value between peers without any intermediary (Khanh and Khang, 2021).

This technology is what powers cryptocurrencies such as Bitcoin, and it offers a vast global platform for conducting transactions and business without relying on a third party to establish trust (Khang et al., 2022b).

FIGURE 16.10 Cybersecurity challenges.

Although it is challenging to anticipate the precise impact of blockchain technology on cybersecurity, experts in this field can offer some informed predictions. As the use and application of blockchain technology in the context of cybersecurity continue to evolve, there will likely be a healthy mix of tension and complementary integrations with traditional, established approaches to security (Tailor et al., 2022).

IoT threats—*IoT* refers to a network of physical devices that are interconnected and accessible through the internet. These devices are identified by a unique identifier (UID) and can transfer data over a network without the need for human-to-human or human-to-computer interaction. However, the firmware and software running on these devices make both consumers and businesses highly vulnerable to cyberattacks.

When IoT devices were designed, cybersecurity and commercial purposes were not the primary considerations. Therefore, every organization must collaborate with cybersecurity professionals to ensure the security of their password policies, session handling, user verification, multifactor authentication, and security protocols. This collaboration will help mitigate the risk associated with IoT devices.

AI expansion—The expanded form of AI is "artificial intelligence." John McCarthy, father of AI defined it as "[t]he science and engineering of making intelligent machines, especially intelligent computer programs."

It is a subfield of computer science dedicated to making robots that can perform human-like tasks and mimic our emotional responses. Speech recognition, learning, planning, problem-solving, and others are all part of AI. Integrating AI into our cybersecurity strategy allows us to better guard and fight our systems from the moment a malicious assault is launched (Hajimahmud et al., 2022).

Serverless Apps Vulnerability—Serverless architecture and apps are those that do not require any servers or other hardware, instead relying on a service like Google Cloud Function or Amazon Web Services Lambda in the cloud (Rani et al., 2023).

Because users use serverless apps locally or off-server on their devices, they leave themselves vulnerable to cyberattacks. Users must take their own safeguards to ensure their own safety when utilizing a serverless application.

The serverless applications do not protect our data in any way. If an attacker acquires access to our data via a vulnerability—such as leaking credentials, a compromised insider, or any other method other than serverless—then the serverless application is of no use.

16.10   CONCLUSION

The integration of AI and ML technologies with cybersecurity has no doubt expanded the horizon of cybersecurity to a great extent, but due to this cybersecurity has to face some challenges like difficulties in manual tracking of any incidents, problems of biasedness, or generating false-positive results (Khang et al., 2023b).

This chapter revealed different emerging cybercrimes and -threats and cybersecurity architecture and discussed about currently available cybersecurity tools and techniques. Moreover, it gives an indication of the future road map of this research area (Jaiswal et al., 2023).

Also, a case study–based discussion in this chapter reveals the real-life cybersecurity scenario, which will help budding researchers find a direction for cybersecurity tools and techniques (Hajimahmud et al., 2023).

The future growth of cybersecurity along with qualitative, as well as quantitative, analysis of different available technologies will be explored to improve this area of research in the near future (Khang et al., 2023c).

REFERENCES

Allen BV. *Brobston & Brobston, Bessemer, for Appellant. Mead, Norman & Fitzpatrick, Birmingham, for Appellee Sellers.* LAWSON, Justice. Archie BAGGETT, Supreme Court of Alabama, January 18, 1962. https://law.justia.com/cases/alabama/supreme-court/1962/137-so-2d-37-1.html

Ansari MF, Dash B, Sharma P, Yathiraju N. "The Impact and Limitations of Artificial Intelligence in Cybersecurity: A Literature Review," *International Journal of Advanced Research in Computer and Communication Engineering*, 2022. https://papers.ssrn.com/sol3/papers.cfm?abstract_id=4323317

Bhambri P, Rani S, Gupta G, Khang A. *Cloud and Fog Computing Platforms for Internet of Things.* CRC Press, 2022. https://doi.org/10.1201/9781003213888

Bob Thomas, The Creeper Worm, the First Computer Virus. 1971. https://www.historyofinformation.com/detail.php?entryid=2860

Carley KM, Cervone G, Agarwal N, Liu H. "Social Cyber-Security," Thomson R, Dancy C, Hyder A, Bisgin H (Eds.) *Social, Cultural, and Behavioral Modeling: SBP-BRiMS 2018. Lecture Notes in Computer Science* (vol. 10899). Springer, 2018. https://doi.org/10.1007/978-3-319-93372-6_42

Cascavilla G, Tamburri DA, Heuvel W. "Cybercrime Threat Intelligence: A Systematic Multi-Vocal Literature Review," *Computers & Security*, 105, 102258, 2021. https://doi.org/10.1016/j.cose.2021.102258

Choi YH. et al. "Using Deep Learning to Solve Computer Security Challenges: A Survey," *Cybersecurity*, 3(15), 2020. https://doi.org/10.1186/s42400-020-00055-5

Dey AK, Gupta GP, Sahu SK. "Hybrid Meta-Heuristic Based Feature Selection Mechanism for Cyber-Attack Detection in IoT-enabled Networks," *Procedia Computer Science*, 218(1), 318–327, 2023. www.sciencedirect.com/science/article/pii/S1877050923000145

Douligeris C, Mitrokotsa A. "DDoS Attacks and Defense Mechanisms: Classification and State-of-the-Art," *Computer Networks*, 44(5), 643–666, 2004. www.sciencedirect.com/science/article/pii/S1389128603004250

Easttom C, Butler W. "A Modified McCumber Cube as a Basis for a Taxonomy of Cyber-Attacks," *2019 IEEE 9th Annual Computing and Communication Workshop and Conference (CCWC)* (pp. 0943–0949). IEEE, January 2019. https://ieeexplore.ieee.org/abstract/document/8666559/

Eunaicy C, Suguna S. "Web Attack Detection Using Deep Learning Models," *Materials Today*, 62(2), 2022. https://doi.org/10.1016/j.matpr.2022.03.348(Preprint)

Hajimahmud VA, Khang A, Hahanov V, Litvinova E, Chumachenko S, Alyar AV. "Autonomous Robots for Smart City: Closer to Augmented Humanity," *AI-Centric Smart City Ecosystems: Technologies, Design and Implementation* (1st Ed.). CRC Press, 2022. https://doi.org/10.1201/9781003252542-7

Hajimahmud VA. et al. (Eds.). "The Role of Data in Business and Production," *AI-Aided IoT Technologies and Applications in the Smart Business and Production*. CRC Press, 2023. https://doi.org/10.1201/9781003392224-2

Jaiswal N, Misra A, Misra PK, Khang A (Eds.). "Role of the Internet of Things (IoT) Technologies in Business and Production," *AI-Aided IoT Technologies and Applications in the Smart Business and Production*. CRC Press, 2023. https://doi.org/10.1201/9781003392224-1

Jayashree M. et al. (Eds.). "Vehicle and Passenger Identification in Public Transportation to Fortify Smart City Indices," *Smart Cities: IoT Technologies, Big Data Solutions, Cloud Platforms, and Cybersecurity Techniques*. CRC Press, 2024. https://doi.org/10.1201/9781003376064-13sensor

Jun Y, Craig A, Shafik W, Sharif L. "Artificial Intelligence Application in Cybersecurity and Cyberdefense," *Wireless Communications and Mobile Computing*, 1–10, 2021. https://search.proquest.com/openview/d0519dc5d5d6896dc213a5753ec2fccf/1?pq-origsite=gscholar&cbl=2034344

Kaur G, Habibi Lashkari Z, Habibi Lashkari A. "Introduction to Cybersecurity," *Understanding Cybersecurity Management in FinTech: Challenges, Strategies, and Trends*, 17–34, 2021. https://link.springer.com/chapter/10.1007/978-3-030-79915-1_2

Khang A. "Material4Studies," *Material of Computer Science, Artificial Intelligence, Data Science, IoT, Blockchain, Cloud, Metaverse, Cybersecurity for Studies*, 2021. www.researchgate.net/publication/370156102_Material4Studies

Khang A, Chowdhury S, Sharma S. *The Data-Driven Blockchain Ecosystem: Fundamentals, Applications, and Emerging Technologies* (1st Ed.). CRC Press, 2022.

Khang A, Hahanov V, Abbas GL, Hajimahmud VA. "Cyber-Physical-Social System and Incident Management," *AI-Centric Smart City Ecosystems: Technologies, Design and Implementation* (1st Ed., vol. 2, p. 15). CRC Press, December 30, 2022b. https://doi.org/10.1201/9781003252542-2

Khang A, Rani S, Gujrati R, Uygun H, Gupta SK (Eds.). *Designing Workforce Management Systems for Industry 4.0: Data-Centric and AI-Enabled Approaches*. CRC Press, 2023a. https://doi.org/10.1201/9781003357070

Khang A, Rani S, Sivaraman AK. *AI-Centric Smart City Ecosystems: Technologies, Design and Implementation* (1st Ed.). CRC Press, 2022a. https://doi.org/10.1201/9781003252542

Khang A, Shah V, Rani S. *AI-Based Technologies and Applications in the Era of the Metaverse* (1st Ed.). IGI Global Press, 2023b. https://doi.org/10.4018/9781668488515

Khanh HH, Khang A. "The Role of Artificial Intelligence in Blockchain Applications," *Reinventing Manufacturing and Business Processes Through Artificial Intelligence* (pp. 20–40). CRC Press, 2021. https://doi.org/10.1201/9781003145011-2

Krumay B, Bernroider EW, Walser R. "Evaluation of Cybersecurity Management Controls and Metrics of Critical Infrastructures: A Literature Review Considering the NIST Cybersecurity Framework," *In Secure IT Systems: 23rd Nordic Conference, NordSec 2018, Oslo, Norway, November 28–30, 2018, Proceedings 23* (pp. 369–384). Springer International Publishing, 2018. https://link.springer.com/chapter/10.1007/978-3-030-03638-6_23

Li Y, Liu Q. "A Comprehensive Review Study of Cyber-Attacks and Cyber Security," *Emerging Trends and Recent Developments, Energy Reports*, 7, 2021. https://doi.org/10.1016/j.egyr.2021.08.126

Lim M, Abdullah A, Jhanjhi M, Khan MK. "Situation-Aware Deep Reinforcement Learning Link Prediction Model for Evolving Criminal Networks," *IEEE Access*, 8(6), 16550–16559, 2020. https://doi.org/10.1109/ACCESS.2019.2961805

Luke J, Khang A, Chandrasekar V, Pravin AR, Sriram K (Eds.). "Smart City Concepts, Models, Technologies and Applications," *Smart Cities: IoT Technologies, Big Data Solutions, Cloud Platforms, and Cybersecurity Techniques* (1st Ed.). CRC Press, 2024. https://doi.org/10.1201/9781003376064-1

Maleh Y, Shojafar M, Alazab M, Romdhani I. (Eds.). *Blockchain for Cybersecurity and Privacy: Architectures, Challenges, and Applications*, 2020. www.google.com/books?hl=en&lr=&id=dC7yDwAAQBAJ&oi=fnd&pg=PP1&dq=Blockchain+for+cybersecurity+and+privacy:+architectures,+challenges,+and+applications&ots=e5LDU6-5QBx&sig=fzLWFxOX5tJifgwIbjPcpXawaPs

Matari Al OM, Helal IM, Mazen SA, Elhennawy S. "Cybersecurity Tools for IS Auditing," *2018 Sixth International Conference on Enterprise Systems (ES)* (pp. 217–223). IEEE, October 2018. https://ieeexplore.ieee.org/abstract/document/8588282/

Mbaziira A, Jones J. "A Text-Based Deception Detection Model for Cybercrime," *International Conference on Technology Management*, 2016. www.researchgate.net/profile/Alex-Mbaziira/publication/307594168_A_Text-based_Deception_Detection_Model_for_Cybercrime/links/5851b1be08aef7d030a19101/A-Text-based-Deception-Detection-Model-for-Cybercrime.pdf

Mbaziira M, Vincent A. *Detecting and Analyzing Cybercrime in Text-Based Communication of Cybercriminal Networks through Computational Linguistic and Psycholinguistic Feature Modeling.* Eric, George Mason University, 2017. https://search.proquest.com/openview/ff03063a39a5e92d4902c3ef7775ec42/1?pq-origsite=gscholar&cbl=18750

Patel H. "The Future of Cybersecurity with Artificial Intelligence (AI) and Machine Learning (ML)," *Preprints*, 2023. https://doi.org/10.20944/preprints202301.0115.v1

Pritiprada P, Satpathy I, Patnaik BCM, Patnaik A, Khang A (Eds.). "Role of the Internet of Things (IoT) in Enhancing the Effectiveness of the Self-Help Groups (SHG) in Smart City," *Smart Cities: IoT Technologies, Big Data Solutions, Cloud Platforms, and Cybersecurity Techniques.* CRC Press, 2024. https://doi.org/10.1201/9781003376064-14

Rani S, Bhambri P, Kataria A, Khang A. "Smart City Ecosystem: Concept, Sustainability, Design Principles and Technologies," *AI-Centric Smart City Ecosystems: Technologies, Design and Implementation* (1st Ed.). CRC Press, 2022. https://doi.org/10.1201/9781003252542-1

Rani S, Bhambri P, Kataria A, Khang A, Sivaraman AK. *Big Data, Cloud Computing and IoT: Tools and Applications* (1st Ed.). Chapman and Hall/CRC Press, 2023. https://doi.org/10.1201/9781003298335https://doi.org/10.1201/9781003269281

Rani S, Chauhan M, Kataria A, Khang A (Eds.). "IoT Equipped Intelligent Distributed Framework for Smart Healthcare Systems," *Networking and Internet Architecture.* CRC Press, 2021. https://doi.org/10.48550/arXiv.2110.04997

Rahman HA. "A Proposed Model for Cybercrime Detection Algorithm Using a Big Data Analytics," *International Journal of Computer Science and Information Security (IJCSIS)*, 18(6), 2020. www.academia.edu/download/64119494/17%20Paper%2001062019%20IJCSIS%20Camera%20Ready%20pp146-153.pdf

Rana G, Khang A, Sharma R, Goel AK, Dubey AK (Eds.). *Reinventing Manufacturing and Business Processes Through Artificial Intelligence.* CRC Press, 2021. https://doi.org/10.1201/9781003145011

Rupa C, Thippa RG, Abidi MH, Ahmari A. "Computational System to Classify Cyber Crime Offenses Using Machine Learning," *Sustainability*, 12(10), 4087, 2020. https://doi.org/10.3390/su12104087

Sakthivel M, Gupta SK, Karras DA, Khang A, Dixit CK, Haralayya B. "Solving Vehicle Routing Problem for Intelligent Systems Using Delaunay Triangulation," *2022 International Conference on Knowledge Engineering and Communication Systems (ICKES)*, 2022. https://ieeexplore.ieee.org/abstract/document/10060807/

Samtani S, Zhao Z, Krishnan R. "Secure Knowledge Management and Cybersecurity in the Era of Artificial Intelligence," *Information Systems Frontiers*, 2023. https://doi.org/10.1007/s10796-023-10372-y

Shieh C-S, Lin W-W, Nguyen T-T, Chen C-H, Horng M-F, Miu D. "Detection of Unknown DDoS Attacks with Deep Learning and Gaussian Mixture Model," *Applied Sciences*, 11(11), 5216, 2021. https://doi.org/10.3390/app11115213

Subhashini R, Khang A (Eds.). "The Role of Internet of Things (IoT) in Smart City Framework," *Smart Cities: IoT Technologies, Big Data Solutions, Cloud Platforms, and Cybersecurity Techniques.* CRC Press, 2024. https://doi.org/10.1201/9781003376064-3

Taddeo M, McCutcheon T, Floridi L. "Trusting Artificial Intelligence in Cybersecurity Is a Double-Edged Sword," *Nature Machine Intelligence*, 1(12), 557–560, 2019. www.nature.com/articles/s42256-019-0109-1

Tailor RK, Ranu Pareek, Khang A (Eds.). "Robot Process Automation in Blockchain," *The Data-Driven Blockchain Ecosystem: Fundamentals, Applications, and Emerging Technologies* (1st Ed., pp. 149–164). CRC Press, 2022. https://doi.org/10.1201/9781003269281-8

Ullah F. et al. "Cyber Security Threats Detection in Internet of Things Using Deep Learning Approach," *IEEE Access*, 7, 124379–124389, 2019. https://ieeexplore.ieee.org/abstract/document/8812669/

Xin Y. et al. "Machine Learning and Deep Learning Methods for Cybersecurity," *IEEE Access*, 6(2), 35365–35381, 2021. www.academia.edu/download/59828182/IRJET-V6I461320190622-57214-128ovl3.pdf

Yampolskiy RV, Spellchecker MS. *Artificial Intelligence Safety and Cybersecurity: A Timeline of AI Failures*, 2016. https://arxiv.org/abs/1610.07997

17 A Smart City Metaverse Using the Internet of Things with Cloud Security

Karthikeyan M. P., Nidhya M. S., Radhamani E., and Ananthi S.

17.1 INTRODUCTION

The word *metaverse* refers to a shared, immersive, and persistent three-dimensional virtual environment with many users that spans a variety of digital platforms and interacts with the actual world in real time while allowing for real-time shopping, work, play, and socializing.

Data are encoded using the encryption technique so that only authorized users may decode it (Mystakidis, 2022). However, encryption by itself is insufficient to protect your data. Data in a database can be protected using several techniques, including access control, data integrity, encryption, and auditing.

Giving your data to a trustworthy custodian, like a data repository, is the most responsible approach to preserving it. Try to keep your research data in a repository that also provides data curation services whenever it is practical to do so (Rani et al., 2021).

17.1.1 BASIC CONCEPTS OF MQTT

Publish/Subscribe: The Message Queue Telemetry Transport (MQTT) protocol uses a concept known as a "Publish/Subscribe," where a publisher publishes messages and users subscribe to topics. Subscribers sign up for specific themes that are relevant to them and, as a result, receive all messages produced on those topics (Alshammari, 2023).

Subscriptions and topics: A publisher in MQTT publishes messages to topics that can be thought of as message subjects. As a result, customers sign up for specific communications by subscribing to themes.

The expressed topic subscriptions might restrict data collection to just that one subject (Zeghida et al., 2023). Topics include two wildcard levels that can be used to get information on a number of related topics.

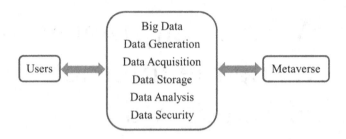

FIGURE 17.1   Big data meets the metaverse.

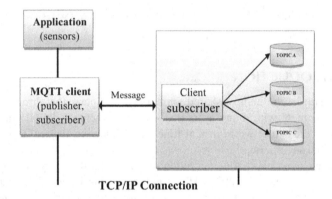

FIGURE 17.2   MQTT architecture.

Quality-of-service (QoS) levels: These are the agreements between two parties to a communication regarding the assurance of data dissemination. Three levels of service quality are supported, detailed in the following.

QoS1 (At most once): This quality level of service does not ensure message delivery and only sends messages a maximum of once.

QoS2 (Exactly once): Using four-way handshaking, the message is conveyed only once under these quality levels of service. The system determines which QoS level to use; for example, if the system requires constant data delivery, it will adjust to QoS2 for data transmission. When a second membership for the same topic is purchased, any saved messages on those topics are sent to the new client (Roldán-Gómez et al., 2022).

Clean sessions and reliable connections: If its value is false, it first connection is taken into account as being permanent. In this task, the delivery of consecutive messages that arrive with the greatest QoS assignment is reserved until the association is reestablished. The use of these flags is not required.

Wills: In the event of an unforeseen disconnect. These are especially useful in systems with security or alarm settings, where managers are alerted right away when a sensor loses contact with the system (Riedel, 2022).

17.1.2 ARTIFICIAL INTELLIGENCE METAVERSE

The metaverse, which is thought to represent the next-generation internet, has drawn a lot of interest from both academia and business where people can access and communicate with others via augmented and virtual reality (AR/VR) technologies.

In essence, it is a universe of endless, interconnected virtual communities where people can communicate via mobile applications to socialize, work, and play (Hwang et al., 2022).

The metaverse is related to the technologies of blockchain, digital twins technology, cloud service and IoT technology, and artificial intelligence (AI) technology, among others.

The architecture of the metaverse is shown in Figure 17.3. In this figure, The metaverse and the physical world are intersected by digital twins technology.

The metaverse is used for hardware and also software platforms and real-time applications in recent technologies (Yang et al., 2022). The architecture ought to connect the real and virtual worlds.

This chapter is divided into five sections. Section 17.1 provided an introduction. Section 17.2 deals with the current method and its limitations of data privacy in the metaverse.

Section 17.3 explores the suggested techniques with the proposed particle swarm optimization (PSO) plus logistic regression (LR) method. The experimental results of testing and their analysis are discussed in Section 17.4, and Section 17.5 provides our conclusion.

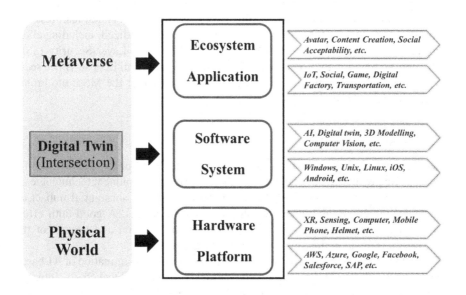

FIGURE 17.3 Architecture of metaverse.

Source: Khang (2021).

17.2 LITERATURE REVIEW

This section discusses the existing or present system that has some limitations in data preservation in metaverse techniques.

17.2.1 THE MQTT PROTOCOL

On top of the transport control protocol, an IoT ecosystem was explored by Dinculeana et al. in 2019. MQTT was approved as a component of OASIS after being standardized by ISO/IEC 20922.

At its core, the publisher subscribes to the communication model used by MQTT, reducing the number of resources required and making this method appropriate for usage in low-bandwidth contexts because clients do not need to be updated.

Updates are sent to MQTT clients by a broker, a server that implements the protocol. Additionally, a server–client system employs the protocol.

Instead of speaking with one another directly, the clients communicate through the broker. Each MQTT message includes a subject that is organized like a tree and a customer with a specific value or instruction and distributes it to all other clients who have subscribed to that specific subject (Khan et al., 2021).

By deciding between three different dependability mechanisms, sometimes referred to as QoS, the protocol is also capable of facilitating dependable transfers.

The MQTT protocol is significantly more suited to contexts with limited resources because it has a lot lower footprint than other protocols like HTTP, as was already mentioned.

While there are many benefits to the MQTT protocol, not all MQTT-based brokers have equivalent or comparable entity authentication or encryption features.

Most of the aspects of the MQTT protocol that are standardized, including client certificates and support for Secure Sockets Layer/Transport Layer Security (SSL/TLS), may be found in the open-source Mosquito program from Eclipse. Since security procedures are required to protect the sent data because the Mosquito broker does not by default (Hindy et al., 2021).

17.2.2 AI METAVERSE

Scientists have explained the term metaverse has been used in this context to refer to a supported by numerous cutting-edge shared virtual environment technologies.

One of these technologies, AI, has been shown to have a substantial impact on increasing immersion and enabling intelligent virtual agents. A good-faith effort to investigate how AI particularly contributed to the creation and growth of the metaverse was made in this survey (Rana et al., 2021).

The primary contributions of this context are an in-depth examination of AI-based techniques in relation to many technological fields (including blockchain, networking, digital twins, natural language processing, machine vision, and neural interface) that have the potential to build metaverse virtual worlds (Luke et al., 2023).

Several important AI-aided applications are also being looked into for possible use in virtual worlds, including those in manufacturing, healthcare, smart cities, and gaming (Lin et al., 2023). Finally, metaverse wrap up their major contribution and lay the groundwork for several potential future AI metaverse research topics.

How the metaverse will influence how consumer research will be researched in the future. In the marketing-focused debate, there has been a lot of discussion about the primary challenges and game-changing prospects for marketers as a result of the possibility of customer engagement with businesses in the metaverse (Kliestik et al., 2022). This study explores the marketing consequences of a fictitious broad adoption of the metaverse, drawing on the expertise of its contributors. They pinpoint fresh lines of inquiry and put forth a fresh paradigm that will benefit academia, industry, and decision-makers alike; value creation and consumer wellness are the results of our future research agenda (Dwivedi et al., 2023).

Aks et al. (2022) take an assessment of blockchain technology for privacy and security in the metaverse. The relationship between the metaverse and the blockchain is established because data are distributed using the blockchain method, and because the data will be sent securely using the data privacy process, the distribution of data will be improved. The universe that the metaverse is now creating is known as the metaverse. Because it creates a separate world, the virtual world will grow quickly. This research is strong because it can identify issues with earlier research because of the method that was utilized, which is innovative based on earlier research (Park et al., 2022). The goal of this research is to understand how the blockchain method and the metaverse are related.

By focusing on data security and privacy, this idea ensures that data dissemination is secure since it employs the blockchain method. The method used shows how to develop a system and test various meta-force and blocking components to have a measurable effect (Khanh and Khang, 2021).

The next-generation internet's metaverse is expected to provide fully immersive and customized experiences for socializing, working, and playing. The development of AI-extended reality (XR) metaverse applications will be largely made possible by improvements in different technologies (Lavdas et al., 2023).

Protecting AI's security in crucial applications like AI-XR metaverse applications is of the utmost importance to prevent unwanted acts that can jeopardize user security and privacy and ultimately endanger users' lives. In particular, a taxonomy of answers to these problems is currently available, and it may be used to create safe, private, reliable, and trustworthy AI-XR applications (Qayyum et al., 2022).

Given the exponential growth of cybercrime, it is obvious that a thorough examination of security in the metaverse based on AI is needed. As a result of a rise in distributed denial-of-service (DDoS) assaults and the theft of user-identifying data, it is crucial to conduct a thorough comprehensive investigation in this area to uncover the Metaverse's vulnerabilities and limitations (Sivasankar, 2022).

According to the findings, user identification, for which biometric techniques are most frequently utilized, is a crucial issue in the works that have been presented.

Although using biometric data is thought to be the safest approach, because it is so special. An AI-based cyber-situation management system should be able to use algorithms to analyze data of any amount.

AI systems may produce erroneous conclusions and false positives when there are few data or occurrences. If organizations do not catch data tampering, it could have devastating results.

Data collection and classification are desirable ideas for developing AI-based metaverse cybersecurity because implicit intelligence systems need big data sets (Pooyandeh et al., 2022).

17.3 SYSTEM DESIGN

This section explains the proposed system algorithm for a smart city metaverse using the IoT with cloud security technology by using PSO with LR algorithms (Khang et al., 2022a).

17.3.1 Sensors

An object that generates an output signal as a sensor detects actual events. Examples of inputs include light, heat, motion, moisture, pressure, and a range of other environmental events. They are utilized in both more commonplace and more industrial applications. Predictive and one or two benefits of using sensors and sensing technology include preventive maintenance. They ensure speedier measurement data transmission while also improving accuracy, which improves asset health and process control.

A smart sensor is a device that gathers data from the environment, analyzes it before sending it, and then uses internal processing capacity to carry out planned actions when specific inputs are acknowledged (Hajimahmud et al., December 30, 2022).

A sensor is a tool used to detect quantities of physical, chemical, or biological qualities and transform them into discernible signals. There are numerous sensors that are easily accessible for practically every industrial application (Hahanov et al., 2022).

17.3.2 IoT and Cloud Security Connected with the Metaverse

The IoT and the metaverse are similar twin technologies that are about to transform how we communicate, collaborate, and make money. They foresee facilitating more active, than passive, internet access for users.

The information provided here explains how the IoT will link the physical world to the metaverse. Integrating the IoT and the metaverse depends heavily on cloud computing. Cloud service providers will support innovation that opens the market to more advanced capabilities.

Developers and organizations may now take advantage of the best data services at their convenience and choice thanks to well-known cloud platforms like Amazon's Amazon Web Services and Microsoft's Azure (Bhambri et al., 2022).

The primary goal of the IoT will be to establish the required connection between the physical world and the internet. Building more sophisticated IoT infrastructure that can readily accommodate the complexities of the digital environment is crucial for empowering the metaverse (Khang et al., June 2023a).

17.3.3 PSO

PSO, a famous method of optimization technique, is based on how swarms move and behave. PSO settles a conflict by utilizing the idea of social interaction. It makes use of a swarm of particles (called agents) that roam the search space in search of the optimum solution.

PSO was chosen for this reason since it has the main benefit of having fewer parameters to tweak. PSO uses particle interaction to find the optimal solution.

Figure 17.4 shows the PSO algorithm function for the aforementioned mathematical expression. The S value is given to the fitness function, and its output is given to the fitness value.

In this structure, the first stage acts as an initialization, and after the few stages, the optimization criteria is checked in the last stage. The optimization criterion "**Yes**" means it gives the optimization solution. Otherwise, "**No**" means it returns back to the evaluation of the fitness function, and this process takes place continuously.

Swarm: A set of particles (S)
Particle: A potential solution as Equation 17.1

$$\text{Position, } Xi = (Xi1, Xi2 \dots Xin) \in Rn \tag{17.1}$$

$$\text{Velocity, } Vi = (Vi1, Vi2 \dots Vin) \in Rn \tag{17.2}$$

Each particle maintains as shown in Figure 17.5.

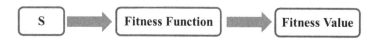

FIGURE 17.4 Particle swarm optimization algorithm function.

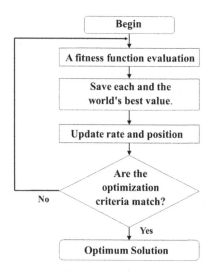

FIGURE 17.5 Basic structure of particle swarm optimization.

Individual best positions are presented in Equations 17.3 and 17.4:

$$Pi = (Pi1, Pi2 \dots Pin) \in Rn \tag{17.3}$$

$$pbesti = f \ (Pi) \tag{17.4}$$

Swarm maintains its global best as shown in Equation 17.5:

$$Pg \in Rn$$

$$gbest = f(Pg) \tag{17.5}$$

Every particle starts out at a different, randomly determined position and velocity. Each particle must maintain its "pbest" or local best position and "gbest" or overall best position among all other particles. The pseudocode for PSO is shown in Figure 17.6.

17.3.4 LR

One of the most fundamental and often used algorithms for solving a classification problem is LR. It is called "logistic regression" because the fundamental methodology is quite similar to that of linear regression.

In a logistic regression, the dependent variable is necessary. Professionals in numerous industries use the logistic regression technique to categorize data for a variety of purposes.

Based on certain dependent variables, the logistic regression machine learning classification technique is used to estimate the likelihood of a given class. Before creating the logistic of the outcome, the logistic regression model essentially computes the sum of the input features (in most cases, there is a bias component).

Initialize Population
for t=1: maximum generation
    for i=1: population size
        Iff $(X_{i,d}(t)) < f(P_i(t))$**then** $P_i(t) = X_{i,d}(t)$
        $f(P_g(t)) = min_i(f(P_i(t)))$
        end
        for d = 1: dimension
        $V_{i,d}(t + 1) = WV_{i,d}(t) + C_1R_1 (P_i - X_{i,d}(t)) + C_2R_2(P_g - X_{i,d}(t))$
        $X_{i,d}(t + 1) = X_{i,d}(t) + V_{i,d}(t + 1)$
        if$V_{i,d}(t + 1) > V_{max}$**then**$V_{i,d}(t + 1) = V_{max}$
        else if$V_{i,d}(t + 1) < V_{min}$**then**$V_{i,d}(t + 1) = V_{min}$
        end
        if$X_{i,d}(t + 1) > X_{max}$**then**$X_{i,d}(t + 1) = X_{max}$
        else if$X_{i,d}(t + 1) < X_{min}$**then**$X_{i,d}(t + 1) = X_{min}$
    end

FIGURE 17.6 Pseudocode of particle swarm optimization.

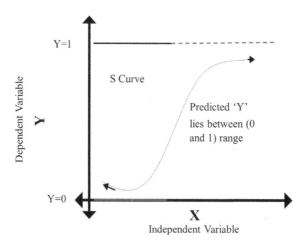

FIGURE 17.7   Logistic regression.

Source: Khang (2021).

Many people who want to be able to better understand their data and forecast trends among its elements will find logistic regression to be helpful.

In Figure 17.7, the numbers "1" and "0" indicate positive and negative classes, respectively. Logistic regression is widely used in binary classification problems, where the outcome variable reveals one of two groups (0 or 1) or (0 or 1).

The following definition applies to the sigmoid function, often known as the activation function for logistic regression as shown in Equation 17.6:

$$f(x) = \frac{1}{1+e^{-x}}, \tag{17.6}$$

where
e is the natural logarithm's base.
"Value" must be transformed into a numerical value.

The following equation is a representation of logistic regression as Equation 17.7:

$$y = \frac{e^{(b0+b1X)}}{1+e^{(b0+b1X)}} \tag{17.7}$$

Here,
x = value of input,
y = output of prediction,
$b0$ = intercept term or bias, and
$b1$ = coefficient for input (x).

Similar to linear regression, this equation predicts the output value after linearly merging the input values using weights or coefficient values.

Input: Training Data
Begin
    For i= 1 to k
        For every occurrence of training data*di*
        Set the regression's goal value to $Z_i = y_i - P(1Id_j)$

$$[P(1Id_j)(1 - P(1Id_j))]$$

        Set the instance weight to zero. *dj* to [(1*Idj*) (1–(*Idj*))]
        Finalize af (j) to the data with class value (Zj) and weight (ωj)
        Classical label decision
      Assign (class label: 1) if *Pid* > 0.5, otherwise (class label: 2)
End

FIGURE 17.8 Pseudocode of logistic regression.

Instead of a numeric value, the output value provided above is a binary value (0 or 1). In contrast to linear regression, the logistic regression pseudocode is shown in Figure 17.8.

17.4 RESULTS AND DISCUSSION

The overall result outperforms the proposed method of PSO with LR is discussed in this section. And it is compared with the existing methods in terms of accuracy, sensitivity, specificity, and *F*-measure metrics.

Convolutional neural network (CNN), particle swarm optimization (PSO), and Lagrangian relaxation (LR) are the existing methods of metaverse data privacy–preserving systems, and the proposed system (PSO+LR) gives the better result. The newly suggested approach makes use of MATLAB 2013 to evaluate the performance metrics (MATLAB, 2023).

The performance evaluation metrics are used to predict how well our trained machine learning models perform. And it helps us determine how well the machine learning algorithm will outperform with a data set that it has never been worked in previously. A confusion matrix performs and determines the output effectively. The working of confusion matrix is explained and shown in Figure 17.9.

Accuracy: The number of correct forecasts divided by the total number of predictions generated by the model yields the accuracy in Table 17.1. The accuracy is calculated by using Equation 17.8:

$$\text{Accuracy} = \frac{TP + TN}{TP + TN + FN + FP} \tag{17.8}$$

The confusion matrix is used to calculate the sensitivity, and it delivers the output result that is explained in previous tabulations and Figure 17.10. The performance output of accuracy for the existing and proposed system is shown in Table 17.1.

The output graph of accuracy for the existing and proposed system is clearly shown in Figure 17.10. From this graph, our proposed PSO + LR system outperforms the other systems with better accuracy.

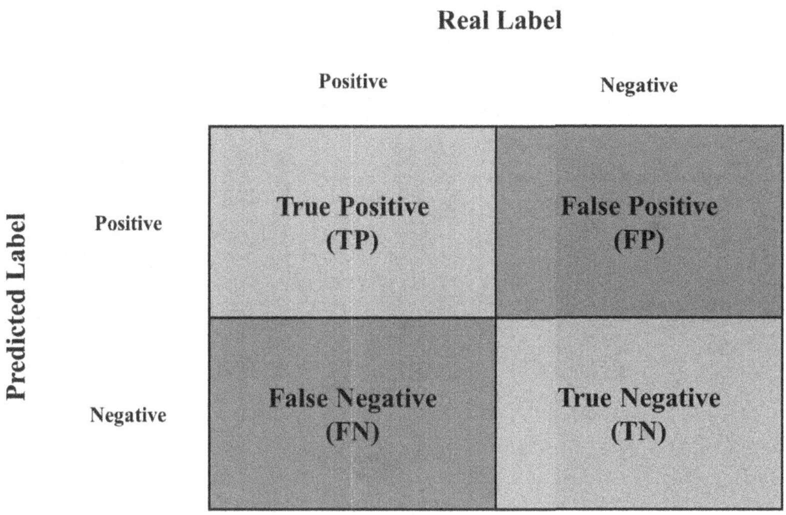

Real Label

FIGURE 17.9   Confusion matrix.

Source: Khang (2021).

TABLE 17.1
Accuracy Results of the Existing and Proposed Algorithm

Algorithm	Result (%)
CNN	76.83
LR	80.12
PSO	85.58
PSO + LR	89.79

Note: CNN = convolutional neural network; LR = logistic regression;
PSO = particle swarm optimization.

Sensitivity: Sensitivity (SN) is calculated as the ratio of the number of correct positive forecasts to the total positives. Equation 17.9 is used for calculating sensitivity.

$$\text{Sensitivity} = \frac{TP}{TP + FN} \tag{17.9}$$

The sensitivity is calculated by using a confusion matrix, and it gives the sensitivity outputs of the existing and proposed systems in Table 17.2.

The sensitivity output graph of the existing and proposed systems is shown in Figure 17.11. The newly proposed PSO + LR algorithm gives the better sensitivity result among all other algorithms.

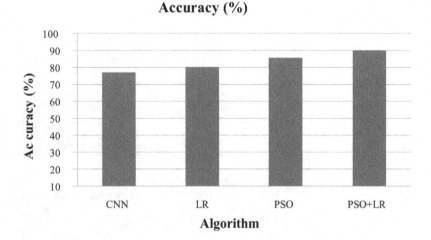

FIGURE 17.10 Accuracy graph for existing and proposed algorithm.

TABLE 17.2

Sensitivity Result of Existing and Proposed Algorithm

Algorithm	Result (%)
CNN	75.54
LR	81.33
PSO	84.27
PSO + LR	90.36

Note: CNN = convolutional neural network; LR = logistic regression; PSO = particle swarm optimization.

Specificity: Calculating specificity (SP) involves dividing the number of accurate negative predictions by the total number of negatives. The specificity is calculated using Equation 17.10.

$$\text{Specificity} = \frac{TN}{(TN + FP)} \qquad (17.10)$$

By using Equation 17.10 with the confusion matrix, the specificity is calculated. And it gives the specificity output results of the existing and proposed systems shown in Table 17.3.

The specificity output graph of the existing and proposed systems is shown in Figure 17.12. The new approach, our proposed PSO + LR algorithm works and gives a better specificity result compared to the other existing algorithms.

F-Measure: The *F*-measure, commonly referred to as the *F*-score, is a mean that balances recall and precision. The *F*-measure is calculated by using Equation 17.11.

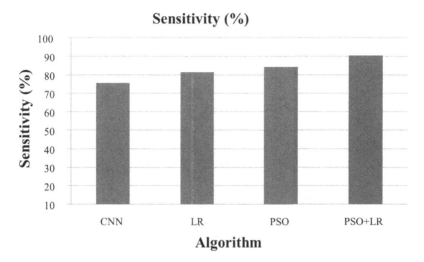

FIGURE 17.11   Sensitivity graph for the existing and proposed algorithm.

TABLE 17.3
Specificity Results of the Existing and Proposed Algorithms

Algorithm	Result (%)
CNN	72.68
LR	77.91
PSO	86.53
PSO + LR	91.42

Note: CNN = convolutional neural network; LR = logistic regression;
PSO = particle swarm optimization.

$$\text{F-measure} = \frac{2 * recall * precision}{recall + precision} \tag{17.11}$$

The confusion matrix is used to determine the *F*-measure value by using Equation 17.11, and the obtained outputs of the existing and proposed systems are shown in Table 17.4.

The *F*-measure output result graph of the existing and proposed systems is shown in Figure 17.13.

The newly constructed PSO + LR algorithm performs well, and it gives a better *F*-measure result than the other existing algorithms.

The overall performance of existing systems is compared with the newly designed hybrid, and the PSO+LR algorithm is determined by using the performance matrix.

The measuring parameters of accuracy, sensitivity, specificity, and *F*-measure values are calculated and their output noted.

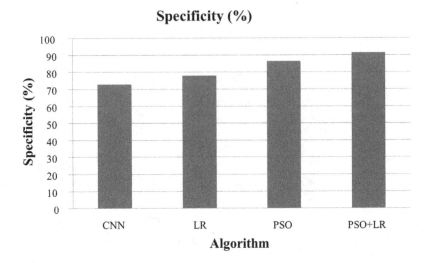

FIGURE 17.12   Specificity graph for the existing and proposed algorithms.

TABLE 17.4
F-Measure Results of the Existing and Proposed Algorithms

Algorithm	Result (%)
CNN	73.85
LR	79.49
PSO	87.68
PSO + LR	92.36

Note: CNN = convolutional neural network; LR = logistic regression;
PSO = particle swarm optimization.

The bar chart graph was plotted from the noted values. Depending on the output results, our proposed PSO + LR algorithm gives better results compared to the existing systems of CNN, LR, and PSO.

17.5 CONCLUSION

The metaverse offers numerous opportunities in order to develop the next generation of internet users. As the metaverse is still developing, it is difficult to assess its potential and assign specific tags to it.

The IoT, by comparison, has been around for years; therefore, working together, they will be extremely beneficial for both concepts where the partnership can pursue further penetration into the metaverse. This study presents a comprehensive analysis of the benefits of the metaverse, its effects on people, its uses, and related technologies (Khang et al., 2022b).

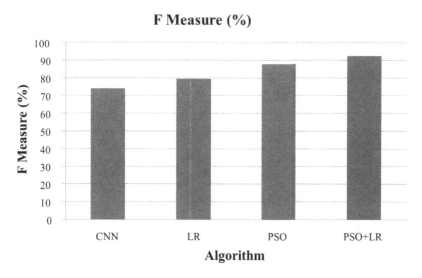

FIGURE 17.13 *F*-measure graph for existing and proposed algorithm.

Tech professionals will be able to manifest the unconventional thinking that is constrained in the conventional setup thanks to the intersection of the IoT and the metaverse. The IoT can be used to reveal new operating capacities. Despite this, we must continue to be concerned about cyber threats that could exacerbate the weaknesses in the metaverse ecosystem. The metaverse used for smart city by IoT and cloud security techniques are processed based on the suggested system. The newly constructed hybrid PSO + LR works well and delivers the very good output results compared to the existing algorithms (Shah et al., 2023).

The suggested hybrid optimization technique for Metaverse ecosystem plays a vital role in privacy and security for a smart city when using IoT and cloud. In-depth discussions are also had about a few significant of open issues and opportunities (Khang et al., 2023b).

At last, we bring this chapter to a close. With this study, we hope to increase the metaverse's allure for businesses and smart-city development. In order to open up new options for themselves within the industry, consumers should take advantage of the expansion and investigate the metaverse IoT.

REFERENCES

Aks SMY, Karmila M, Givan B, Hendratna G, Setiawan HS, Putra AS, Herawaty MT. "A Review of Blockchain for Security Data Privacy with Metaverse," *2022 International Conference on ICT for Smart Society (ICISS)* (pp. 1–5). IEEE, August 2022. https://iee-explore.ieee.org/abstract/document/9915055/

Alshammari HH. "The Internet of Things Healthcare Monitoring System Based on MQTT Protocol," *Alexandria Engineering Journal*, 69, 275–287, 2023. www.sciencedirect.com/science/article/pii/S1110016823000881

Bhambri P, Rani S, Gupta G, Khang A. *Cloud and Fog Computing Platforms for Internet of Things*. CRC Press, 2022. https://doi.org/10.1201/9781003213888

Dwivedi YK, Hughes L, Wang Y, Alalwan AA, Ahn SJ, Balakrishnan J, Wirtz J. "Metaverse Marketing: How the Metaverse Will Shape the Future of Consumer Research and Practice," *Psychology & Marketing*, 40(4), 750–776, 2023. https://onlinelibrary.wiley.com/doi/abs/10.1002/mar.21767

Gorzałczany MB, Rudziński F. "Intrusion Detection in Internet of Things with MQTT Protocol-An Accurate and Interpretable Genetic-Fuzzy Rule-Based Solution," *IEEE Internet of Things Journal*, 9(24), 24843–24855, 2022. https://ieeexplore.ieee.org/abstract/document/9844142/

Hahanov V, Khang A, Litvinova E, Chumachenko S, Hajimahmud VA, Alyar VA. "The Key Assistant of Smart City—Sensors and Tools," *AI-Centric Smart City Ecosystems: Technologies, Design and Implementation* (1st Ed., vol. 17, p. 10). CRC Press, December 30, 2022. https://doi.org/10.1201/9781003252542-17

Hajimahmud VA, Khang A, Hahanov V, Litvinova E, Chumachenko S, Alyar VA. "Autonomous Robots for Smart City: Closer to Augmented Humanity," *AI-Centric Smart City Ecosystems: Technologies, Design and Implementation* (1st Ed., vol. 7, p. 12). CRC Press, December 30, 2022. https://doi.org/10.1201/9781003252542-7

Hindy H, Bayne E, Bures M, Atkinson R, Tachtatzis C, Bellekens X. "Machine Learning Based IoT Intrusion Detection System: An MQTT Case Study (MQTT-IoT-IDS2020 Dataset)," *Selected Papers from the 12th International Networking Conference: INC 2020* (pp. 73–84). Springer International Publishing, January 2021. https://link.springer.com/chapter/10.1007/978-3-030-64758-2_6

Hwang GJ, Chien SY. "Definition, Roles, and Potential Research Issues of the Metaverse in Education: An Artificial Intelligence Perspective," *Computers and Education: Artificial Intelligence*, 3, 100082, 2022. www.sciencedirect.com/science/article/pii/S2666920X22000376

Khan MA, Khan MA, Jan SU, Ahmad J, Jamal SS, Shah AA, Buchanan WJ. "A Deep Learning-Based Intrusion Detection System for MQTT Enabled IoT," *Sensors*, 21(21), 7016, 2021. www.mdpi.com/1325806

Khang A. "Material4Studies," *Material of Computer Science, Artificial Intelligence, Data Science, IoT, Blockchain, Cloud, Metaverse, Cybersecurity for Studies*, 2021. www.researchgate.net/publication/370156102_Material4Studies

Khang A, Gupta SK, Rani S, Karras DA (Eds.). *Smart Cities: IoT Technologies, Big Data Solutions, Cloud Platforms, and Cybersecurity Techniques*. CRC Press, 2023b. https://doi.org/10.1201/9781003376064

Khang A, Hahanov V, Abbas GL, Hajimahmud VA. "Cyber-Physical-Social System and İncident Management," *AI-Centric Smart City Ecosystems: Technologies, Design and Implementation* (1st Ed., vol. 2, p. 15). CRC Press, December 30, 2022a. https://doi.org/10.1201/9781003252542-2

Khang A, Rani S, Sivaraman AK. *AI-Centric Smart City Ecosystems: Technologies, Design and Implementation* (1st Ed.). CRC Press, December 30, 2022b. https://doi.org/10.1201/9781003252542

Khang A, Shah V, Rani S. *AI-Based Technologies and Applications in the Era of the Metaverse* (1st Ed.). IGI Global Press, June 2023a. https://doi.org/10.4018/9781668488515

Khanh HH, Khang A. "The Role of Artificial Intelligence in Blockchain Applications," *Reinventing Manufacturing and Business Processes Through Artificial Intelligence* (vol. 2, pp. 20–40). CRC Press, 2021. https://doi.org/10.1201/9781003145011-2

Kliestik T, Novak A, Lăzăroiu G. "Live Shopping in the Metaverse: Visual and Spatial Analytics, Cognitive Artificial Intelligence Techniques and Algorithms, and Immersive Digital Simulations," *Linguistic and Philosophical Investigations*, 21, 187–202, 2022. www.ceeol.com/search/article-detail?id=1045822

Lavdas AA, Mehaffy MW, Salingaros NA. "AI, the Beauty of Places, and the Metaverse: Beyond 'Geometrical Fundamentalism'," *Architectural Intelligence*, 2(1), 8, 2023. https://link.springer.com/article/10.1007/s44223-023-00026-z

Lin Y, Du H, Niyato D, Nie J, Zhang J, Cheng Y, Yang Z. "Blockchain-Aided Secure Semantic Communication for AI-Generated Content in Metaverse," *IEEE Open Journal of the Computer Society*, 4, 72–83, 2023. https://ieeexplore.ieee.org/abstract/document/10079087/

Luke J, Khang A, Chandrasekar V, Pravin AR, Sriram K (Eds.). "Smart City Concepts, Models, Technologies and Applications," *Smart Cities: IoT Technologies, Big Data Solutions, Cloud Platforms, and Cybersecurity Techniques* (1st Ed.). CRC Press, 2023. https://doi.org/10.1201/9781003376064-1

MATLAB. *Take Your Ideas beyond Research to Production*, 2023. https://www.mathworks.com/products/matlab.html

Mystakidis S. "Metaverse," *Encyclopedia*, 2(1), 486–497, 2022. www.mdpi.com/1492130

Park WH, Siddiqui IF, Qureshi NMF. "AI-Enabled Grouping Bridgehead to Secure Penetration Topics of Metaverse," *Computers, Materials and Continua*, 73(3), 5609–5624, 2022. https://scholarworks.bwise.kr/skku/handle/2021.sw.skku/99856

Pooyandeh M, Han KJ, Sohn I. "Cybersecurity in the AI-Based Metaverse: A Survey," *Applied Sciences*, 12(24), 12993, 2022. www.mdpi.com/article/10.3390/app122412993

Qayyum A, Butt MA, Ali H, Usman M, Halabi O, Al-Fuqaha A, Qadir J. *Secure and Trustworthy Artificial Intelligence-Extended Reality (AI-XR) for Metaverses*, 2022, arXiv preprint arXiv:2210.13289. https://arxiv.org/abs/2210.13289

Rana G, Khang A, Sharma R, Goel AK, Dubey AK (Eds.). *Reinventing Manufacturing and Business Processes Through Artificial Intelligence*. CRC Press, 2021. https://doi.org/10.1201/9781003145011

Rani S, Chauhan M, Kataria A, Khang A (Eds.). "IoT Equipped Intelligent Distributed Framework for Smart Healthcare Systems," *Networking and Internet Architecture* (vol. 2, p. 30). CRC Press, 2021. https://doi.org/10.48550/arXiv.2110.04997

Riedel E. "MQTT Protocol for SME Foundries: Potential as an Entry Point into Industry 4.0, Process Transparency and Sustainability," *Procedia CIRP*, 105, 601–606, 2022. www.sciencedirect.com/science/article/pii/S2212827122001019

Roldán-Gómez J, Carrillo-Mondéjar J, Castelo Gómez JM, Ruiz-Villafranca S. "Security Analysis of the MQTT-SN Protocol for the Internet of Things," *Applied Sciences*, 12(21), 10991, 2022. www.mdpi.com/1915730

Shah V, Jani S, Khang A (Eds.). "Automotive IoT: Accelerating the Automobile Industry's Long-Term Sustainability in Smart City Development Strategy," *Smart Cities: IoT Technologies, Big Data Solutions, Cloud Platforms, and Cybersecurity Techniques*. CRC Press, 2023. https://doi.org/10.1201/9781003376064-9

Sivasankar GA. "Study of Blockchain Technology, AI and Digital Networking in Metaverse," *IRE Journals*, 5(8), 110–115, 2022. www.irejournals.com/formatedpaper/1703198.pdf

Yang Q, Zhao Y, Huang H, Xiong Z, Kang J, Zheng Z. "Fusing Blockchain and AI with Metaverse: A Survey," *IEEE Open Journal of the Computer Society*, 3, 122–136, 2022. https://ieeexplore.ieee.org/abstract/document/9815155/

Zeghida H, Boulaiche M, Chikh R. "Securing MQTT Protocol for IoT Environment Using IDS Based on Ensemble Learning," *International Journal of Information Security*, 1–12, 2023. https://link.springer.com/article/10.1007/s10207-023-00681-3

18 Internet of Things–Integrated Cloud and Data Solutions

Sanskar Verma, Vrinda Vishnoi, Rajat Verma, Namrata Dhanda, and Anil Kumar

18.1 INTRODUCTION

The Internet of Things (IoT) has revolutionized the way people collaborate with the physical world. With the internet's capability to link billions of devices and things, real-time data collection and sharing are made possible (Conti et al., 2018). This has created an unprecedented opportunity to gain insights and make data-driven decisions based on the information generated by these connected devices (Feki et al., 2022).

An IoT-integrated cloud and data solution is a system that leverages the power of the IoT by connecting physical devices to the cloud, where the data created by these devices can be collected, processed, and stored. It allows businesses to assemble and analyze gigantic amounts of data from numerous sources, resulting in insightful data that can be utilized to enhance operations, create new products, and spur innovation (Zhan et al., 2017).

The cloud infrastructure is the backbone of the IoT-integrated cloud and data solution, providing the necessary assessing and storage resources required to host and run the applications and services that make up the system. This infrastructure is typically accessed over the internet and allows organizations to scale their resources as needed without the need to invest in expensive on-premises hardware.

Data processing and analytics are critical components of an IoT-integrated cloud and data solution. These tools allow organizations to extract insights from the data generated by IoT devices and apply machine learning algorithms, statistical inspection, and other moves to gain new insights (Saggi et al., 2018).

IoT-integrated cloud and data solutions can be used in various domains like healthcare, smart cities, retail, and manufacturing where the data launched by IoT devices can be used to optimize operations, reduce costs, and improve the overall user experience.

In summary, IoT-integrated cloud and data solutions allow organizations to connect physical devices to the cloud, where the data generated by these devices can be assembled, analyzed, and stored. This provides a powerful means for organizations to gain insights, improve operations, and drive innovation, making it a key technology for the digital transmutation of business and industry (Berman et al., 2012). The IoT is highlighted in Figure 18.1.

DOI: 10.1201/9781003392224-18

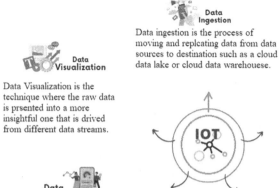
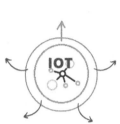

Data Ingestion

Data ingestion is the process of moving and replcating data from data sources to destination such as a cloud data lake or cloud data warehouese.

Data Visualization

Data Visualization is the technique where the raw data is prsented into a more insightful one that is drived from different data streams.

Data Processing

Data processing is the collection and manupulation of items of data to procedure meaningful information

Data Analysis

Data Analytics is the process of systematically applying statistical and / or logical techniques to describe and illustrate, condense and recap, and evaluate data.

Data Transmission

The collected data by IoT nodes will be transferred to a gaetway server which may aggregate data and send it to a cloud platform fir further data processing

FIGURE 18.1   Illustration showing the Internet of Things.

In Figure 18.1, the major components of IoT have highlighted the architecture that tells about the data ingestion–data can be as simple as humidity by collecting from the resources. The data need to be analyzed once they are dispatched to the cloud, data visualization. It is necessary to process data to make it insightful for users.

The user has the interface to check in to the system and gain a better understanding that aids in forecasting data for future occurrences; data analysis and prediction are conducted using historical data. Data transmission is conducted to the cloud through the doorway to ensure data security protocols are used during data processing. Data processing on the IoT platform involves obtaining, manipulating, or categorizing data from the cloud.

18.2   MATERIAL AND METHODOLOGY

18.2.1   Problem Statement

The inability to gather, analyze, and analyze substantial volumes of data from IoT devices in a way that is effective, affordable, and scalable is the issue that IoT-integrated cloud and data solutions seek to address.

One of the main predicaments in implementing an IoT-integrated cloud and data solution is dealing with the sheer volume and velocity of data that is launched by IoT devices. These devices can generate a humongous amount of data, which can be difficult and expensive to store and process using traditional on-premises hardware and software.

The variety of data that IoT devices create presents another difficulty. These data may be quite heterogeneous, with various data kinds arriving from many sources and in various forms. This can make it difficult to integrate and analyze these data and may require specialized data processing and analytics tools.

The security of the data generated by IoT devices is a major concern. These devices might be vulnerable to a range of cyber threats, including malware, denial-of-service attacks, and hacking. An IoT-integrated cloud and data solution must be successful in protecting the data and guaranteeing its integrity and confidentiality (Hahanov et al., 2022).

The ability to draw insights from the data produced by IoT devices and utilize those insights to improve operations, foster innovation, and create new products and services is another issue (Khang et al., 2023c).

The successful adoption of IoT for providing these services faces several difficulties. In summary, the problem that IoT-integrated cloud and data solutions aim to address is the capability to assemble, rectify, and evaluate substantial amounts of information from IoT devices in a way that is efficient, cost-effective, and scalable while ensuring the security of the data and being able to wrest valuable insights from it.

18.2.2 CLOUD AND ITS REQUIREMENTS IN THE IoT

IoT is the technological tool that makes the dream of a fully interconnected world a reality. The IoT domain is anticipated to provide services like security, scalability, effective resource use, and high-performance metrics to realize this future vision.

An IoT cloud is a sizable network that supports IoT devices and applications. This includes the auxiliary hardware, servers, and storage needed for instantaneous processes and processing (Rao et al., 2012). In many industries, the combination of cloud computing with IoT may be quite beneficial. This will enable the cloud to connect to actual physical items since IoT makes it possible to link heterogeneous objects.

By providing the required infrastructure, platform, and software services for the gathering, storage, and evaluation of the data released by IoT devices, cloud computing impersonates a key role in the IoT (Darwish et al., 2017). The key requirements for cloud-based IoT systems include scalability, security, data processing and analysis, integration, flexibility, real-time processing, high availability, reliability, and cost-effectiveness (Suciu et al., 2013).

Scalability enables organizations to adjust resources in real time, security ensures the protection of data and devices, data processing enables insights extraction and real-time processing allows decision-making, integration allows cross-system data sharing, flexibility enables a wide range of IoT use cases, while availability, reliability, and cost-effectiveness ensure smooth and continuous operation at a reasonable cost. Cloud provides the necessary means of control and flexibility to the IoT solution (Farahzadi et al., 2017).

18.2.3 REQUIREMENTS FOR CLOUD-BASED IoT ARCHITECTURE

The requirements and description of IoT are shown in Table 18.1 for easy understanding.

This table outlines the various requirements that a cloud-based IoT architecture must meet to be effective. The requirements include scalability, security, remote

TABLE 18.1

Key Requirements for Cloud-Based IoT Architecture

Requirement	Description
Scalability	The ability to quickly scale up or down as necessary to manage a large number of devices and a large amount of data.
Security	Attacks concerning the internet are protected by strong security measures like encryption of data and verification for users.
Remote Management	The ability to remotely manage and monitor IoT devices to detect and respond to security threats more quickly.
Data Processing and Analysis	Ability to process, analyze, and store enormous amounts of data generated by Internet of Things (IoT) devices.
Integration	The ability to integrate multiple IoT devices and systems, as well as integration of IoT data with other data sources.
Flexibility	Support for many platforms, devices, and protocols is required to support a variety of IoT use cases.
Real-Time Processing	Real-time data handling and processing are essential for instantaneous administration and making choices.

management, data rectifying and analysis, integration, flexibility, and real-time processing (Verma et al., 2023).

Scalability ensures that a substantial number of devices and a substantial volume of information can be handled, and security offers strong security measures like data encryption and user authentication to fend against online dangers (Junaid et al., 2023).

Faster detection and response to security risks are made possible by remote management, and data created by IoT devices may be processed, examined, and stored via data rectification and analysis (Shukla et al., 2023).

Multiple IoT systems and devices can be combined through integration, and IoT data can be combined with data from other sources (Khang et al., 2022a).

Flexibility to support various protocols, platforms, and devices to enable a variety of IoT use cases, and to validate real-time decision-making and control, real-time processing enables handling and processing of data in actual time (Khang et al., 2023a). All these requirements must be met for a cloud-based IoT architecture to be effective and functional (Costa et al., 2020).

The cloud-based IoT design should also offer high availability, cost-effectiveness, and dependability in addition to these objectives. Many cloud service providers, including Google Cloud Platform, Amazon Web Services, and connectivity services, provide IoT-specific services, such as analytics, data processing, and data storage, which assist organizations in meeting the requirements for their IoT implementation and speed up their IoT projects.

Notably, to maximize IoT, a well-designed and optimized architecture that meets the particular requirements of the IoT application are required. Understanding the needs for data flow, data processing, networks, devices, and security is all-important to accomplish this (Sisinni et al., 2018).

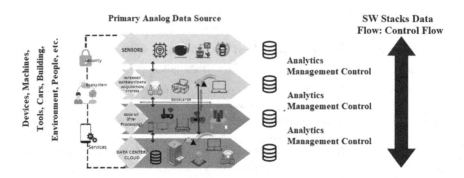

FIGURE 18.2   The four phases of IoT solutions formation.

Organizations may select the appropriate cloud services and create a tailored solution to match their unique demands once they are understood (Henderson et al., 1985). The four phases of IoT Solutions Formation are shown in Figure 18.2.

The four stages of IoT solutions architecture are typically considered to be the following:

- **Device/sensor layer:** This is the layout of the IoT solution, which includes the sensors, devices, and actuators that collect and transmit data (Khang et al., 2023b).
- **Network layer:** This part of the stack is in charge of the infrastructure and protocols utilized for data transmission, as well as the connectivity and communication between the devices (Verma et al., 2020).
- **Data layer:** The data handling, storage, and access layer is in charge of channeling the information the devices have acquired for use in evaluation and decision-making (Rani et al., 2021).
- **Application layer:** The IoT solution's top layer is called the application layer, where data are utilized to create apps and analytics that provide value to businesses (Bhambri et al., 2022).

18.3   RELATIONSHIP OF IOT-INTEGRATED CLOUD AND DATA WITH OTHER SYSTEMS

18.3.1   AN IOT-INTEGRATED CLOUD AND DATA SOLUTION

An IoT-integrated cloud and data solution can work with other systems in an assortment of ways, reliant on the specific requirements of the application.

- Cloud and data solutions for the IoT are regularly connected with other systems via APIs (application programming interfaces). The IoT-integrated cloud and data solution may interface with other systems like enterprise resource planning (ERP) systems, and manufacturing execution systems (MESs) since it enables various systems to talk with one another and share data.

- Another way to integrate an IoT-integrated cloud and data solution with other systems is through the use of middleware (Symeonaki et al., 2020). Middleware acts as a bridge between distinct systems, allowing them to communicate and share data. This can include message brokers, integration brokers, and event-driven architecture (Khriji et al., 2021).
- The IoT-integrated cloud and data solution can also be integrated with other systems by using data analytics tools that enable the data to be visualized, reported, dash-boarding, and used for the decision-making process. Edge computing, which allows data processing to take place closer to IoT devices at the network's edge rather than in the cloud, can also be incorporated into the system (Atlam et al., 2018). This can help minimize latency and enhance the responsiveness of the system (Blumenthal, 2013).

In conclusion, there are several methods to interface an IoT-integrated cloud and data solution with other systems, such by using APIs, middleware, data analytics tools, and edge computing (Kim et al., 2019).

Understanding the requirements for data flow, data rectifying, networks, devices, and security can help you select the integration strategies that will best serve the unique needs of your IoT application.

18.3.2 Reactive Machine Learning Models in IoT-Integrated Scenario

Reactive machine learning (ML) models are a type of ML model that can make predictions or decisions in real time based on streaming data (Akbar et al., 2017). They are well suited for use in IoT-integrated systems because they can process and evaluate data from IoT devices in real time.

In IoT-integrated systems, reactive ML models can be used for a variety of tasks:

- **Anomaly detection:** identifying unusual patterns or behaviors in sensor data that may indicate a malfunction or other issue
- **Predictive maintenance:** analyzing sensor data to predict when equipment or machines may require maintenance or repair
- **Smart home automation:** analyzing sensor data from smart home devices to control lighting, temperature, and other settings
- **Industrial IoT:** analyzing sensor data from industrial equipment to optimize operations and improve efficiency

Depending on the particular needs of the IoT-integrated system, reactive ML models can be deployed at the edge or in a cloud context (Shumba et al., 2022). While edge deployment lowers latency and bandwidth needs, cloud deployment enables the use of more powerful resources.

In summary, reactive ML models are a type of ML model that can make predictions or decisions in real time based on streaming data. They are perfect for use in systems that include the IoT since they can manage and analyze data from IoT devices in real time. These models may be used for a variety of activities, including anomaly detection, predictive maintenance, smart home automation, and industrial

IoT. Depending on the particular needs of the system, they can be positioned in the cloud or at the edge (Huang et al., 2021).

18.4 IOT DEVICES PROVISION AND ONBOARDING

Configuring and setting up IoT devices so they can connect to and communicate with an IoT network and platform are known as IoT device provisioning and onboarding (Szabo et al., 2021).

The device must first be registered with the network, the required software and firmware must be installed, and security configurations must be made. Assigning a distinctive identification to the device, such as an Internet Protocol (IP) address or device identifier (ID), as well as defining the communication protocols and security settings that will be used to send data are all part of the provisioning process (Vollmer et al., 2014).

The process of integrating the device into the larger IoT ecosystem, including the software and systems that will be used to manage, analyze, and act on the data provided by the device, is referred to as onboarding. This may include integrating the device with other cloud platforms, data analytics tools, and business systems (Lee et al., 2019).

There are numerous ways to accomplish IoT devices' provisioning and onboarding:

- **Over-the-air (OTA) updates:** OTA updates enable the configuration and installation of software and firmware without having physical access to the device, enabling remote provisioning and onboarding of devices.
- **Manual provisioning:** Devices can be provisioned and onboarded manually by connecting to the device directly and configuring settings and software through a user interface.
- **Automatic provisioning:** By utilizing specialized software that scans the device, verifies it, and assigns it an identity, security settings, and communication protocols, some devices can be automatically provisioned and onboarded.

Overall, provisioning and onboarding are crucial steps in the deployment of an IoT solution, as they ensure that devices are properly configured and integrated into the overall system, enabling them to collect and transmit data effectively (Stock et al., 2014).

18.4.1 IoT Data Management and Storage

The process of acquiring, arranging, and storing the large amount of data produced by IoT devices and sensors is referred to as IoT data management and storage (Verma et al., 2021). The following are the essential components of IoT data handling and storage:

- **Data collection:** It is the efficient and safe collection of data from various IoT gadgets and devices (Verma et al., 2022).

- **Data storage:** It is a strategy for storing substantial volumes of IoT data in a versatile and expandable way, either on-premises or in the cloud (Rani et al., 2023).
- **Data analytics:** The IoT data are examined to glean insightful information and guide judgment.
- **Data visualization:** IoT data are made visually appealing to make data analysis easier and give stakeholders insights that are simple to understand (Khanh and Khang, 2021)
- **Data management:** Organizing and managing IoT data are organized and managed, which includes data integration, data archiving, and data quality control (Wolfert et al., 2017).
- **Data security:** IoT data are kept secure and shielded from unauthorized access or alteration.
- **Data privacy:** Data privacy includes ensuring IoT data privacy and following data protection laws.

In conclusion, IoT handling of data and storage is a crucial part of IoT technology that enables businesses to transform raw IoT data into insights that influence business outcomes (Wolfert et al., 2017).

The procedure of systematizing, storing, and securely managing data is referred to as data management (Poltavtseva et al., 2019).

The technology and methods utilized to store electronic information are referred to as data storage (Corallo et al., 2022). These two ideas go hand in hand since secure and effective data storage is a prerequisite for good data management. The purpose of data management and storage is to make sure that data is trustworthy, accessible, and usable to support company operations and decision-making (Ginige et al., 2007).

18.4.2 FRAMEWORK DEPICTION

Six layers of stacking make up the recommended IoT data management system, two of which feature sublayers and complementary (Elkheir et al., 2013). Figure 18.3 shows how the lookup/orchestration process is viewed as an extra activity that, while technically not a part of the data life cycle, is described together with the structure of layers that closely resemble the stages of the IoT data life cycle (Vermesam et al., 2022).

IoT sensors, smart devices (data production items), and modules for in-network processing, as well as information collection/real-time aggregation, are all included in the "Things" layer (processing, gathering; Sandeepa et al., 2022).

The communication layer supports the conveyance of demands, questions, information, and outcomes (Notaro et al., 2021). The data/sources twin layers take care of finding and cataloguing data sources as well as storing and indexing gathered data (data sources; Chen et al., 2014).

The data layer also manages the processing of data and queries for regional, independent data depository sites (interpreting, screening, prior processing; Sarramia et al., 2022).

The association layer offers the data repository integration and abstraction required for global analysis requests (preprocessing, collaboration, merging) by exploiting the

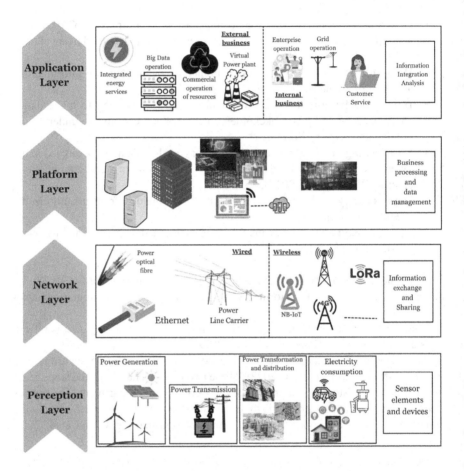

FIGURE 18.3 Internet of Things data life cycle.

metadata provided in the data sources layer to enable both real-time source integrations and geographically targeted queries (Chakravarthi, 2021).

The query layer works with the federation layer and the complementing transactions layer to manage the complexities of processing requests and to optimize (Elkheir et al., 2013).

18.5   IOT DATA WAREHOUSING VERSUS DATA LAKES

A collection of interconnected machines, sensors, and gadgets known as the IoT that collect and share data (Gupta et al., 2020). Making informed choices, increasing efficiency, and creating new business models can all be done with the help of these data (Lee, 2019).

However, it could be difficult to manage and analyze the massive volumes of data that IoT devices generate (Jin et al., 2013). Data lakes and data warehouses are useful in this situation (Aissi et al., 2021)

18.5.1 Data Warehouses

A central data depository known as business intelligence (BI) functions including reporting, research, and data mining are supported by a data warehouse. Data are gathered from vivid citations, formatted uniformly, and organized for archiving (Kondo et al., 2010). To obtain insights into business performance, it is simple to search, analyze, and visualize this structured data (Sieber, 1991).

Data storage in the IoT entails gathering data from various devices and sensors, converting it to a standardized format, and archiving it in a central location (Emeakaroha et al., 2015). Then, with the help of these data, decisions can be made more effectively by spotting patterns and optimizing efficiency (Tang et al., 2019). The data warehouse is shown in Figure 18.4.

Data are collected from various sources, converted into a uniform format, and archived in an orderly manner (Anisimov, 2003). These data can then be easily searched, analyzed, and visualized to gain insights into business performance (Kandogan et al., 2015).

18.5.2 Data Lakes

A significant, consolidated depository of both unstructured and structured data that enables big data analytics (Liu et al., 2020). Data lakes, as opposed to data warehouses, keep data in their original format, providing more flexibility and quicker processing times (Khine et al., 2018). Text, pictures, videos, and sensor data can all be stored in data pools (Constantiou et al., 2014).

Data lakes are used in the IoT context to store and handle the huge volumes of unorganized and semi-generated data produced by IoT devices (Monino, 2016). New patterns and relationships that might not be obvious in structured data can be found using these data (Sutton et al., 1995) as Figure 18.5 illustrates the concept of data lakes.

To gain insights and utility from these data, a variety of tools and technologies can be used to process them (Alohakoon and Yu, 2016). The ideal place to store and

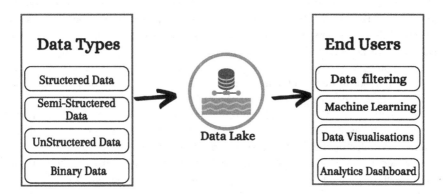

FIGURE 18.4 Data warehousing.

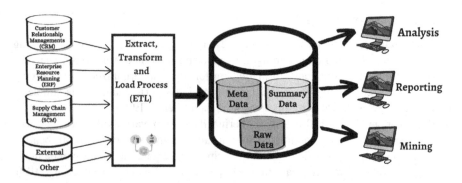

FIGURE 18.5 Image depicting data lakes.

handle the huge volumes of semistructured and unorganized data created by IoT devices is in data lakes.

18.5.3 THE DISTINCTION BETWEEN DATA WAREHOUSES AND DATA LAKES

Although the terms *data lakes* and *data warehouses* are frequently used interchangeably, they are not the same (Ramchand et al., 2022). The various structures between them are one of the biggest differences (i.e., processed vs. raw data; Firouzi et al., 2020).

Data lakes hold raw or unprocessed data, while data warehouses store data that have been processed and filtered. To be more precise, structured and organized data are delivered to a data lake whereas raw and unstructured information is first sorted into a single schema and then stored in the warehouse. In the database, analysis is done on the cleaned-up data. Instead, in a data pool, data are picked and arranged as and when they are required (Munappy et al., 2019).

An affordable option for keeping a data warehouse contains data that have been analyzed. By comparison, data lakes are superior to data warehouses in terms of capacity and are perfect for ML applications and the study of raw, unprocessed data.

The goal or use is another important distinction (Nambiar et al., 2022). The data lake's intended function is not stated, and in a perfect world, it may be utilized for any purpose. But there will not be a waste of the allocated storage space because the processed data that enter data warehouses are regularly employed for certain objectives. There is no need for specialized knowledge in order to use filtered or processed data; all you need to know is how data are shown (e.g., in graphs, spreadsheets, tables of data, and slideshows; Heckmann et al., 2019). Therefore, any company or person can use data warehouses (Corbellini et al., 2017).

Contrarily, it is relatively challenging to analyze data lakes without experience with raw data; as a result, data scientists with the necessary expertise or tools are required to understand them for particular business use (Elshawi et al., 2018).

Another feature that sets data lakes apart from data warehouses is the utility or accessibility of data repositories. A data lake's architecture is flexible because it lacks a formal framework (Filar et al., 1996). Instead, a data warehouse's construction ensures that no outside particles enter it, and manipulating one is expensive. It is also very secure thanks to this function (Carciochi et al., 2017).

18.6 IOT EDGE COMPUTING

The "edge" of the network or a data center, which is closest to the source of the data, is where data processing and analysis are carried out under the IoT Edge computing paradigm (Giannoutakis et al., 2020). As a result, network traffic and latency are reduced while thinking and decision-making are fastened up.

18.6.1 EDGE COMPUTING

Edge computing in the IoT context entails the deployment of edge devices, which are compact computing units placed near the sensors and IoT devices (Papageorgiou et al., 2015). Data filtering, aggregation, preprocessing, and analysis are just a few of the duties that these devices are capable of performing. They can also offer real-time insights and actions based on the data they process (Badidi et al., 2019).

These are a few of the primary benefits of IoT peripheral computing:

- **Reduced latency:** Edge computing reduces latency and allows real-time decision-making by processing data closer to the source. Optimizing bandwidth means only pertinent data are transmitted to the cloud or information hub by filtering and aggregating them at the periphery, which lowers network traffic and bandwidth usage.
- **Enhanced data privacy:** Edge computing can increase data security by decreasing the need to transfer sensitive data to the cloud or data center.
- **Reliability:** By working and analyzing at the edge, edge devices can continue to work and offer insights even if the cloud or data center is down.

18.6.2 EDGE COMPUTING ARCHITECTURE

The broad adoption of the edge computing paradigm was also made feasible by the significant growth of IoT applications (Naveen et al., 2019) in that it decreases bandwidth and reaction times; edge computing architecture is similar to that of cloud computing, but it is more effective at processing and storing data close to the origin (Escamilla-Ambrosio et al., 2016).

The edge computing architecture offers the structural basis for installing computer components and applications in Figure 18.6.

The IoT system's beating heart is the enormous volume of data that IoT devices generate (Alwarafy et al., 2020). These data will be collected and stored in the edge network or edge equipment before being reviewed (Samuel et al., 2018).

For faster results and more network bandwidth, edge computing systems are placed close to consumers or IoT devices (Premsankar et al., 2018). The following will be accomplished through edge computing:

- Organizing IoT gadgets and collecting data (Atlam et al., 2018)
- Ensuring that the data from the sensor to the server are highly secure
- Implementing hands-free gadget boarding
- Data intake, gathering, storing, and analysis at the periphery

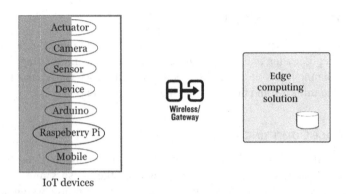

IoT devices

FIGURE 18.6   Image illustrating edge computing architecture.

18.7   CLOUD COMPUTING

Businesses may leverage the cloud computing infrastructure, to maximize the return on their investment in information technology (IT) software and hardware (Buyya et al., 2009). Applications that have been expanded to be accessed online fall under the umbrella of cloud computing (Aazam et al., 2014).

Web applications and web services are hosted in the cloud by colossal data centers and despotic servers (Larumbe et al., 2013). Both a platform and a particular kind of application are included with the phrase "cloud computing" (S. Dash and Pani, 2016).

Infrastructure as a service (IaaS), platform as a service (PaaS), and software as a service (SaaS) are the three kinds of cloud computing services (Mahmood, 2011). IaaS provides customers with virtualized computing resources including storage facilities, servers, and networking in contrast to PaaS, which offers a platform for developing, testing, and deploying applications (Hashizume et al., 2012). SaaS provides software applications such as email, customer contact management, and productivity tools through the internet (Akande et al., 2014).

18.8   EDGE COMPUTING VERSUS CLOUD COMPUTING

Although it cannot completely replace cloud computing, many extremely substantial IoT applications benefit from edge computing (Sultan, 2010). The distinction between edge and cloud computing is discussed in Table 18.2.

No matter how much data are created or how quickly it is processed, cloud computing offers a wide range of services and is currently prevalent in many applications. Even though it only offers a few services, real-time applications can employ edge computing.

18.9   CONCLUSION

IoT device integration with cloud computing and data management systems has become a potent tool for allowing data-driven decision-making and enhancing corporate operations, to sum up.

TABLE 18.2

Distinctions between Edge and Cloud Computing

Features	Edge Computing	Cloud Computing
Region	Nodes and systems are distributed but they are close to the end user.	It might be anywhere in the world.
Reaction Period	The response time is brief.	The response time is rapid.
Processing and Judgment	At local edge device	Remote Cloud
Connection	Real-time connection	Constraints with bandwidth and internet access
Dimension of Storage	Diminutive	Humongous
Device Diverseness	Intensely Supported	Restricted Support
The Capability for Processing	Average	Lofty

The advantages and drawbacks of IoT-integrated cloud and data solutions have been examined in this study. These solutions' capacity to assemble and analyze enormous amounts of data in real time, facilitate predictive analytics, and improve operational efficiency are among their key advantages (Khang et al., 2022b).

IoT-integrated cloud and data solutions are one of the many study fields that are now being investigated by academics and business professionals.

1. **Platforms for edge computing:** As IoT systems produce more data that must be analyzed in real time, edge computing platforms are becoming increasingly crucial.
2. **Platforms for IoT that are cloud-native**: These platforms can offer the scalability and flexibility required to enable large-scale IoT deployments since they are built to operate in cloud environments.
3. **Standardization of data interchange formats and protocols:** Because IoT systems use a variety of devices, platforms, and suppliers, maintaining compatibility and interoperability can be difficult.
4. **Dashboards and intuitive interfaces:** IoT systems create a lot of data, which can be daunting for consumers. User-friendly dashboards and interfaces can help.
5. **Enhancing data privacy and security:** In IoT-integrated cloud and data solutions, data privacy and security are crucial challenges.

These are only a few potential directions for IoT-integrated cloud and data solutions. People may anticipate seeing more creative solutions that will aid in addressing the difficulties and opportunities given by this technology as IoT continues to develop and becomes more widespread in everyday life.

Overall, the chapter indicates that cloud and data solutions with IoT integration have the potential to alter enterprises in a variety of industries. Building strong, secure, and scalable data management systems that can accommodate the rising volumes of data produced by IoT devices will be a priority as more businesses use these solutions.

The needs and difficulties of establishing cloud and data solutions with IoT integration are covered in this chapter. Dealing with the immense volume and diversity of data produced by IoT devices is one of the key issues.

The capacity to use data insights to enhance operations and develop new goods and services is another important worry about the security of the data produced by IoT devices. The infrastructure, platform, and software services required for the collection, storage, and analysis of the data produced by IoT devices are provided by cloud computing, which is essential to the IoT.

Scalability, security, data processing and analysis, integration, flexibility, real-time processing, high availability, dependability, and cost-effectiveness are the main criteria for cloud-based IoT systems. IoT-specific services from cloud providers like Google Cloud Platform and Amazon Web Services enable businesses to satisfy IoT implementation requirements and move forward with IoT initiatives more quickly.

Additionally, managing and analyzing the substantial amounts of data launched by IoT devices is made possible by data warehouses and data lakes.

REFERENCES

Aazam M, Khan I, Alsaffar AA, Huh E-N. "Cloud of Things: Integrating Internet of Things and Cloud Computing and the Issues Involved," *Proceedings of 2014 11th International Bhurban Conference on Applied Sciences & Technology (IBCAST) Islamabad, Pakistan*, January 14–18, 2014. https://doi.org/10.1109/ibcast.2014.6778179

Aissi El MEM. "Data Lake Versus Data Warehouse Architecture: A Comparative Study," *Lecture Notes in Electrical Engineering*, 201–210, 2021. https://doi.org/10.1007/978-981-33-6893-4_19

Akande AO, Van Belle J-P. "Cloud Computing in Higher Education: A Snapshot of Software as a Service," *2014 IEEE 6th International Conference on Adaptive Science & Technology (ICAST)*, 2014. https://doi.org/10.1109/icastech.2014.7068111

Akbar A, Khan A, Carrez F, Moessner K. "Predictive Analytics for Complex IoT Data Streams," *IEEE Internet of Things Journal*, 4(5), 1571–1582, 2017. https://doi.org/10.1109/jiot.2017.2712672

Alahakoon D, Yu X. "Smart Electricity Meter Data Intelligence for Future Energy Systems: A Survey," *IEEE Transactions on Industrial Informatics*, 12(1), 425–436, 2016. https://doi.org/10.1109/tii.2015.2414355

Anisimov AA. "Review of the Data Warehouse Toolkit," *ACM SIGMOD Record*, 32(3), 101–102, 2003. https://doi.org/10.1145/945721.945741

Alwarafy A, Al-Thelaya KA, Abdallah M, Schneider J, Hamdi M. "A Survey on Security and Privacy Issues in Edge-Computing-Assisted Internet of Things," *IEEE Internet of Things Journal*, 8(6), 4004–4022, 2021. https://doi.org/10.1109/jiot.2020.3015432

Atlam H, Walters R, Wills G. "Fog Computing and the Internet of Things: A Review," *Big Data and Cognitive Computing*, 2(2), 10, 2018. https://doi.org/10.3390/bdcc2020010

Badidi E, Moumane K. "Enhancing the Processing of Healthcare Data Streams using Fog Computing," *2019 IEEE Symposium on Computers and Communications (ISCC)*, 2019. https://doi.org/10.1109/iscc47284.2019.8969736

Berman SJ. "Digital Transformation: Opportunities to Create New Business Models," *Strategy & Leadership*, 40(2), 16–24, 2012. https://doi.org/10.1108/10878571211209314

Bhambri P, Rani S, Gupta G, Khang A. *Cloud and Fog Computing Platforms for Internet of Things*. CRC Press, 2022. https://doi.org/10.1201/9781003213888

Blumenthal SH. "Medium Earth Orbit Ka Band Satellite Communications System," *MILCOM 2013-2013 IEEE Military Communications Conference*, 2013. https://doi.org/10.1109/milcom.2013.54

Buyya R, Pandey S, Vecchiola C. "Cloudbus Toolkit for Market-Oriented Cloud Computing," *Lecture Notes in Computer Science*, 24–44, 2009. https://doi.org/10.1007/978-3-642-10665-1_4

Carciochi RA, D'Alessandro LG, Vauchel P, Rodriguez MM, Nolasco SM, Dimitrov K. "Valorization of Agrifood By-Products by Extracting Valuable Bioactive Compounds Using Green Processes," *Ingredients Extraction by Physicochemical Methods in Food*, 191–228, 2017. https://doi.org/10.1016/b978-0-12-811521-3.00004-1

Chakravarthi VS. "IoT Database Management and Analytics," *Internet of Things and M2M Communication Technologies*, 207–229, 2021. https://doi.org/10.1007/978-3-030-79272-5_13

Constantiou ID, Kallinikos J. "New Games, New Rules: Big Data and the Changing Context of Strategy," *Journal of Information Technology*, 30(1), 44–57, 2015. https://doi.org/10.1057/jit.2014.17

Conti M, Dehghantanha A, Franke K, Watson S. "Internet of Things Security and Forensics: Challenges and Opportunities," *Future Generation Computer Systems*, 78, 544–546, 2018. https://doi.org/10.1016/j.future.2017.07.060

Corallo A, Crespino AM, Lazoi M, Lezzi M. "Model-Based Big Data Analytics-as-a-Service Framework in Smart Manufacturing: A Case Study," *Robotics and Computer-Integrated Manufacturing*, 76, 102331, 2022. https://doi.org/10.1016/j.rcim.2022.102331

Corbellini A, Mateos C, Zunino A, Godoy D, Schiaffino S. "Persisting Big-Data: The NoSQL Landscape," *Information Systems*, 63, pp. 1–23, 2017. https://doi.org/10.1016/j.is.2016.07.009

Costa FS. "FASTEN IIoT: An Open Real-Time Platform for Vertical, Horizontal and End-To-End Integration," *Sensors*, 20(19), 5499, 2020. https://doi.org/10.3390/s20195499

Darwish A, Hassanien AE, Elhoseny M, Sangaiah AK, Muhammad K. "The Impact of the Hybrid Platform of Internet of Things and Cloud Computing on Healthcare Systems: Opportunities, Challenges, and Open Problems," *Journal of Ambient Intelligence and Humanized Computing*, 10(10), 4151–4166, 2017. https://doi.org/10.1007/s12652-017-0659-1

Dash S, Pani SK. "E-Governance Paradigm Using Cloud Infrastructure: Benefits and Challenges," *Procedia Computer Science*, 85, 843–855, 2016. https://doi.org/10.1016/j.procs.2016.05.274

Elkheir AM, Hayajneh M, Ali N. "Data Management for the Internet of Things: Design Primitives and Solution," *Sensors*, 13(11), 15582–15612, 2013. https://doi.org/10.3390/s131115582

Elshawi R, Sakr S, Talia D, Trunfio P. "Big Data Systems Meet Machine Learning Challenges: Towards Big Data Science as a Service," *Big Data Research*, 14, 1–11, 2018. https://doi.org/10.1016/j.bdr.2018.04.004

Emeakaroha VC, Cafferkey N, Healy P, Morrison JP. "A Cloud-Based IoT Data Gathering and Processing Platform," *2015 3rd International Conference on Future Internet of Things and Cloud*, 2015. https://doi.org/10.1109/ficloud.2015.53

Escamilla-Ambrosio PJ, Rodríguez-Mota A, Aguirre-Anaya E, Acosta-Bermejo R, Salinas-Rosales M. "Distributing Computing in the Internet of Things: Cloud, Fog and Edge Computing Overview," *NEO 2016*, 87–115, 2017. https://doi.org/10.1007/978-3-319-64063-1_4

Farahzadi A, Shams P, Rezazadeh J, Farahbakhsh R. "Middleware Technologies for Cloud of Things: A Survey," *Digital Communications and Networks*, 4(3), 176–188, 2018. https://doi.org/10.1016/j.dcan.2017.04.005

Feki M. "Big Data Analytics Driven Supply Chain Transformation," *Research Anthology on Big Data Analytics, Architectures, and Applications*, 1413–1432, 2022. https://doi. org/10.4018/978-1-6684-3662-2.ch068

Filar JA, Zapert R. "Uncertainty Analysis of a Greenhouse Effect Model," *Operations Research and Environmental Management*, 101–118, 1996. https://doi.org/10.1007/ 978-94-009-0129-2_5

Firouzi F, Farahani B. "Architecting IoT Cloud," *Intelligent Internet of Things*, 173–241, 2020. https://doi.org/10.1007/978-3-030-30367-9_4

Giannoutakis KM, Spanopoulos-Karalexidis M, Filelis Papadopoulos CK, Tzovaras D. "Next Generation Cloud Architectures," *The Cloud-to-Thing Continuum*, 23–39, 2020. https:// doi.org/10.1007/978-3-030-41110-7_2

Ginige JA, Ginige A. "Challenges and Solutions in Process Automation in Tertiary Education Institutes: An Australian Case Study," *6th IEEE/ACIS International Conference on Computer and Information Science (ICIS 2007)*, 2007. https://doi.org/10.1109/ icis.2007.74

Gupta A, Srivastava A, Anand R, Tomažič T. "Business Application Analytics and the Internet of Things: The Connecting Link," *New Age Analytics*, 249–273, 2020. https://doi. org/10.1201/9781003007210-10

Hahanov V, Khang A, Litvinova E, Chumachenko S, Hajimahmud VA, Alyar VA. "The Key Assistant of Smart City—Sensors and Tools," *AI-Centric Smart City Ecosystems: Technologies, Design and Implementation* (1st Ed., vol. 17, p. 10). CRC Press, 2022. https:// doi.org/10.1201/9781003252542-17

Hashizume K, Fernandez EB, Larrondo-Petrie MM. "A Pattern for Software-as-a-Service in Clouds," *2012 ASE/IEEE International Conference on BioMedical Computing (BioMedCom)*, 2012. https://doi.org/10.1109/biomedcom.2012.29

Heckmann L, de Figueroa B. "A Modeling Approach for Bioinformatics Workflows," *Lecture Notes in Business Information Processing*, 167–183, 2019. https://doi. org/10.1007/978-3-030-35151-9_11

Henderson JC, Schilling DA. "Design and Implementation of Decision Support Systems in the Public Sector," *MIS Quarterly*, 9(2), 157, 1985. https://doi.org/10.2307/249116

Huang H, Yang L, Wang Y, Xu X, Lu Y. "Digital Twin-Driven Online Anomaly Detection for an Automation System Based on Edge Intelligence," *Journal of Manufacturing Systems*, 59, 138–150, 2021. https://doi.org/10.1016/j.jmsy.2021.02.010

Junaid A, Nawaz A, Usmani MF, Verma R, Dhanda N. "Analyzing the Performance of a DAPP Using Blockchain 3.0," *2023 13th International Conference on Cloud Computing, Data Science & Engineering (Confluence)*, 2023. https://doi.org/10.1109/ confluence56041.2023.10048887

Kandogan E. "LabBook: Metadata-Driven Social Collaborative Data Analysis," *2015 IEEE International Conference on Big Data (Big Data)*, 2015. https://doi.org/10.1109/ bigdata.2015.7363784

Khang A, Chowdhury S, Sharma S. *The Data-Driven Blockchain Ecosystem: Fundamentals, Applications, and Emerging Technologies* (1st Ed.). CRC Press, 2022a. https://doi. org/10.1201/9781003269281

Khang A, Gupta SK, Rani S, Karras DA (Eds.). *Smart Cities: IoT Technologies, Big Data Solutions, Cloud Platforms, and Cybersecurity Techniques*. CRC Press, 2023a. https:// doi.org/10.1201/9781003376064

Khang A, Gupta SK, Rani S, Karras DA (Eds.). *Smart Cities: IoT Technologies, Big Data Solutions, Cloud Platforms, and Cybersecurity Techniques*. CRC Press, 2023b. https:// doi.org/10.1201/9781003376064

Khang A, Gupta SK, Shah V, Misra A (Eds.). *AI-Aided IoT Technologies and Applications in the Smart Business and Production*. CRC Press, 2023c. https://doi. org/10.1201/9781003392224

Khang A, Ragimova NA, Hajimahmud VA, Alyar VA. "Advanced Technologies and Data Management in the Smart Healthcare System," *AI-Centric Smart City Ecosystems: Technologies, Design and Implementation* (1st Ed., vol. 16, p. 10), CRC Press, 2022b. https://doi.org/10.1201/9781003252542-16

Khanh HH, Khang A. "The Role of Artificial Intelligence in Blockchain Applications," *Reinventing Manufacturing and Business Processes Through Artificial Intelligence*, 2, 20–40, 2021. https://doi.org/10.1201/9781003145011-2

Khine PP, Wang ZS. "Data Lake: A New Ideology in Big Data Era," *ITM Web of Conferences*, 17, 03025, 2018. https://doi.org/10.1051/itmconf/20181703025

Khriji S, Benbelgacem Y, Chéour R, Houssaini DE, Kanoun O. "Design and Implementation of a Cloud-Based Event-Driven Architecture for Real-Time Data Processing in Wireless Sensor Networks," *The Journal of Supercomputing*, 78(3), 3374–3401, 2021. https://doi.org/10.1007/s11227-021-03955-6

Kim M, Lee J, Jeong J. "Open Source Based Industrial IoT Platforms for Smart Factory: Concept, Comparison and Challenges," *Computational Science and Its Applications—ICCSA 2019*, 105–120, 2019. https://doi.org/10.1007/978-3-030-24311-1_8

Kondo D, Javadi B, Iosup A, Epema D. "The Failure Trace Archive: Enabling Comparative Analysis of Failures in Diverse Distributed Systems." *2010 10th IEEE/ACM International Conference on Cluster, Cloud and Grid Computing*, 2010. https://doi.org/10.1109/ccgrid.2010.71

Larumbe F, Sanso B. "A Tabu Search Algorithm for the Location of Data Centers and Software Components in Green Cloud Computing Networks," *IEEE Transactions on Cloud Computing*, 1(1), 22–35, 2013. https://doi.org/10.1109/tcc.2013.2

Lee C, Fumagalli A. "Internet of Things Security—Multilayered Method For End to End Data Communications Over Cellular Networks," *2019 IEEE 5th World Forum on Internet of Things (WF-IoT)*, 2019. https://doi.org/10.1109/wf-iot.2019.8767227

Lee I. "The Internet of Things for Enterprises: An Ecosystem, Architecture, and IoT Service Business Model," *Internet of Things*, 7, 100078, 2019. https://doi.org/10.1016/j.iot.2019.100078

Liu R, Isah H, Zulkernine F. "A Big Data Lake for Multilevel Streaming Analytics." *2020 1st International Conference on Big Data Analytics and Practices (IBDAP)*, 2020. https://doi.org/10.1109/ibdap50342.2020.9245460

Mahmood Z. "Cloud Computing for Enterprise Architectures: Concepts, Principles and Approaches," *Computer Communications and Networks*, 3–19, 2011. https://doi.org/10.1007/978-1-4471-2236-4_1

Monino JL. "Data Value, Big Data Analytics, and Decision-Making," *Journal of the Knowledge Economy*, 12(1), 256–267, 2016. https://doi.org/10.1007/s13132-016-0396-2

Munappy A, Bosch J, Olsson HH, Arpteg A, Brinne B. "Data Management Challenges for Deep Learning." *2019 45th Euromicro Conference on Software Engineering and Advanced Applications (SEAA)*, 2019. https://doi.org/10.1109/seaa.2019.00030

Nambiar A, Mundra D. "An Overview of Data Warehouse and Data Lake in Modern Enterprise Data Management," *Big Data and Cognitive Computing*, 6(4), 132, 2022. https://doi.org/10.3390/bdcc6040132

Naveen S, Kounte MR. "Key Technologies and Challenges in IoT Edge Computing," *2019 Third International Conference on I-SMAC (IoT in Social, Mobile, Analytics and Cloud) (I-SMAC)*, 2019. https://doi.org/10.1109/i-smac47947.2019.9032541

Notaro P, Cardoso J, Gerndt M. "A Survey of AIOps Methods for Failure Management," *ACM Transactions on Intelligent Systems and Technology*, 12(6), 1–45, 2021. https://doi.org/10.1145/3483424

Papageorgiou A, Cheng B, Kovacs E. "Real-Time Data Reduction at the Network Edge of Internet-of-Things Systems," *2015 11th International Conference on Network and Service Management (CNSM)*, 2015. https://doi.org/10.1109/cnsm.2015.7367373

Premsankar G, Di Francesco M, Taleb T. "Edge Computing for the Internet of Things: A Case Study," *IEEE Internet of Things Journal*, 5(2), 1275–1284, 2018. https://doi.org/10.1109/jiot.2018.2805263

Ramchand S, Mahmood T. "Big Data Architectures for Data Lakes: A Systematic Literature Review," *2022 IEEE 46th Annual Computers, Software, and Applications Conference (COMPSAC)*, 2022. https://doi.org/10.1109/compsac54236.2022.00179

Rani S, Bhambri P, Kataria A, Khang A, Sivaraman AK. *Big Data, Cloud Computing and IoT: Tools and Applications* (1st Ed.). Chapman and Hall/CRC Press, 2023. https://doi.org/10.1201/9781003298335

Rani S, Chauhan M, Kataria A, Khang A (Eds.). "IoT Equipped Intelligent Distributed Framework for Smart Healthcare Systems," *Networking and Internet Architecture*, V(2), 30, 2021. https://doi.org/10.48550/arXiv.2110.04997

Rao BBP, Saluia P, Sharma N, Mittal A, Sharma SV. "Cloud Computing for Internet of Things & Sensing Based Applications," *2012 Sixth International Conference on Sensing Technology (ICST)*, 2012. https://doi.org/10.1109/icsenst.2012.6461705

Saggi MK, Jain S. "A Survey Towards an Integration of Big Data Analytics to Big Insights for Value-Creation," *Information Processing & Management*, 54(5), 758–790, 2018. https://doi.org/10.1016/j.ipm.2018.01.010

Sandeepa C, Siniarski B, Kourtellis N, Wang S, Liyanage M. "A Survey on Privacy for B5G/6G: New Privacy Challenges, and Research Directions," *Journal of Industrial Information Integration*, 30, 100405, 2022. https://doi.org/10.1016/j.jii.2022.100405

Sarramia D, Claude A, Ogereau F, Mezhoud J, Mailhot G. "CEBA: A Data Lake for Data Sharing and Environmental Monitoring," *Sensors*, 22(7), 2733, 2022. https://doi.org/10.3390/s22072733

Shukla U, Dhanda N, Verma R. *Augmented Reality Product Showcase E-Commerce Application*, February 16, 2023. SSRN. https://ssrn.com/abstract=4361319 or http://dx.doi.org/10.2139/ssrn.4361319

Shumba A-T, Montanaro T, Sergi I, Fachechi L, De Vittorio M, Patrono L. "Leveraging IoT-Aware Technologies and AI Techniques for Real-Time Critical Healthcare Applications," *Sensors*, 22(19), 7675, 2022. https://doi.org/10.3390/s22197675

Sieber JE. "Openness in the Social Sciences: Sharing Data," *Ethics & Behavior*, 1(2), 69–86, 1991. https://doi.org/10.1207/s15327019eb0102_1

Sisinni E, Saifullah A, Han S, Jennehag U, Gidlund M. "Industrial Internet of Things: Challenges, Opportunities, and Directions," *IEEE Transactions on Industrial Informatics*, 14(11), 4724–4734, 2018. https://doi.org/10.1109/tii.2018.2852491

Stock D, Stöhr M, Rauschecker U, Bauernhansl T. "Cloud-Based Platform to Facilitate Access to Manufacturing IT," *Procedia CIRP*, 25, 320–328, 2014. https://doi.org/10.1016/j.procir.2014.10.045

Suciu G, Vulpe A, Halunga S, Fratu O, Todoran G, Suciu V. "Smart Cities Built on Resilient Cloud Computing and Secure Internet of Things," *2013 19th International Conference on Control Systems and Computer Science*, 2013. https://doi.org/10.1109/cscs.2013.58

Sultan N. "Cloud Computing for Education: A New Dawn?" *International Journal of Information Management*, 30(2), 109–116, 2010. https://doi.org/10.1016/j.ijinfomgt.2009.09.004

Sutton RI, Staw BM. "What Theory Is Not," *Administrative Science Quarterly*, 40(3), 371, 1995. https://doi.org/10.2307/2393788

Symeonaki E, Arvanitis K, Piromalis D. "A Context-Aware Middleware Cloud Approach for Integrating Precision Farming Facilities into the IoT toward Agriculture 4.0," *Applied Sciences*, 10(3), 813, 2020. https://doi.org/10.3390/app10030813

Szabo G. "Assessment of the Efficiency of 5G Network Exposure for the Industrial Internet of Things," *2021 IEEE Conference on Standards for Communications and Networking (CSCN)*, 2021. https://doi.org/10.1109/cscn53733.2021.9686079

Tang S, Shelden DR, Eastman CM, Pishdad-Bozorgi P, Gao X. "A Review of Building Information Modeling (BIM) and the Internet of Things (IoT) Devices Integration: Present Status and Future Trends," *Automation in Construction*, 101, 127–139, 2019. https://doi.org/10.1016/j.autcon.2019.01.020

Verma R, Dhanda N. "Application of Supply Chain Management in Blockchain and IoT—A Generic Use Case," *2023 13th International Conference on Cloud Computing, Data Science & Engineering (Confluence)*, 2023. https://doi.org/10.1109/confluence56041.2023.10048815

Verma R, Dhanda N, Nagar V. "Addressing the Issues & Challenges of Internet of Things Using Blockchain Technology," *International Journal of Advanced Science and Technology*, 29(5), 10074–10082, 2020. http://sersc.org/journals/index.php/IJAST/article/view/19491

Verma R, Dhanda N, Nagar V. "Security Concerns in IoT Systems and Its Blockchain Solutions," *Cyber Intelligence and Information Retrieval*, 485–495, 2021. https://doi.org/10.1007/978-981-16-4284-5_42

Verma R, Dhanda N, Nagar V. "Application of Truffle Suite in a Blockchain Environment," *Proceedings of Third International Conference on Computing, Communications, and Cyber-Security*, 693–702, 2022. https://doi.org/10.1007/978-981-19-1142-2_54

Vollmer T, Manic M. "Cyber-Physical System Security with Deceptive Virtual Hosts for Industrial Control Networks," *IEEE Transactions on Industrial Informatics*, 10(2), 1337–1347, 2014. https://doi.org/10.1109/tii.2014.2304633

Walker C, Alrehamy H. "Personal Data Lake with Data Gravity Pull." *2015 IEEE Fifth International Conference on Big Data and Cloud Computing*, 2015. https://doi.org/10.1109/bdcloud.2015.62

Wolfert S, Ge L, Verdouw C, Bogaardt M-J. "Big Data in Smart Farming—A Review," *Agricultural Systems*, 153, 69–80, 2017. https://doi.org/10.1016/j.agsy.2017.01.023

Zhan Y, Tan KH, Ji G, Chung L, Tseng M. "A Big Data Framework for Facilitating Product Innovation Processes," *Business Process Management Journal*, 23(3), 518–536, 2017. https://doi.org/10.1108/bpmj-11-2015-0157

19 Public Service Strategy Empowered for Internet of Things Technologies and Its Challenges

*Shashi Kant Gupta, Sunil Kumar Vohra,
Olena Hrybiuk, and Arvind Kumar Shukla*

19.1 INTRODUCTION

To better the lives of its citizens, each portion of the complicated ecosystem necessitates extensive planning, such as the creation of a data-based network for transportation, significant usage of energy, and enforcement of the law. In addition, decisions in many cases may be politically motivated, so the resulting information is often subjective and analogous. Therefore, we rely on digital data to get things done.

With the help of Internet of Things (IoT) gadgets and big data tools, we can generate smart data with bidirectional information flow, smart system interaction with our smartphones, smart devices, gadgets, and, most significantly, machine-to-machine communication (Wei et al., 2017).

For example, Harbin, a Chinese city of 11 million, was forced to close in 2013 due to hazardous air, and the capital city of Beijing was also hit (Perera et al., 2014). Nonetheless, 70% of the global gross domestic product is still dependent on urbanization.

A nervous system with cutting-edge technology, leveraging the IoT should be established in smart cities as a substitute for siloed approaches to individual functions such as energy, transportation, food, water, shelter, and so on. Everything, from public health to food to education to employment to political stability, can be kept running thanks to this centralized location (Rana et al., 2021).

The smart city initiative aims to bring the benefits of big data and the IoT to our everyday lives through technological innovation and a concentrated effort in collecting and analyzing relevant data (Khanh and Khang, 2021).

19.2 IMPACT

More and more organizations, both public and private, are beginning to see the benefits of blockchain technology. Peer-to-peer (P2P) platforms for the decentralized exchange of data, assets, and digital goods are made possible by these

DOI: 10.1201/9781003392224-19

innovations. The use of blockchain technology has the potential to greatly enhance the effectiveness of regulatory controls in a wide range of economic sectors (Atzori et al., 2012).

Along with other emerging technologies like artificial intelligence (AI), autonomous vehicles, and cloud computing, to name a few, blockchain has the potential to disrupt many business sectors and our society. This is because the current Fourth Industrial Revolution is characterized by the merging of technologies that blur the boundaries between the physical and the digital.

The potential business uses are too vast to even begin to explore here. Actually, it would be more enlightening to investigate how these technologies will deliver efficient and cost-effective solutions to markets around the world.

19.2.1 SMART CITY

Computing advances have given the internet, social environment, smartphones, and, most recently, the IoT; now, we are seeing this concept of making the digital shift to our cities (Khang et al., 2023a).

Massive people are leaving rural areas for cities in search of employment, higher standards of living, and more opportunities in the digital and global economies (Macke et al., 2018).

The concept of a methodologically driven, associated city, commonly referred to as a "Smart City," has emerged as a means of supporting such a movement by providing new approaches to managing existing resources while also planning for and making the most of potential future ones as Figure 19.1.

Since people are moving to cities and urban areas at an alarming rate, they need to expand in all ways to accommodate their residents and strike a healthy balance between the city's economic, social, and environmental needs (Khang, 2021).

The city's primary goal in becoming a "Smart City" to enhance the life's quality for its residents (Mattern et al., 2010). Therefore, the city must establish critical features on its outskirts and incorporate specific essential components.

Simply put, the IoT is a network of interconnected devices and machines that can be monitored and controlled from a central location (Stoianov et al., 2007). These

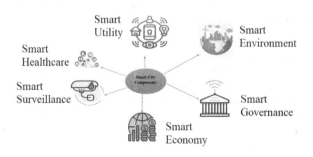

FIGURE 19.1 IoT-based smart components.

devices can see, communicate with, and share data because they will each contain significant internet protocol (IP) addresses.

In the case of home and industrial automation (Hajimahmud et al., 2022), for example, IoT allows traditionally separate systems like

- fire alarms,
- motion detectors,
- access control,
- surveillance cameras,
- heating,
- ventilation,
- air-conditioning units, and
- energy management control panels, among others.

The IoT lays the groundwork for a smart city by allowing all devices to monitor and react to their surroundings. Automated networks and the identification of frequency, machine, and various other embedded devices are connected on a network, making it difficult to pinpoint the precise scale of the IoT (Khang et al., 2022a).

However, it is predicted that the various existing associated devices in a city will reach billions within the next few years, far outpacing the growth of the internet itself. The most critical aspects of the IoT are sensor networks and cloud computing.

The monitoring, transmitting, analyzing, and recording of various conditions can be accomplished through a sensor network (Mahmood et al., 2011). Each sensor node has its dedicated transducer, microcontroller, transceiver, and power source, allowing for distributed sensing.

The transceiver serves as both a transmitter and a receiver, relaying the central computer's command to send the previously recorded output to the latter for analysis. The sensor relies on either external power from the electrical grid or its internal battery to function.

Data from associated devices are gathered on a central server, another critical feature of the IoT. Information is stored in the cloud and made available as an application, typically hosted as software as a service (SaaS) (Horsman, 2016), which users are allowed from various locations around the globe using smartphones, tablets, laptops, smart watches, and so on. Regarding data, cloud computing serves as a service that we can use for various purposes.

19.2.2 DATA FORMAT

Information is needed to generate and sustain messages for all the devices that make up the IoT's network. Extensible Markup Language (XML) and Efficient XML Interchange Format (XSIF) are widely used to exchange data (EXI).

However, the situation becomes more complicated when the XML files grow too large for the onboard devices to process. Therefore, the chapter's authors propose using the EXI format because it solves these problems, and any constrained device can read and understand it, turning it into a general-purpose IoT node.

19.3 APPLICATION AND TRANSPORT LAYERS

Hypertext Transfer Protocol (HTTP) is the de facto standard for transporting application-layer internet traffic over the Transmission Control Protocol (TCP). However, HTTP's complexity and TCP's lack of scalability with limited devices render HTTP unfit for an IoT environment (Zanella et al., 2014).

Overcoming this issue and providing a trustworthy solution is the restricted protocol which is communicated over user datagram protocol (UDP), which is also easily interoperable with HTTP.

19.4 NETWORK LAYER

Recently, the World Wide Web Consortium (W3C) announced that the most widely used addressing technology, Internet Protocol Version 4 (IPv4), has used up all of its address blocks (Gomez et al., 2020).

Fortunately, the Internet Protocol Version 6 (IPv6) was utilized to give a 128-bit address field, which we can use to give each node in an IoT network a distinct identifier. However, this work suggests adopting the IPv6 over low-energy effective personal area system (6LoWPAN) standard, which is based on the IPv6 and UDP header and operates over low-power constrained networks because IPv6 is not significantly capable of all restricted nodes (Govil et al., 2008).

In the future, we will evaluate a nation's economic growth and standing based on the success of its efforts to build smart cities sustainably. Smart and urban mobility, the use of renewable items for tracking the network, and so on will all be made possible by IoT devices and big data technology in the future city. Thus, it is safe to say that we and our cities are part of a globally interconnected network (Rani et al., 2022).

However, making progress toward the vision of a smart city is not without its difficulties, and failure to do so could have dire consequences for the well-being of its residents and the stability of the country's government and economy. Identified the following as some of the most pressing issues at hand:

- A smart city will be brimming with IoT-enabled devices and systems that operate on a wide range of platforms. The data generated by these sources will arrive simultaneously across several networks in various formats (Hahanov et al., 2022). Connecting either city to a more extensive network would pose the most difficulty.
- The security of the vast amounts of data stored on IoT devices and transmitted to a centralized server (the cloud) via a network connection poses the greatest threat to the smart city concept, which is inextricably linked to the IoT and big data. Attack vectors multiply proportionally with the growth in data volume and monetary value. Attackers and hackers can easily steal sensitive data, and they can alter or destroy data as well (Hajimahmud et al., 2023a).
- One set of obstacles facing smart city development is the absence of industry standards for the IoT and big data (Rani et al., 2021). Moreover, there

exist false standards to follow certain most important features of IoT, such as handling data, adapting to the actual values, transmitting, storing, analyzing, and, most importantly, securing the information.

19.5   MIDDLEWARE FOR SERVICE MANAGEMENT AND DEVELOPMENT

In order to provide coherent and intelligent services, cyber-physical systems (CPSs) integrate computational services with physical systems. *Middleware* is a broad term for software that connects many services to carry out coherent operations.

As a result, run-time environments on computing devices and operating systems both function as middleware. The environment to build, deploy, execute, and manage services for CPS/IoT is represented in this chapter by middleware.

Zhang et al. (2008) suggested an application in 2008, a reconfigurable real-time middleware with periodic events was proposed for distributed CPS. The adaptable workflows in CPS were the focus of this effort. A variety of distributed CPS must manage recurring events with various needs.

Real-time middleware, such real-time Common Object Request Broker Architecture (CORBA), has shown promise as a platform for time-constrained distributed systems, but it lacks the flexible configuration methods needed to control end-to-end timing for a range of distinct CPS with both aperiodic and periodic events.

This work's primary contribution is the design, implementation, and performance assessment of the first adjustable component middleware services for admission control and load balancing of periodic and periodic event handling in distributed CPS.

The following functionalities are supported by it: recognizing devices that are heterogeneous in association with the user's setting. The second step entails setting up and converting hardware into service component form.

The third step involves modifying and distributing application programs to get the desired outcome. The final step is to carry out everything mentioned earlier using sensors that can be accessed remotely.

Another objective is to build high-level apps using the WuKong (Zhang et al., 2008) middleware as the cornerstone of next-age IoT application development. The middleware WuKong comes with a lightweight Java virtual machine (JVM). The JVM platform enables programmers to extend the hardware's built-in functionality by adding behavior particular to a certain application.

Intelligent middleware and a framework for constructing flow-based IoT applications are the two main parts of WuKong middleware, which bridges the gap between developing and maintaining IoT applications.

Virtual middleware is what WuKong middleware is referred to as VMW. There are two factors at play here.

First, as sensor networks spread, it is possible that applications will need to use sensors made by various manufacturers and interact via various network protocols. Applications can operate on heterogeneous networks of sensors thanks to a virtual sensor.

Second, the virtual device sensors provide higher phase primitives relevant to IoT applications, the procedure of reprogramming a device will be lower in cost when the network concludes to recognize the system by employing a virtual machine design.

19.6 OPEN ISSUES ON MIDDLEWARE SCHEME FOR SMART BUILDINGS

The middleware described earlier offers a variety of IoT applications for development and the runtime environment. Several difficulties still need to be resolved.

19.6.1 PRIVACY MECHANISM

The IoT for smart cities and smart buildings collects user behavior with or without identifying information. Critical issues for the residents of buildings and cities include the collection of data that are owned, publicized, preserved, and used.

To meet the privacy criteria set forth by developers or users, the middleware itself should include a way to preserve privacy and regulate data flows. CPS/IoT systems must implement the middleware-defined privacy control policy.

As a result, a machine-understandable illustration for controlling privacy is lacking and has to be created so that it is adhered to by all system participants' devices (Luke et al., 2024).

19.6.2 ACCESS MECHANISM

Contrary to privacy controls, access control policies outline which services and gadgets are allowed access to data. The data gathered in smart cities and smart buildings include geographic and temporal information. As a result, when authorizing access to data, access management should be considered not only for services but also for data collection.

19.6.3 SCALABILITY

There may be thousands or even millions of devices and services installed in a city or a building. With complex network connectivity between devices and complex communication interactions between service components, it is difficult to develop a single service of this size. Therefore, it is preferable to create a single middleware service and deploy it to all the devices in the city and building. This is because similar sorts of devices are best served by a single middleware service.

According to a set of rules, the middleware should connect the connections between the services. One illustration is a smoke detector that is clever enough to suggest the best escape route based on data gathered by other device detectors in similar buildings. Services administration is difficult to manage services used by tens of thousands or even millions of devices in a single system.

Additionally, in order to perform the same or similar functions, it is quite likely that the equipment will be various. A wide range of service management technologies are available to fully control computing systems, including network protocols that permit remote management and storage systems that capture events and actions on the monitored devices. The aforementioned functionalities are not available on every CPS/IoT device.

Other gadgets have a limited amount of energy, while some have a restricted window of time during which they can be used. Through CPS/IoT middleware for smart buildings and smart cities, the services that have been delivered to the devices should be able to be monitored and managed remotely (Khang et al., 2023a).

19.7   DATA QUALITY CHALLENGES

19.7.1   CAUSES OF INVALID DATA

The variable's value in computing systems is constant until it is changed or eliminated. Finding the times when a variable is being changed is simple for systems with a single task or process. The execution of concurrent tasks makes it challenging to determine the stance of duration at which a parameter is being altered in systems with numerous tasks, processes, and threads. A variable can be created in CPS/IoT to provide information about the physical item status, which is subject to modification at any time and without prior notification (Tailor et al., 2022).

As a result, the variable's data value may suddenly cease to be valid at any time. In this section of the study, we examine the factors that lead to invalid data in mission-critical CPS/IoT systems, classify the factors, and describe the characteristics of the data.

19.7.2   REPRESENTATION OF QUALITY DATA

It is not easy to display the data quantity due to the many reasons why data are invalid. The presentation of temporal data in procedure critical is the focus of this section. The definition of the variables will include a few additional factors.

The bare minimum arrival interval, valid interval, and criticality of the data, as well as the bare minimum accuracy requirement of the data, are a few examples. There is no requirement to use a sensor at one-tenth of one-hundredth of a degree to determine the external temperature. It will be utilized to determine whether currently available data or the various data collected match the need.

19.7.3   DATA QUALITY ASSURANCE FOR COMPUTING NETWORK

The quality of sampled data can occasionally be enhanced, as stated earlier. The section's goal is to develop a framework for minimizing sampling and non-sampling mistakes in order to ensure the accuracy of the data.

In order to prevent failure, there are two possible strategies: handling the failure and avoiding it. By raising the sampling rates or adding more sensors, the quality of the data can frequently be increased.

The challenge is choosing the right approach so that it will not use up too many system resources. It also prompts queries about scheduling sampling or performing a preliminary calibration of the sensor.

By raising the sampling rates or adding more sensors, the quality of the data can frequently be increased. The challenge is choosing the right approach so that it will not use up too many system resources. It also prompts queries about scheduling sampling or performing a preliminary calibration of the sensor (Hahanov et al., 2022).

19.8 ISSUE ON COMPUTATION STRATEGY FOR THE IOT

The appropriate computing techniques for CPS/IoT schemes heavily depend on the needs and application areas. It is unlikely that a single computation model will be used to carry out all the system's services for a system with the scope of cities and smart spaces. Consequently, the difficult problems include identifying the computing network in different estimation methods (Subhashini et al., 2024).

19.8.1 SERVICE IMPEDANCE: PRIVATE AND PUBLIC PROVISIONS

Applications for the IoT include sensing data, clever decision logic, and actuation. Logic based on intelligence can be used on local-based computing or remote systems, while actuation and sensed information are performed on physical systems. The primary focus is to work on logics that are the backbone of IoT-based applications.

Public and private services are the two categories into which intelligent decision logics can be divided. By comparing their (1) accessibility, (2) access method, and (3) decision logics data sources, public and private facilities can be separated.

The three principles and two types of decision services are defined in the paragraphs that follow. The time periods for accessing the services are referred to as the decision logics' availability. Public services are housed on constantly accessible computing strategies and are reachable with the right authentication mechanisms.

Cloud-based services, such as Amazon Echo and Smart Things, are examples of public services. Private services, by comparison, are housed on computers with restricted resources and are not always accessible. In order to save electricity, resource-limited computing devices may occasionally be installed on-site and put to sleep or hibernate as shown in Figure 19.2.

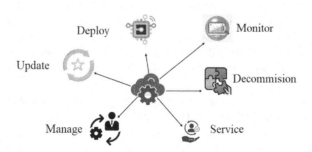

FIGURE 19.2   Life cycle of the Internet of Things.

Source: Khang (2021).

Decision logic's access model makes reference to the accessibility interface's openness. Access to public services is possible through an open communication interface. Any service component can access the public services provided they have access to the protocol name, service addresses, and authentication configuration.

In contrast, only predefined, constrained communication interfaces can be used to access private services. A wired interface is frequently used to access the services. Decision logic's sources of data are referred to as data sources. Using public information like traffic flow, transit schedules, and weather predictions, public services make decisions.

The three aforementioned characteristics separate amenities into private and public systems. Public services that offer high possibilities, open contact, and service provided in public data are by their very nature deployed in the cloud. Nevertheless, deploying private services in the cloud might not be simple.

Additionally, private services are typically tailored for certain users or applications, whereas public services are typically associated with a large total number of users. Deploying private services to the cloud is therefore not practical.

19.8.2 IoT-Based Fault Tolerance System

To locate anomalies on various computing devices and find the network so that these faults can be substituted during the process of recovery, the aim of fault handling in the IoT is to employ a central supervisor. Periodically, the controller polls the devices to find out how they are doing (Khang, 2023b).

A gadget may have a defect if any unusual behavior is noticed or if it does not respond. Based on the services that have been deployed on that device, the central controller will then determine which services might be at the problem. This straightforward approach may run into scaling problems. First, the controller must manage more fault tolerance combinations when more devices are added to a network.

Finally, the controller and devices nearby may experience bottlenecks as a result of all monitoring-related signals that are sent to it. In addition to wasting communication resources and energy, this will also have an impact on how well these devices' running active applications perform.

We, therefore, separate IoT-based clusters for handling rather use a single controller for all of them. Only each cluster is monitored independently. For efficiency, a manager, or cluster head (CH), oversees each cluster.

The backup data for every device in an organization is stored in a CH. Certain devices in the group will notice a failure when it occurs and report it to CH. Then CH will substitute backup services for any malfunctioning ones.

An alert-based communication behavior of the IoT may be preferred in many situations to reduce transmission costs. Event activated refers to the fact that a message is only issued when a large deviation in gathered value is identified. This has the drawback of making it impossible for a receiving scheme to determine whether failure on the sender side if no message is received (Khang et al., 2024).

Periodic beacons are frequently employed in event-triggered communication to inform the receiver that the transmitter is in good condition and to aid in fault detection. When transmitting beacon signals, we employ the loop architecture to reduce communication costs and distribute a CH's workload.

The star topology anticipates that CH will transmit and receive each beacon or back message. A loop topology, by comparison, divides the work among all of the cluster's devices. No requirement to report back is present until the existing beacon can be transmitted via the loop. As a result, a loop topology cluster's monitoring packet usage is drastically reduced.

The loop topology's principal drawback is that any flaw could prevent the propagation of subsequent beacon messages. Two regulators are employed to estimate the period of beacon for every device in order to handle this: one for regulating the speed of the message that is sent by the beacon and the other for a time-out period.

The transmission and receiving times of various devices may vary depending on the requirements of the various applications. The receiver must cause the computing to alert CH about the irregular reception if there is no message from the beacon for the duration.

19.9 CHALLENGES

The country's economic prosperity and prestige will be assessed in the coming days by smart city developments.

Cities will be inundated that consist of various techniques related to big data and IoT systems—smart and urban mobility, smart buildings operating smart meters, security monitoring, the running of elevators, smart traffic, the use and tracking of renewable products, and so on. As a result, it is acceptable to claim that we and our cities reside in a networked world (Khang et al., 2023c).

To achieve the objective of constructing a smart city, there are numerous obstacles to overcome, and doing so could have disastrous effects on people, society, the government, and even the development and economics of the entire nation (Khang et al., 2023b).

Several of the pressing issues that must be resolved right away, according to the chapter's authors, follow:

- The IoT-enabled systems and devices that make up a smart city will run on a number of operating systems, and the data produced by these systems and devices will be distributed through a variety of networks in varied varieties, velocities, and volumes. Integrating either city into a network would be the most difficult task.
- The biggest issue facing the concept of a "smart city," which is primarily related to security in the big data environment (Bhambri et al., 2022). IoT devices contain a lot of data of different kinds, and when they are associated to a system, the data are sent to a central server (the cloud), which makes them even more vulnerable. As data volume and value increase over time, so, too, are the targets for assaults, which are multiplied by the same number (Liu et al., 2022). Attackers and hackers have the ability to steal all sensitive data, possibly alter it, and even jeopardize its integrity.

- Another set of obstacles to the development of smart cities is the absence of values in the big data and IoT fields. There are numerous concepts, but not many of them have been chosen to become standards just yet. As a result, there are no genuine standards to adhere to for key essential IoT and big data activities, such as handling data, adapting to real values, and securely transmitting, storing, and analyzing information (Abri et al., 2022).

Since the microservice paradigm permits the reuse of software components and building blocks, the adoption of microservice-oriented architectures contributes to the defeat of the monolithic platform approach (Nagarajan et al., 2022).

Additionally, microservice designs are flexible and can more readily leverage external services (if required for outsourcing some processing or gaining access to additional services or applications; Blackstock et al., 2014).

They can be easily modified, for example, to handle practically all IoT and communication protocols as well as data-driven and event-driven push modalities. The introduction of more intelligent business intelligence and data analytics frameworks, as well as more interactive visual analysis tools, is advancing in that direction (Bhat et al., 2022).

Additionally, platforms for IoT-enabled smart cities are developing toward cross-organizational and multitenancy IoT platforms and applications. Because the infrastructures are shared by several operators, this enables the construction of sizable infrastructures that can support numerous enterprises, improve scalability, and lower infrastructure costs (Khang et al., 2023b).

This feature attempts to balance and overcome the efforts required to construct specifically tailored platforms for each city or each unique situation, which is economically wasteful (Boyinbode et al., 2011). It is strongly related to the reuse of components in smart city frameworks.

The introduction and use of cutting-edge network technologies, such as 5G, are some other potential future directions. Networking and device solutions technological advancements are significant.

In fact, adopting the newest network technologies, like 5G, along with more efficient construction methods and technologies that adhere to the net-zero-energy infrastructure paradigm, can result in building solutions that are strive for net-zero carbon emissions (Radanliev et al., 2022).

19.10 CONCLUSION

The growth of all the elements and sectors surrounding and related to the smart city is necessary to make it a reality. A city's design, planning, and development depend heavily on a number of essential smart city features and requirements, such as smart utility, smart living, smart mobility, and smart government.

Despite the fact that the intelligent city has extremely variable resource needs, it offers great benefits to the population and society. The fact that industry analysts believe that smart cities will account for roughly USD 40 billion in the market by 2016 shows that the major businesses and the government realize the value of urban cities and are taking the necessary steps to execute it (Jaiswal et al., 2023).

The IoT provides the essential hardware and software for efficient city management and an elevated quality of life, acting as the brains of the smart city concept. The brain of the smart city is big data, which claims to generate insight from the data received from IoT-enabled devices and other data sources (Hajimahmud et al., 2023b).

However, to achieve their objective of creating smart cities, the city must use the developments of IoT and cutting-edge big data techniques (Rani et al., 2023). This will allow the city to be smarter, exhibit progress, be more competitive, and have sustainable growth. Security, privacy, data management, and analysis are just a few of the many issues and unknowns that still need to be resolved (Khang et al., 2022d).

REFERENCES

Abri Al KAS, Jabeur N, Yasar A, El-Hansali Y. "New Cyber Physical System Architecture for the Management of Driving Behavior Within the Context of Connected Vehicles," *Computing and Informatics*, 41(2), 527–549, 2022. https://doi.org/10.31577/cai_2022_2_527

Atzori L, Iera A, Morabito G, Nitti M. "The Social Internet of Things (SIOT)-When Social Networks Meet the Internet of Things: Concept, Architecture and Network Characterization," *Computer Networks*, 56(16), 3594–36018, 2012. https://doi.org/10.1016/j.comnet.2012.07.010

Bhambri P, Rani S, Gupta G, Khang A. *Cloud and Fog Computing Platforms for Internet of Things*. CRC Press, 2022. ISBN: 978-1-032-101507. https://doi.org/10.1201/9781003213888

Bhat JR, AlQahtani SA, Nekovee M. "FinTech Enablers, Use Cases, and Role of Future Internet of Things," *Journal of King Saud University-Computer and Information Sciences*, 2022. https://doi.org/10.1016/j.jksuci.2022.018.033

Blackstock M, Lea R. "Toward a Distributed Data Flow Platform for the Web of Things (Distributed Node-RED)," *Proceedings of the 5th International Workshop on Web of Things, WoT'14*, 34–39, 2014. https://doi.org/10.1145/2684432.2684439

Boyinbode O, Le H, Takizawa M. "A Survey on Clustering Algorithms for Wireless Sensor Networks," *International Journal of Space-Based and Situated Computing*, 1(2–3), 130–136, 2011.

Gomez C, Minaburo A, Toutain L, Barthel D, Zuniga JC. "IPv6 Over LPWANs: Connecting Low Power Wide Area Networks to the Internet (of Things)," *IEEE Wireless Communications*, 27(1), 206–213, 2020. https://doi.org/10.1109/MWC.001.1900215

Govil J, Kaur N, Kaur H. "An Examination of IPv4 and IPv6 Networks: Constraints and Various Transition Mechanisms," *IEEE SoutheastCon* (pp. 178–185). IEEE, 2008. https://doi.org/10.1109/SECON.20018.4494282

Hahanov V, Khang A, Litvinova E., Chumachenko S, Hajimahmud VA, Alyar VA. "The Key Assistant of Smart City—Sensors and Tools," *AI-Centric Smart City Ecosystems: Technologies, Design and Implementation* (1st Ed.). CRC Press, 2022. https://doi.org/10.1201/9781003252542-17

Hajimahmud VA, Khang A, Gupta SK, Babasaheb J, Morris G. *AI-Centric Modelling and Analytics: Concepts, Designs, Technologies, and Applications* (1st Ed.). CRC Press, 2023a. https://doi.org/10.1201/9781003400110

Hajimahmud HV, Khang A, Hahanov V, Litvinova E, Chumachenko S, Alyar VA (Eds.). "Autonomous Robots for Smart City: Closer to Augmented Humanity," *AI-Centric Smart City Ecosystems: Technologies, Design and Implementation* (1st Ed.). CRC Press, 2022. https://doi.org/10.1201/9781003252542-7

Hajimahmud VA. et al. (Eds.). "The Role of Data in Business and Production," *AI-Aided IoT Technologies and Applications in the Smart Business and Production*. CRC Press, 2023b. https://doi.org/10.1201/9781003392224-2

Horsman G. "Unmanned Aerial Vehicles: A Preliminary Analysis of Forensic Challenges," *Digital Investigation*, 16, 1–11, 2016. https://doi.org/10.1016/j.diin.2015.11.002

Jaiswal N, Misra A, Misra PK, Khang A (Eds.). "Role of the Internet of Things (IoT) Technologies in Business and Production," *AI-Aided IoT Technologies and Applications in the Smart Business and Production*. CRC Press, 2023. https://doi.org/10.1201/9781003392224-1

Khang A. "Material4Studies," *Material of Computer Science, Artificial Intelligence, Data Science, IoT, Blockchain, Cloud, Metaverse, Cybersecurity for Studies*, 2021. www.researchgate.net/publication/370156102_Material4Studies. Last visit 2023.

Khang A (Ed.). *AI-Oriented Competency Framework for Talent Management in the Digital Economy: Models, Technologies, Applications, and Implementation*. CRC Press, 2023. https://doi.org/10.1201/9781003440901https://doi.org/10.1504/IJSSC.2011.040339

Khang A, Abdullayev V, Hahanov V, Shah V. *Advanced IoT Technologies and Applications in the Industry 4.0 Digital Economy* (1st Ed.). CRC Press, 2024. https://doi.org/10.1201/978-1-003-43426-9

Khang A, Gupta SK, Rani S, Karras DA (Eds.). *Smart Cities: IoT Technologies, Big Data Solutions, Cloud Platforms, and Cybersecurity Techniques* (1st Ed.). CRC Press, 2023a. https://doi.org/10.1201/9781003376064

Khang A, Gupta SK, Shah V, Misra A (Eds.). *AI-Aided IoT Technologies and Applications in the Smart Business and Production* (1st Ed.). CRC Press, 2023b. https://doi.org/10.1201/9781003392224

Khang A, Hahanov V, Abbas GL, Hajimahmud VA. "Cyber-Physical-Social System and Incident Management," *AI-Centric Smart City Ecosystems: Technologies, Design and Implementation* (1st Ed.). CRC Press, 2022b. https://doi.org/10.1201/9781003252542-2

Khang A, Rani S, Gujrati R, Uygun H, Gupta SK (Eds.). *Designing Workforce Management Systems for Industry 4.0: Data-Centric and AI-Enabled Approaches* (1st Ed.). CRC Press, 2023c. https://doi.org/10.1201/99781003357070

Khang A, Rani S, Sivaraman AK. *AI-Centric Smart City Ecosystems: Technologies, Design and Implementation* (1st Ed.). CRC Press, 2022a. https://doi.org/10.1201/9781003252542

Khang A, Shah V, Rani S. *AI-Based Technologies and Applications in the Era of the Metaverse* (1st Ed.). IGI Global Press, 2023d. https://doi.org/10.4018/9781668488515

Khanh HH, Khang A. "The Role of Artificial Intelligence in Blockchain Applications," *Reinventing Manufacturing and Business Processes through Artificial Intelligence* (pp. 20–40). CRC Press, 2021. https://doi.org/10.1201/9781003145011-2

Liu Y, Yu Z, Wang J, Guo B, Su J, Liao J. "Crowd-Manager: An Ontology-Based Interaction and Management Middleware for Heterogeneous Mobile Crowd Sensing," *IEEE Transactions on Mobile Computing*, 2022. https://doi.org/10.1109/TMC.2022.3199787

Luke J, Khang A, Chandrasekar V, Pravin AR, Sriram K (Eds.). "Smart City Concepts, Models, Technologies and Applications," *Smart Cities: IoT Technologies, Big Data Solutions, Cloud Platforms, and Cybersecurity Techniques* (1st Ed.). CRC Press, 2024. https://doi.org/10.1201/9781003376064-1

Macke J, Casagrande RM, Sarate JAR, Silva KA. "Smart City and Quality of Life: Citizens' Perception in a Brazilian Case Study," *Journal of Cleaner Production*, 182, 717–726, 2018. https://doi.org/10.1016/j.jclepro.20118.02.078

Mahmood Z. "Cloud Computing for Enterprise Architectures: Concepts, Principles and Approaches," *Computer Communications and Networks* (pp. 3–19). Springer, 2011. https://doi.org/10.1007/978-1-4471-2236-4_1

Mattern F, Floerkemeier C. "From the Internet of Computers to the Internet of Things," *Lecture Notes in Computer Science* (pp. 242–259). Springer, 2010. https://doi.org/10.1007/9783-642-17226-7_15

Nagarajan MS, Anandhan P, Muthukumaran V, Uma K, Kumaran U. "Security Framework for IoT and Deep Belief Network-Based Healthcare System Using Blockchain Technology," *International Journal of Electronic Business*, 17(3), 226–243, 2022. https://doi.org/10.1504/IJEB.2022.124324

Perera C, Liu CH, Jayawardena S, Chen M. "A Survey on Internet of Things from Industrial Market Perspective," *IEEE Access*, 2, 1660–1679, 2014. https://doi.org/10.1109/ACCESS.2015.2389854

Radanliev P, De Roure D, Nicolescu R, Huth M, Santos O. "Digital Twins: Artificial Intelligence and the IoT Cyber-Physical Systems in Industry 4.0," *International Journal of Intelligent Robotics and Applications*, 6(1), 171–185, 2022. https://doi.org/10.1007/s41315-021-00180-5

Rana G, Khang A, Sharma R, Goel AK, Dubey AK (Eds.). *Reinventing Manufacturing and Business Processes Through Artificial Intelligence*. CRC Press, 2021. https://doi.org/10.1201/9781003145011

Rani S, Bhambri P, Kataria A, Khang A. "Smart City Ecosystem: Concept, Sustainability, Design Principles and Technologies," *AI-Centric Smart City Ecosystems: Technologies, Design and Implementation* (1st Ed.). CRC Press, 2022. https://doi.org/10.1201/9781003252542-1

Rani S, Bhambri P, Kataria A, Khang A, Sivaraman AK (Eds.). *Big Data, Cloud Computing and IoT: Tools and Applications* (1st Ed.). Chapman and Hall/CRC Press, 2023. https://doi.org/10.1201/9781003298335

Rani S, Chauhan M, Kataria A, Khang A (Eds.). "IoT Equipped Intelligent Distributed Framework for Smart Healthcare Systems," *Networking and Internet Architecture*. CRC Press, 2021. https://doi.org/10.48550/arXiv.2110.04997

Stoianov I, Nachman L, Madden S, Tokmouline TP. "Wireless Sensor Network for Pipeline Monitoring," *Proceedings of the 6th International Conference on Information Processing in Sensor Networks*, 264–273, 2007. https://doi.org/10.1145/1236360.1236396

Subhashini R, Khang, A (Eds.). "The Role of Internet of Things (IoT) in Smart City Framework," *Smart Cities: IoT Technologies, Big Data Solutions, Cloud Platforms, and Cybersecurity Techniques* (1st Ed.). CRC Press, 2024. https://doi.org/10.1201/9781003376064-3

Tailor RK, Pareek R, Khang A (Eds.). "Robot Process Automation in Blockchain," *The Data-Driven Blockchain Ecosystem: Fundamentals, Applications, and Emerging Technologies* (1st Ed., pp. 149–164). CRC Press, 2022. https://doi.org/10.1201/9781003269281-8

Wei J, Zhan W, Guo X, Marinova D. "Public Attention to the Great Smog Event: A Case Study of the 2013 Smog Event in Harbin, China," *Nature Hazards*, 89(2), 923–9318, 2017. https://doi.org/10.1007/s11069017-3000-6

Zanella A, Bui N, Castellani A, Vangelista L, Zorzi M. "Internet of Things for Smart Cities," *IEEE Internet Things Journal*, 1(1), 22–32, 2014. https://doi.org/10.1109/JIOT.2014.2306328

Zhang Y, Gill C, Lu C. "Reconfigurable Real-Time Middleware for Distributed Cyber-Physical Systems with Aperiodic Events," *The 28th International Conference on Distributed Computing Systems, ICDCS'08*, 581–5818, 2008. https://doi.org/10.1109/ICDCS.20018.96

Index

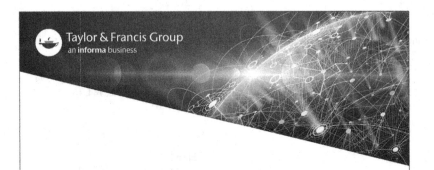

Printed in the United States
by Baker & Taylor Publisher Services